Genetic Mosaics
and Cell Differentiation

Edited by W. J. Gehring

With Contributions by

H. J. Becker   A. García-Bellido   R. L. Gardner
J. C. Hall   K. Illmensee   W. Janning
A. Mc Laren   J. R. Merriam   P. Ripoll
C. Tokunaga   E. Wieschaus

With 75 Figures

Springer-Verlag Berlin Heidelberg GmbH 1978

Professor Dr. WALTER J. GEHRING
Biozentrum der Universität
Abteilung Zellbiologie
Klingelbergstraße 70
CH-4056 Basel, Switzerland

ISBN 978-3-662-21952-2   ISBN 978-3-540-35803-9 (eBook)
DOI 10.1007/978-3-540-35803-9

Library of Congress Cataloging in Publication Data. Main entry under title: Genetic mosaics and cell differentiation. (Results and problems in cell differentiation; v. 9) Includes bibliographies and index. 1. Mosaicism. 2. Cell differentiation. I. Becker, H. J. II. Gehring, Walter J., 1939–   . III. Series. QH607.R4 vol. 9 [QH445.7] 574.8'761s [591.8'761]. 78-13470.

2131/3130-543210

Results and Problems in
Cell Differentiation

A Series of Topical Volumes in Developmental Biology

9

Preface

The relationship of cell lineage and differentiation is one of the most intriguing problems in developmental biology. In most higher organisms, the analysis of the cell lineage has to rely on indirect methods. Only in the most suitable systems, like the nematodes, can the pattern of cell division be determined by direct observation under the microscope. In cases where this is not possible, the fate of the cells has to be examined by using cell markers. Most suitable for this purpose are genetic markers, provided that they do not interfere with the developmental pathway to be studied. However, suitable genetic markers and techniques for generating genetic mosaics are available in a few organisms only. Therefore, this volume is largely concerned with *Drosophila* and the mouse, which have been studied most extensively.

In 1929 STURTEVANT introduced the analysis of gynandromorphs into developmental genetics. However, this important contribution remained largely unnoticed until the late sixties, when the potential of this technique for determining embryonic fate maps and the number of primordial cells was exploited, and the methodology extended to the mapping of mutational foci. Mitotic recombination was demostrated by STERN in 1936, who realized the potential of this technique for the analysis of cell lineage and pattern formation. The usefulness of this method was substantially increased by the discovery of FRIESEN that mitotic recombination can be induced by x-irradiation. This allows the induction of mosaicism at any stage of development.

When these elegant genetic techniques cannot be used, chimaeras can be generated by combining genetically marked cells from different donors into one embryo. These techniques have been developed by TARKOWSKI, MINTZ, and GARDNER for the mouse and other mammals. They are discussed extensively in this volume.

Genetic mosaics and chimaeras allow us to determine the number and location of progenitor cells in the embryo and their subsequent cell lineage. In addition, the autonomy of cell differentiation can be studied. If cells differentiate autonomously in chimeras or genetic mosaics, their developmental pathway is determined by intrinsic properties of the cells themselves, whereas nonautonomy indicates that cellular interactions are involved. The extent to which cell differentiation is determined by cell lineage or by cellular interactions is still a matter of debate. An attempt was made to have various points of view represented in this volume, so that the reader can use his own judgement.

I would like to thank the contributors for their collaboration in preparing this book, which illustrates nicely the potential of genetic mosaics in the analysis of development.

Summer 1978   WALTER J. GEHRING

Contents

Drosophila Chimeras and the Problem of Determination

By K. ILLMENSEE. With 7 Figures

Estimating Primordial Cell Numbers in Drosophila Imaginal Discs and Histoblasts

By J. R. MERRIAM. With 9 Figures

Cell Lineage Relationships in the Drosophila Embryo

By E. WIESCHAUS. With 8 Figures

Cell Lineage and Differentiation in Drosophila

By A. García-Bellido and P. Ripoll. With 7 Figures

Genetic Mosaic Studies of Pattern Formation in Drosophila melanogaster, with Special Reference to the Prepattern Hypothesis

By C. Tokunaga. With 17 Figures

Contributors

BECKER, HANS J., Institut für Allgemeine Biologie, Schwarzspanierstraße 17, 1090 Wien, Austria

GARCÍA-BELLIDO, ANTONIO, Centro de Biologia Molecular, Facultad de Ciencias, Universidad Autónoma de Madrid, Canto Blanco, Madrid-34, Spain

GARDNER, RICHARD L., Department of Zoology, University of Oxford, South Parks Road, Oxford OX1 3PS, Great Britain

HALL, JEFFREY C., Department of Biology, Brandeis University, Waltham, MA 02154, USA

ILLMENSEE, KARL, Département de Biologie Animale, 154, Route de Malagnou, 1224 Genève, Switzerland

JANNING, WILFRIED, Zoologisches Institut der Westfälischen Wilhelms-Universität, Badestraße 9, 4400 Münster, FRG

McLAREN, ANNE, MRC Mammalian Development Unit, University College London, Wolfson House, 4 Stephenson Way, London NW1 2HE, Great Britain

MERRIAM, JOHN R., University of California, Department of Biology, 405 Hilgard Avenue, Los Angeles, CA 90024, USA

RIPOLL, PEDRO, Centro de Biologia Molecular, Facultad de Ciencias, Universidad Autónoma de Madrid, Canto Blanco, Madrid-34, Spain

TOKUNAGA, CHIYOKO, University of California, Department of Molecular Biology, Wendell M. Stanley Hall, Berkeley, CA 94720, USA

WIESCHAUS, ERIC, Europäisches Laboratorium für Molekularbiologie (EMBL), Meyerhofstraße, 6900 Heidelberg, FRG

Gynandromorph Fate Maps in Drosophila *

WILFRIED JANNING

Zoologisches Institut der Universität, Münster, FRG

I. Introductory Remarks

Genetic mosaics are organisms that are built up by at least two genotypically different cell types. Gynandromorphs are genetic mosaics that consist of both male and female cells. If one of the two X chromosomes of a *Drosophila* embryo is lost, an XX//XO gynander [1] will develop with female XX and male (mutant) XO cells. By various methods XXY//XY gynanders or pure female XXY//XX or pure male XY//XO mosaics can also be obtained.

The aim of this article is to show how analyses of gynandromorphs lead to the establishment of blastoderm fate maps and thereby to an expansion of our knowledge of embryology and development. Information will be given on the mapping method, its assumptions, "weak points," and limitations. Little will be said about the application of gynandromorph fate maps to localize behavioral, lethal, or biochemical foci because this field is covered by other authors in this volume.

II. Production of Gynandromorphs

Besides transplantation of nuclei or cells there are at present four genetic techniques available for producing mosaic individuals in *Drosophila*:

1. Mitotic recombination.
2. Chromosome loss due to unstable ring-X chromosomes.
3. Mutants that cause chromosome loss during early cleavage mitoses.
4. X-ray-induced chromosome loss.

With the three latter methods gynandromorphs can be generated. The somatically unstable ring chromosomes used are *R(1)2*, *In(1)w^{vC}* (Catcheside and Lea,

---

* This article is dedicated to the memory of Prof. Dr. Ernst Hadorn.
[1] In this article the separation of genotypes within single genetic mosaics will be indicated by double oblique strokes as proposed by Williamson and Kaplan (1976).

1945) and *R(1)5 A* (Pasztor, 1971), whereas the known chromosome-loss mutants are *claret* of *Drosophila simulans* (Sturtevant, 1929), *claret-nondisjunctional* (ca^{nd}, Lewis and Gencarella, 1952), *mitotic loss inducer* (*mit*, Gelbart, 1974), and *paternal loss* (*pal*, Baker, 1975). Patterson and Stone (1938) analyzed gynandromorphs that were found among the progeny of irradiated males or females.

For detecting the genotypically different cells only those mutants can be used that express their phenotype autonomously in individual cells. There are several mutants of this type which permit the scoring of cuticular structures and pigmented eye cells, Malpighian tubules, and testis sheaths (e.g., *yellow, forked, singed, white, chocolate;* for description of the mutants see Lindsley and Grell,1968). More recently, marker systems were developed with which mosaicism in internal noncuticular parts of larvae and adults is detectable (aldehyde oxidase: Janning, 1972, 1974a; *prune:* Falk et al., 1973; acid phosphatase, glucose-6-phosphate dehydrogenase, 6-phosphogluconate dehydrogenase: Kankel and Hall, 1976). An extensive description of techniques to produce gynandromorphs is given by Hall et al. (1976).

III. Principles of Fate Mapping

In his study on *claret*-gynandromorphs of *Drosophila simulans* Sturtevant (1929) calculated the frequencies with which two parts (landmarks) in the adult were of different phenotype or sex and concluded:

"The numbers in these tables furnish rough indexes of the remoteness of relationship of the parts concerned. For clearly those parts that are more often different must be parts that more often come from separate cleavage nuclei. Conversely stated, the smaller the number of times separated, the more closely related are the parts."

This is the fundamental idea that led to the construction of gynandromorph fate maps and to an extensive analysis of these mosaics. Sturtevant himself established the first fate map (Fig. 1), which showed the embryonic relationships of precursor cells for the imaginal tergites and sternites.

Fourty years later Garcia-Bellido and Merriam (1969a) took up Sturtevant's idea. They obtained his data from 379 mosaics of *Drosophila simulans*, calculated the times of genotypic separation between all pairs of 75 selected landmarks, and found they could indeed construct a self-consistent fate map (Fig. 2).

In the construction of fate maps there are many difficulties, some of which cannot be overcome. The orientation of the spindle in the first zygotic division along any possible axis in three dimensions suggests a sphere with the segregational mitosis in its center. The probability of the cleavage plane bisecting two points on the three-dimensional surface of the sphere reflects the angle between the two points from the center, or the arc distance between two points on the surface (Garcia-Bellido and Merriam, 1969a). Since the *Drosophila* egg is not a sphere but a geometrically more complex structure, the transformation of genotypic separation probabilities to a two-dimensional map are even more complicated.

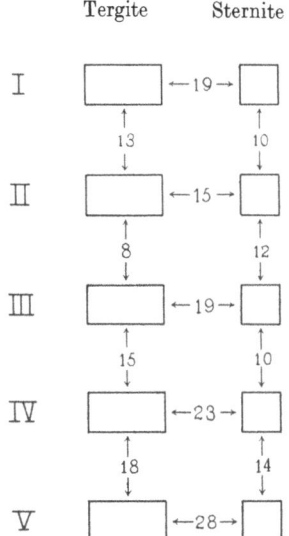

Fig. 1. First gynandromorph fate map established by Sturtevant (1929). The numbers indicate times the two sclerites indicated were different in 96 *claret* gynandromorphs of *Drosophila simulans*

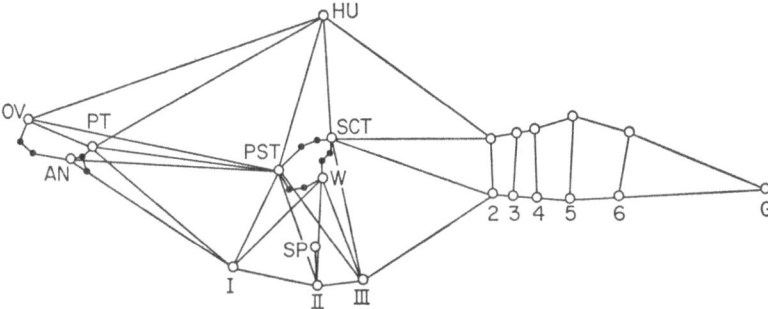

Fig. 2. Gynandromorph fate map of *Drosophila simulans* established by Garcia-Bellido and Merriam (1969a). *AN*, antenna; *G*, genitalia; *HU*, humeral bristles; *OV*, outer vertical bristle; *PST*, presutural bristle; *PT*, postorbital bristles; *SCT*, scutellar bristles; *SP*, sternopleural bristles; *W*, wing; *I, II, III*, legs; *2, 3, 4, 5, 6*, sternites

In the following it is attempted to illustrate some of these complications by considering the theoretical example shown in Figure 3. On the left of this figure a *Drosophila* embryo in the blastoderm stage is drawn with ten dividing lines between male and female cells, whereby each dividing line represents that of one gynandromorph. On the surface of the egg there are 4 points (A, B, C, and D) representing four blastoderm regions, which will develop into different adult structures. If the frequencies of genotypic separation between all pairs of points are established (AB=3, AC=6, BC=5, AD=7, BD=6, CD=5) the numbers can be used to construct the relative positions of the points concerned. The closer two

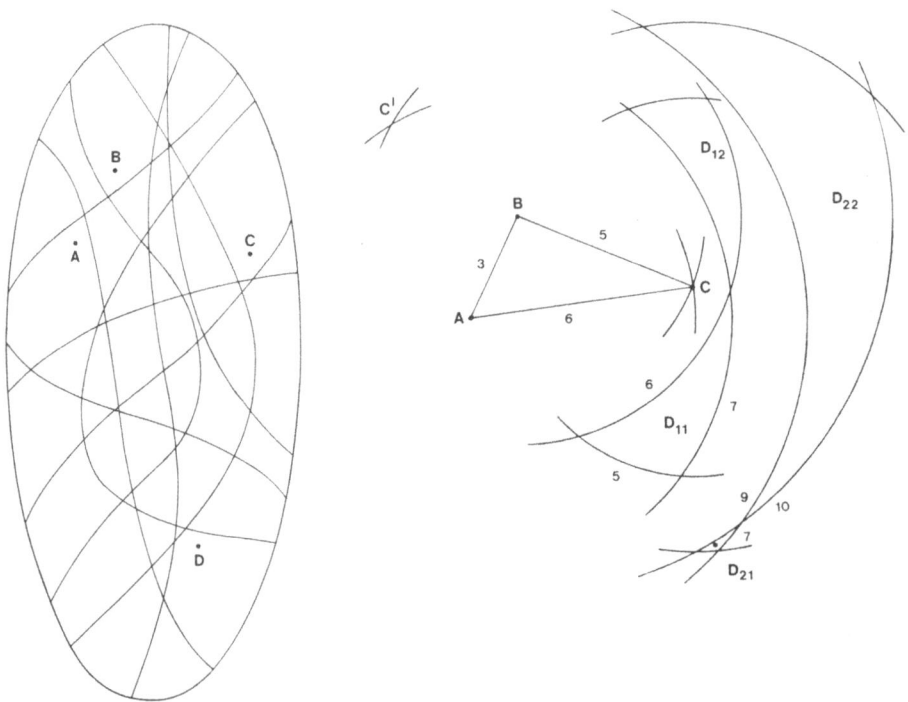

Fig. 3. Demonstration of the principles of fate mapping. *Left*: Projection of ten gynandrom-orph dividing lines on a theoretical *Drosophila* embryo with four surface points (*A*, *B*, *C*, *D*). *Right*: Construction of the corresponding fate map. See text for explanation

points are, the better is the correspondence between the arc distance and the surface distance between them. In this case one would begin with A and B and map C by triangulation (C′ is only the mirror image of C). Difficulties arise if D is localized. With the numbers shown in our example, the position of D cannot be established unequivocally, because it is impossible to distinguish between positions D_{11} and D_{12}. To distinguish between these two possibilities additional points are required. This difficulty is not only due to the relatively large distance between A, B, C on the one hand and D on the other, but it is due to the convolutions of dividing lines on the blastoderm surface. If the dividing line passes twice (or any even number of times) between two points, they are scored as having the same genotype, which leads to an underestimate of the arc distance. It is clearly seen from the figure that if we consider the convolutions, e.g., the AD distance increases from 7 to 9 and D can now be mapped at position D_{21}.

Since the convolutions of the dividing line are difficult to recognize in adult gynandromorphs, it is essential to use short distances for the construction of fate maps. Also, for a short distance the arc distance approximates a straight line most closely. Therefore, the general strategy of constructing fate maps as established by Garcia-Bellido and Merriam (1969a) has been to generate submaps of the head, the thorax, and the abdomen using short distances, and positioning these submaps by a few triangulations (Fig. 2).

IV. Assumptions Underlying the Construction of Fate Maps

Gynandromorph fate mapping is based on some crucial assumptions that were first pointed out by Garcia-Bellido and Merriam (1969a).

A. The Orientation of the Mitotic Spindle of the Cleavage Division at Which a Genotypically Different Clone is Initiated is Random with Respect to the Axes of the Egg

The distribution pattern of genotypes in gynandromorphs varies to a large extent. From the small sample of gynanders in Figure 4 it can be seen that the dividing lines tend to follow the longitudinal midline and intersegmental boundaries; nevertheless all gynanders are different from each other. In general the patches are large and of only one genotype, except the Malpighian tubules, which frequently show a pattern of alternating small cell clusters of different genotypes (Janning, 1974a, 1975).

Since the mosaic areas are large and uniform in genotype, it is concluded (1) that clone initiation takes place at very early developmental stages, i.e., during cleavage, and (2) that there is little mixing of nuclei during the syncytial divisions and their migration to the surface. The cells of sister clones tend to stay together. Clone borders are remarkably stable in these early developmental stages. Therefore, the different dividing lines originate from the time of clone initiation because "... *cleavage of the egg is indeterminate. There is no definite pattern among the mosaics, which can only mean that the cleavage nuclei are distributed to the blastoderm differently in different embryos.*" (Sturtevant, 1929). Histological observations by Parks (1936) showed indeed that there is no preferential orientation of the spindle axis of the first zygotic division.

It is not known how many types of gynandromorphs with different mosaic patterns exist. To try to answer this question we compared the patterns of 735 adult gynandromorphs (the material from Janning 1974b, 1976) with respect to 36 bilaterally symmetric cuticular and five internal landmarks, i.e., a total of 77 structures. We found 723 patterns different from each other with only one single perfect left–right gynander; 713 patterns occurred once, nine patterns each twice, and one pattern four times (with the dividing line between head and thorax). Since it should be random which of the two genotypically different clones is on one or the other side of the dividing line, complementary patterns should exist. In the 735 gynanders we found only three complementary patterns. This means that in 735 gynanders there are at least 720 different dividing lines. This is surprising because the large mosaic patches in gynanders often include a whole body segment or even more. Therefore we looked at only one part of a segment, the notum with 11 bilaterally symmetric bristles, i.e., 22 landmarks. In 334 gynanders the notum was found to be uniform in genotype but in the remaining 401 cases we found 139 different patterns of distributions of the two genotypes with 21 complementary patterns. This means that there are 118 different mosaic dividing lines through the notum. Figure 5 shows 28 of these dividing lines. It can be

Fig. 4. Outlines of a small sample of gynandromorphs generated by loss of the unstable ring-X chromosome *R(1)2, In(1)w^{rC}*. XO areas are *shaded*

seen that there are also mirror image patterns and that the longitudinal midline is preferred as mosaic border. This is to be expected because the imaginal cells for the differentiation of the bilaterally symmetric cuticular structures are separated in very early developmental stages, and left and right structures of each segment are not joined together before metamorphosis.

From the results of this pattern analysis it seems impossible to estimate the maximum number of different gynanders that could be found. This becomes clear if one takes into account that here only 77 landmarks were considered, whereas the number of scorable structures is much higher, including internal parts that can be marked by enzymes (Janning, 1972, 1974a, 1976; Kankel and Hall, 1976).

Despite the fact that all authors scored only a relatively small random sample of gynander types, they arrived at very similar results, for example, with respect to the cell-lineage relationships shown in fate maps. This can only mean that the

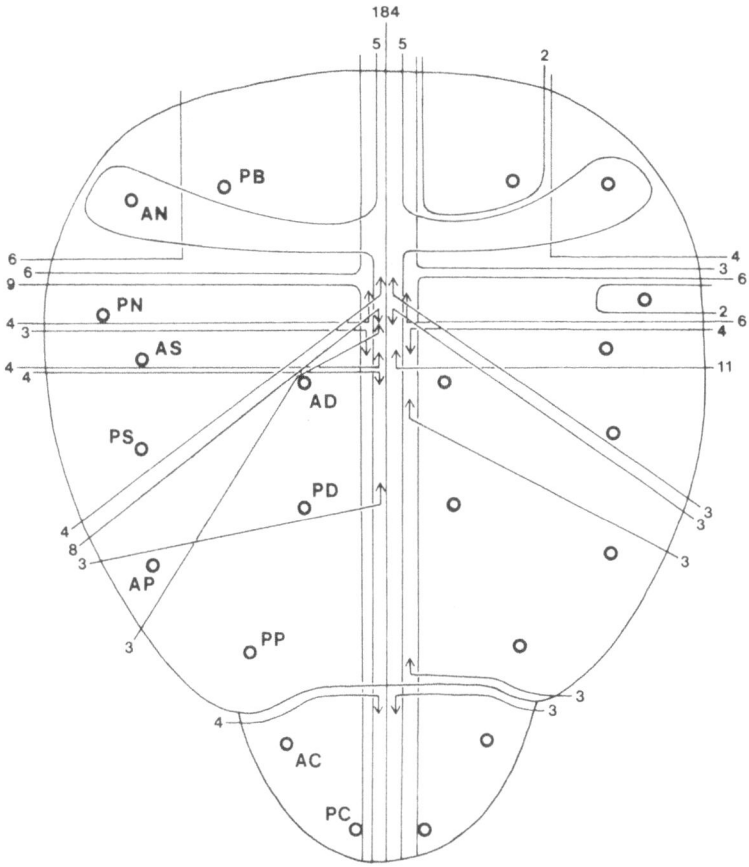

Fig. 5. Dividing lines in the notum of gynandromorphs produced by loss of the $R(1)2$, $In(1)w^{vC}$ chromosome. *Numbers* indicate the frequencies with which the dividing lines were found in 401 mosaic nota (out of 735 gynanders scored). *PB*, presutural bristle; *AN*, *PN*, anterior and posterior notopleural bristle; *AS*, *PS*, anterior and posterior supraalar bristle; *AP*, *PP*, anterior and posterior postalar bristle; *AD*, *PD*, anterior and posterior dorsocentral bristle; *AC*, *PC*, anterior and posterior scutellar bristle

particular dividing line in single individuals is not decisive for these conclusions. Certainly the exact dividing line observed in an individual is fixed late in development when for example single imaginal cells become determined to differentiate defined bristles. This also means that *different apparent dividing lines in the adult can result from identical dividing lines in the blastoderm.*

B. The Two Genotypically Different Clones Have Equal Developmental Parameters

With all methods used so far for the production of gynandromorphs, the time of clone initiation is beyond the control of the experimenter. Considering the sample of gynandromorphs in Figure 4 one could assume that the time of clone initiation, e.g., by loss of the unstable ring-X chromosome, is variable. The fractions of the areas of either genotype vary to a great extent. However, it was found by several authors (e.g., Sturtevant, 1929, for *claret* of *Drosophila simulans*; Hall et al., 1976, for the ring chromosomes $R(1)2$, $In(1)w^{vC}$, and $R(1)5A$) that the frequency distributions of proportions of cuticle surface that are XO in genotype have a maximum at 0.5, but with wide deviations to both sides.

When single landmarks rather than entire gynanders are classified, the frequency of a landmark being XO in the whole set of gynanders will be about 0.5. Since the probabilities of different structures being monosomic (XO) are comparable, they give an indication of the time of initiation of clones. This led to the hypothesis that, for example, elimination of the ring-X chromosomes takes place during the first cleavage division (on possible mechanisms of chromosome elimination see Hall et al., 1976). Equal frequencies of appearance in any adult structure for both clone types suggest equal viability and scorability for both genotypes. However, in most studies the probabilities for structures to be of mutant (male) type are somewhat lower (Garcia-Bellido and Merriam, 1969a; Ripoll, 1972; Janning, 1974b, 1976; Kankel and Hall, 1976), probably due to the marker alleles used, to small growth rate differences between the two cell types, or to the criteria for selecting gynandromorphs (e.g., in larval gynanders, Janning, 1976). However, it cannot be excluded with certainty that in single individuals initiation of mutant clones occurs at later developmental stages or repeatedly during several cleavage mitoses. Ripoll (1972) found in ring/rod-X-chromosome flies, which were classified as nongynanders, some small XO-patches with a single bristle or very few trichomes. This suggests a second but very late stage of ring-X loss.

For *claret* gynandromorphs a mean probability of 0.45 that structures are XO was found by Hall et al. (1976) and in *pal* gynanders this value ranges from 0.3 to 0.4 (Baker, 1975; Kankel and Hall, 1976), whereas in *mit* gynanders this probability is 0.1 (Gelbart, 1974), suggesting a mean time of chromosome loss at the third or fourth cleavage mitosis.

Under the assumption that mutant clones in gynandromorphs arise from chromosome loss in the very early cleavage stages, the variable proportions of mutant patches in adult gynanders (Fig. 4), are easily explained, since only a fraction of cleavage or blastoderm nuclei is represented by clonal descendants in the adult cuticle.

C. At an Early Developmental Stage the Position of a Cell is Correlated with Its Developmental Fate

The gynandromorph mapping procedure assumes that there is a developmental stage in which the fate of a cell is decided largely by its location. This does not necessarily require that cells in this stage are determined to form a specific structure. Arguments that it is most likely the blastoderm stage will be given below.

V. Calculation of Distances

The distances between landmarks on a fate map are proportional to the frequencies of genotypic separation. The distances are given in *sturt units* as proposed by Hotta and Benzer (1972) in memory of Alfred H. Sturtevant. One sturt is equivalent to a probability of 1% that, among all gynanders scored, the two structures in question will be of different genotypes. The method to calculate sturt distances is shown in Table 1. Some of the landmarks used in gynandromorph studies do not originate from a single cell and therefore may be mosaic for both genotypes. The division of the numbers of mosaic landmarks in equal parts to either genotype does not depend on the probability p for landmarks being of one of the two genotypes.

The application of the method to calculate distances [formula (1) in Table 1] is limited to mosaics with p values of 0.5. In mosaics made by mitotic recombination or by chromosomes that tend to be lost at later stages of development, the mutant patches may be much smaller, i.e., the p value decreases. For this reason Gelbart (1974) suggested the *sturtoid unit* [formula (2) in Table 1]. This unit normalizes the distances observed to the probability that any landmark is of mutant type. In other words, the calculation of sturtoid distances does not depend on a certain p value.

In both methods the theoretical maximum distance is 100 sturts or sturtoids, respectively. The empirical maxima do not reach these values but maxima from 61 sturts (Garcia-Bellido and Merriam, 1969a) to 47 sturts (Janning, 1976) and from 96 sturtoids (Gelbart, 1974) to 58 sturtoids (Kankel and Hall, 1976) have been found. In the sturt calculation a distance of 100 sturts between two landmarks means that they will be genotypically separated in every gynandromorph.

Table 1. Calculation of distances (D). a' to i' are the observed fractions of gynandromorphs of that combination. The symbols ♀ and ♂ are used for simplification, since most studies use mosaics of this type. The application of the method does not depend on gynanders but on genetic mosaics. i' is omitted from the calculation because this fraction is almost always only a very few percent. M = mosaic

Structure 2	Structure 1		
	♀	♂	M
♀	a'	b'	e'
♂	c'	d'	f'
M	g'	h'	i'

$\rightarrow$

Structure 2	Structure 1	
	♀	♂
♀	$a' + \dfrac{e'}{2} + \dfrac{g'}{2} = a$	$b' + \dfrac{e'}{2} + \dfrac{h'}{2} = b$
♂	$c' + \dfrac{f'}{2} + \dfrac{g'}{2} = c$	$d' + \dfrac{f'}{2} + \dfrac{h'}{2} = d$

(1) $D = \dfrac{b+c}{a+b+c+d} \times 100$ [sturt] (Hotta and Benzer, 1972)

(2) $D = \dfrac{b+c}{(b+d)+(c+d)} \times 100$ [sturtoid] (Gelbart, 1974)

Since this is not the case, because the dividing lines are convoluted and may pass an even number of times between the precursor cells of the two landmarks, the maximum of 100 sturts cannot be found (resembling the implications of even numbers of crossovers in genetic mapping).

In the sturtoid calculation, the conditions for the maximum distance of 100 sturtoids are quite different. Although the quotient of formula (2) (Table 1) must also equal 1, such a value would mean that there should be no combinations of the two landmarks in which both are of male genotype. This condition depends heavily on the value of p (for the male genotype). With decreasing values of p the maximum size of mutant clones also decreases. Two landmarks, which are far apart in the blastoderm, will never both be of the mutant genotype if p is low enough, and if initiation of clones takes place only once during development. In fact, Gelbart (1974) found a maximum distance of 96 sturtoids in his study on *mit* gynandromorphs. In these mosaics p is 0.1, corresponding to a clone initiation at the third or fourth cleavage division.

VI. History of Fate Maps

Soon after the gynandromorph fate map of *Drosophila simulans* was established by Garcia-Bellido and Merriam (1969a), fate mapping was extended to *Drosophila melanogaster* (Garcia-Bellido and Merriam, 1969b; Hotta and Benzer, 1972, 1973; Ripoll, 1972; Bryant and Zornetzer, 1973; Falk et al., 1973; Gelbart, 1974; Janning, 1974b, 1975, 1976; Baker, 1975; Garcia-Bellido and Ferrús, 1975; Nissani, 1975, 1977; Flanagan, 1976, 1977; Gehring et al., 1976; Hall et al., 1976; Kankel and Hall, 1976; Wieschaus and Gehring, 1976b; Portin, 1977). Not only in *Drosophila* but also in two other species, where gynandromorphs are available, fate maps were constructed: in the honey bee *Apis mellifica* (Milne, 1976), and the wasp *Habrobracon juglandis* (Petters, 1976, 1977). Gynandromorphs have also been used to localize behavioral foci (see Hall, this volume).

In Figure 6 the *D. melanogaster* fate map of imaginal structures shows the relative positions of the cuticular and of some internal organs. As was found by all authors, calculated map distances for bilaterally symmetric landmarks give the same results on the left and right sides.

Since it was shown that aldehyde oxidase (AO) activity is cell-autonomous, *maroonlike (mal)* could be used as a marker gene (Janning, 1972). Flies of the *mal* genotype lack AO, xanthine dehydrogenase (XDH), and pyridoxal oxidase (PO) activities. This is essential for mapping the alimentary system. The inner genitalia, gonads, and Malpighian tubules could be mapped in the same fashion, since they also exhibit AO activity. They can, however, in addition be mapped by their morphology and/or by the use of pigmentation genes such as *white* or *chocolate*. In mapping the gonads, there are some difficulties, since they are composed of germ cells and mesoderm. Gehring et al. (1976) analyzed the two components separately in normal and "transformed" (pure male) gynandromorphs and found indeed that the germ cells mapped as the most posterior structure (as was found

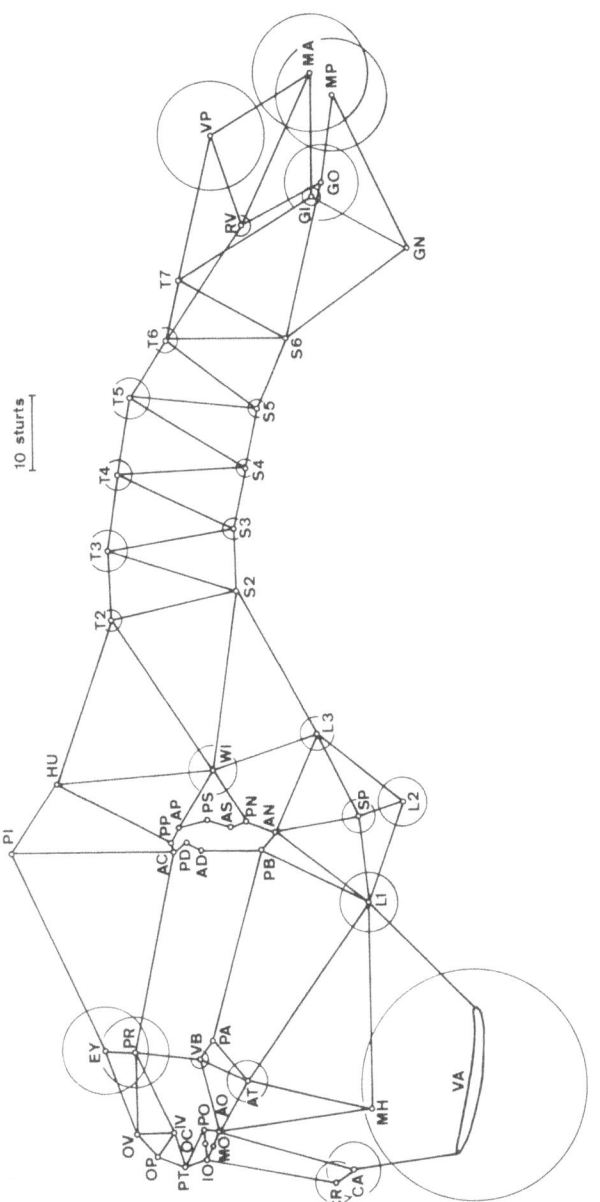

Fig. 6. Gynandromorph fate map of presumptive adult structures in *Drosophila melanogaster*. XX//XO gynanders were produced by loss of the $R(1)2, In(1)w^{vC}$ chromosome. The diameters of the circles are proportional to frequencies of mosaicism within the corresponding landmarks. *AO, MO, PO,* anterior, middle, and posterior orbital bristle; *AT,* antenna; *CA,* cardia; *CR,* crop; *EY,* eye; *GN,* gonads; *GO, GI,* outer and inner genitalia; *HU,* humeral bristles; *IO,* interocellar bristles; *IV, OV,* inner and outer vertical bristle; *L1, 2, 3,* legs; *MA, MP,* anterior and posterior Malpighian tubules; *MH,* larval mouth hooks; *OC,* ocellar bristle; *OP,* occipital bristles; *PA,* maxillary palpus; *PI,* proboscis; *PR,* postorbital bristles; *PT,* postvertical bristle; *RV,* rectal valve; *S2-6,* sternites; *T2-7,* tergites; *VA, VP,* anterior and posterior ventriculus; *VB,* vibrissae; *WI,* wing. Abbreviations for notum bristles see Figure 5

also by Nissani, 1977) and the gonadal mesoderm in a position more ventrally and anteriorly than GN in Figure 6.

The representation of map sites as points is accurate only for structures that derive from a single cell, e.g., bristles, which are ideal landmarks (but see below p. 24). Within larger landmarks the frequency of mosaicism can become so high that only tentative regions can be indicated. For example, the anterior ventriculus (VA in Fig. 6) was mosaic in 61% of all gynanders scored (Janning, 1974b) and therefore was mapped as a stripe in the anterior ventral region. In this case it cannot be excluded that the high frequency of mosaicism results from two separated anlagen on the left and right sides, which fuse later in development.

The diameters of the circles surrounding some map sites are proportional to the frequencies of mosaicism within the corresponding landmarks. These frequencies tell us something about *size and shape of a primordium*. It was pointed out first by Hotta and Benzer (1972) and was further developed by Wieschaus (1974), that frequencies of mosaicism(f) within landmarks are proportional to the diameters (d) of the corresponding primordia rather than to their areas; i.e., $f = k \cdot d$. Obviously k equals 1 if the primordium is rod-shaped, since then the frequency of mosaicism is identical to the distance in map units. For circular primordia k was found to be $\pi/2$, and in general, the frequency of mosaicism is one-half the circumference of the primordium (Wieschaus, 1974). Moreover, Wieschaus and Gehring (1976b) found experimental evidence for these theoretical considerations. In four thoracic primordia (the three legs and the wing) they measured the greatest intradisc distances between landmark bristles, and calculated the expected frequencies of mosaicism if the primordia were rod-shaped, circular, or elliptical. For all four primordia they found the best correspondence between expected and calculated frequencies if a circular shape is assumed. It can further be concluded that if a structure is mosaic twice as often as another, it arises from four times the primordial area, i.e., the number of precursor cells (Wieschaus, 1974).

A further progress in mapping imaginal structures was achieved by Kankel and Hall (1976). They made use of three enzymes that allowed them to mark various internal noncuticular tissues: acid phosphatase (*Acph-1*, 3-101.1), glucose-6-phosphate dehydrogenase (G6PD; genetic locus *Zw*, 1-63), and 6-phosphogluconate dehydrogenase (6PGD; *Pgd*, 1-0.6). With the dehydrogenases oesophagus, stomodeal valve, and ventriculus as well as the abdominal oenocytes could be mapped. The map sites of the latter were found to be dorsal of the tergites. Since the *Acph-1* locus is on the third chromosome, it was linked to the X chromosome by a translocation. The fate map derived by Kankel and Hall (1976) with the acid phosphatase system and the use of the *paternal loss* (*pal*; Baker, 1975) mutation for early chromosome loss, is shown in Figure 7. Besides cuticular landmarks, the central nervous system, alimentary structures (oesophagus, cardia, ventriculus) and the salivary glands were mapped. First of all, the remarkable similarity of the relative positions of cuticular landmark sites between this map and the map in Figure 6 should be pointed out, since the techniques for producing mosaics were very different (unstable ring-X chromosome and *pal*). This seems to be a general phenomenon. All fate maps derived so far are nearly identical, independent of the methods used for generating mosaic flies [*claret* of *Drosophila simulans* (Garcia-Bellido and Merriam, 1969a), $R(1)2$, $In(1)w^{vC}$ (Hotta and Benzer, 1972), *mit*

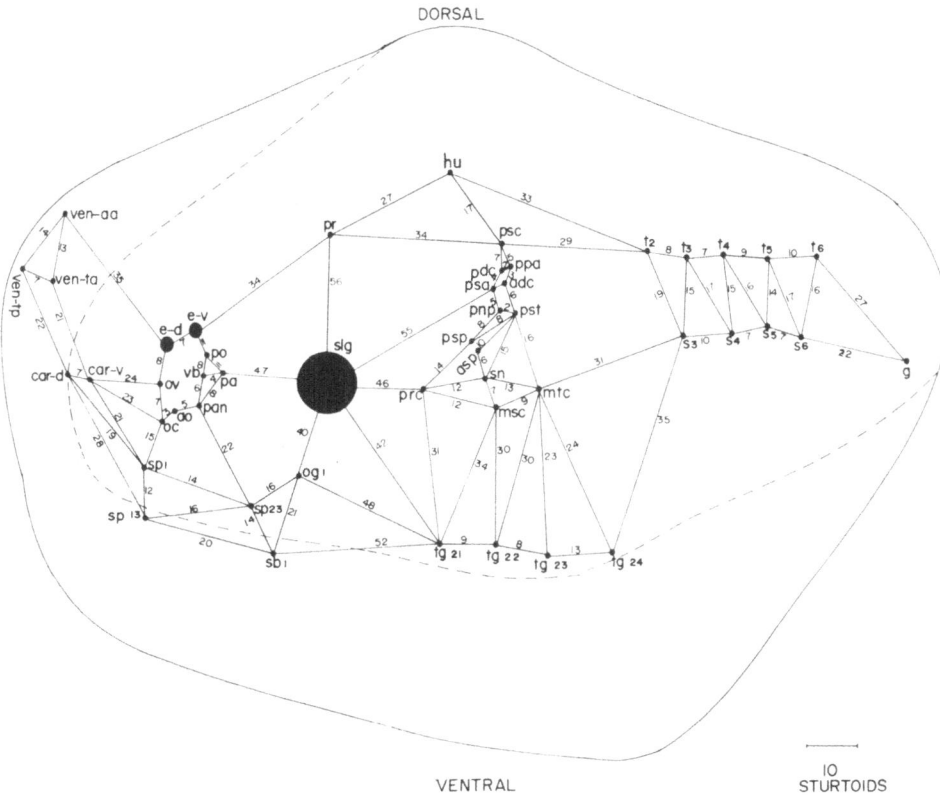

Fig. 7. Gynandromorph fate map of presumptive adult structures constructed by Kankel and Hall (1976; Fig. 10), who used the mutant *paternal loss (pal)* for chromosome loss. *adc, pdc*, anterior and posterior dorsocentral bristle; *ao, po*, anterior and posterior orbital bristle; *asp, psp*, anterior and posterior sternpleural bristle; *car-d, car-v*, dorsal and ventral cardia; *e-d, e-v*, dorsal and ventral eye facet; *g*, external genitalia; *hu*, humeral bristles; *oc*, ocellar bristle; *og*, optic ganglia cortex sites; *ov*, outer vertical bristle; *pa*, maxillary palpus; *pan*, second segment of antenna; *pnp*, posterior notopleural bristle; *ppa*, posterior postalar bristle; *pr*, proboscis; *prc, msc, mtc*, pro-, meso- and metathoracic coxal seta; *psa*, posterior supraalar bristle; *psc*, posterior scutellar bristle; *pst*, presutural bristle; *s3-s6*, sternites; *slg*, salivary gland; *sn*, sternal bristle; *sp, sb*, supra- and suboesophageal ganglion cortex sites; *t2-t6*, tergites; *tg*, thoracic ganglia cortex sites; *vb*, vibrissae; *ven-aa*, anterior abdominal ventriculus; *ven-ta, ven-tp*, anterior and posterior thoracic ventriculus

(Gelbart, 1974), *pal* (Baker, 1975)], and also independent of the sexual constitution of the cells, since pure male (XY//XO), pure female (XXY//XXO), and gynandromorph (XX//XO) mosaics gave the same results (Baker, 1975). The similarity of all fate maps lends support to the validity of the mapping method itself.

As can be seen from Figure 7, the brain ganglia map ventral to the head cuticle region and the thoracic ganglion ventral to the thoracic imaginal discs. The map sites of the alimentary structures are apparently different from the locations in the w^{vC} map (Fig. 6). In the dehydrogenases systems as in the acid phosphatase system, all these structures map anterior and more dorsal to the head, whereas in the aldehyde oxidase system the anterior alimentary structures map ventral to the head cuticle and the posterior ventriculus in a posterior abdominal region. The reasons for these differences remain unclear.

The fate map of Figure 7 is surrounded by the tentative midline of the embryo. The map distances to the midline are calculated by halving the distances between identical structures on the left and right body sides. If this is done for cuticular structures only, the dashed line (Fig. 7) would exclude some of the internal structures due to partly large distances between left and right. Including the bilaterally symmetric ganglia as landmarks the solid line can be drawn. Obvious difficulties with handling large distances can be overcome in part by use of a "mapping function" (see Sec. VII).

The fate maps presented so far showed map sites of adult structures only, with the exception of the larval mouth hooks in Figure 6 (MH). Mapping of larval structures in combination with adult organs using the same gynander individuals is possible by a technique first suggested by Hotta and Benzer (1973). They collected larvae in the third instar, scored the phenotypes of mouth hooks and abdominal setae (marker allele *yellow*), Malpighian tubules *(chocolate)*, and gonads (morphology), and reared them separately until emergence. Map sites of the larval abdominal segments were found to be ventral to the sternites, gonads and Malpighian tubules mapped in the most posterior regions, and the mouth hooks ventral to the imaginal head cuticle.

Since it was shown that most of the internal larval organs exhibit AO activity, a more detailed analysis of larval structures was possible (Janning, 1973, 1975, 1976). The conspicuous feature of the fate map of larval structures (Fig. 8) is the great similarity with the imaginal fate maps (Figs. 2, 6, 7). This is not surprising because it should make no difference in which developmental stages gynandromorphs are scored. Besides larval organs that will differentiate into imaginal structures such as the imaginal discs or the Malpighian tubules, typical larval structures could be mapped, e.g., ring and salivary glands, alimentary structures and the fat body. Since in no case ideal landmarks, i.e., single cells, could be scored, the map sites are tentative only. The larval hypoderm, which also exhibits AO activity, could not be examined exactly, because of interference with the adhering musculature. Most recently Szabad et al. (1976) developed a special technique, which makes scoring of larval hypoderm, histoblasts, and larval oenocytes possible.

Similarities between fate maps of imaginal and larval structures are emphasized by fate maps that only contain the adult structures on the one hand and their corresponding primordia in the larva on the other. These maps are obviously nearly identical (Fig. 9), though difficulties in constructing them arose as a consequence of large map distances (for details see Janning, 1976). Certainly, it would be most useful to have a composite fate map with sites of all structures of the different developmental stages. A first attempt in this direction was made by superimposing the fate maps of imaginal and larval structures (Figs. 6 and 8). Since the adult abdominal segments were mapped rather exactly (tergites and sternites), whereas mapping of the larval hypoderm was not possible, superimposition of the larval map on the adult map was performed with anterior (head,

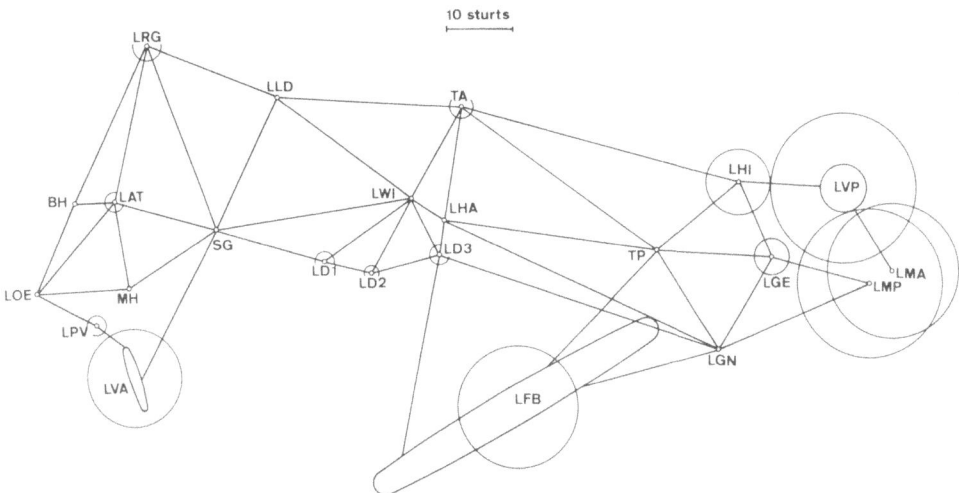

Fig. 8. Gynandromorph fate map of presumptive larval structures. Gynanders were produced by loss of the $R(1)2$, $In(1)w^{vC}$ chromosome. The diameters of the circles are proportional to frequencies of mosaicism within the corresponding landmarks. BH, brain hemisphere. Imaginal discs: LAT, antennal; $LD1$, 2, 3, leg; LGE, genital; LHA, haltere; LLD, labial; LWI, wing; LFB, fat body; LGN, gonad; LHI, hindgut; LMA, LMP, anterior and posterior Malpighian tubules; LOE, oesophagus; LPV, proventriculus; LRG, ring gland; LVA, LVP, anterior and posterior ventriculus; MH, mouth hooks; SG, salivary gland; TA, TP, anterior and posterior spiracles

thorax) and posterior-most regions separately. The result is given in Figure 10, which shows good agreement between the sites of larval and adult structures. This was expected for imaginal discs and the corresponding cuticular structures. The map sites of larval and adult alimentary structures lie close together, which suggests that they develop from adjacent regions of the early embryo. This seems reasonable, since the larval intestine contains nests and rings of imaginal cells that will differentiate into the adult intestine (Bodenstein, 1950). Moreover, the composite map indicates the regions of the larval fat body, the ring and salivary glands, and the larval brain. The sites of the two latter structures agree well with the regions of adult head ganglion and salivary glands found by Kankel and Hall (1976; see Fig. 7).

Many of the structures that were mapped in other studies are missing in the composite map. This is so because it is difficult to superimpose fate maps that were constructed with different methods of mosaic production and/or different marker systems. It is mainly the marker system used that may change map distances in different studies because it determines the selection of gynandromorphs. If, for example, only marker alleles for cuticular structures of the adult were used, all gynandromorphs that exhibit only internal mosaicism were overlooked. Additional mosaics can be found by application of enzyme marker systems, although these are restricted to specific internal structures depending on the distribution of the enzyme's activity in wild-type individuals. Some gynandro-

Imago

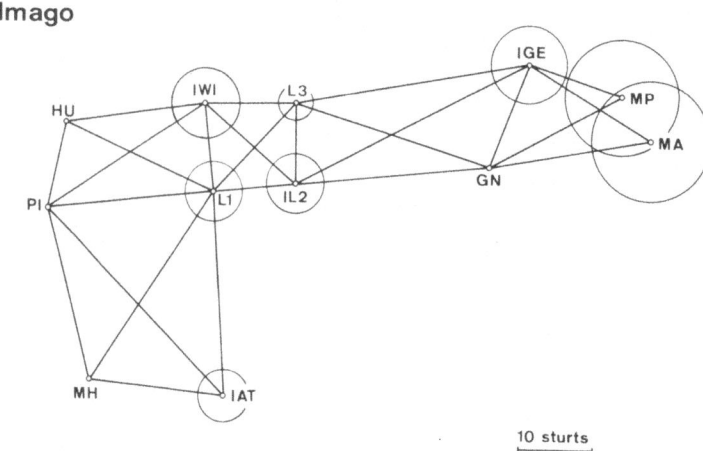

10 sturts

Larva

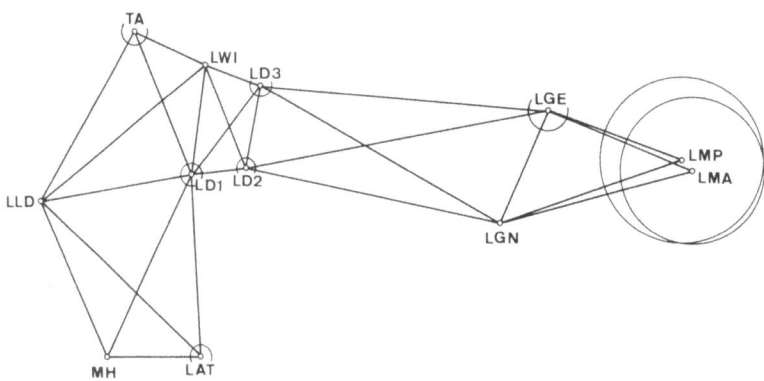

Fig. 9. Fate maps of imago and larva including only structures which are either identical in both developmental stages, or where adult structures are represented by their primordia in the larva. The diameters of the circles are proportional to frequencies of mosaicism within the corresponding landmark. *IAT, IGE, IL2, IWI*, adult derivatives of the antennal, genital, second leg and wing imaginal discs, respectively. Other abbreviations as in Figures 6 and 8

morphs remain "hidden," but they may account for only a few percent (e.g., Hotta and Benzer, 1973; Janning, 1974b; Kankel and Hall, 1976) and will not significantly change the relationships among structures on the fate map.

A more detailed mapping of large landmarks is conceivable (Ripoll, 1972; Wieschaus and Gehring, 1976b). However, with this kind of analysis the intrinsic limitations of gynandromorph fate mapping will become apparent. It was pointed out by Hotta and Benzer (1972) that map distances measured between two closely related points will be expanded if a site on the map represents the center of a distribution function for the probability that cells in this region will eventually give rise to a particular structure. For the visualization of this "map expansion,"

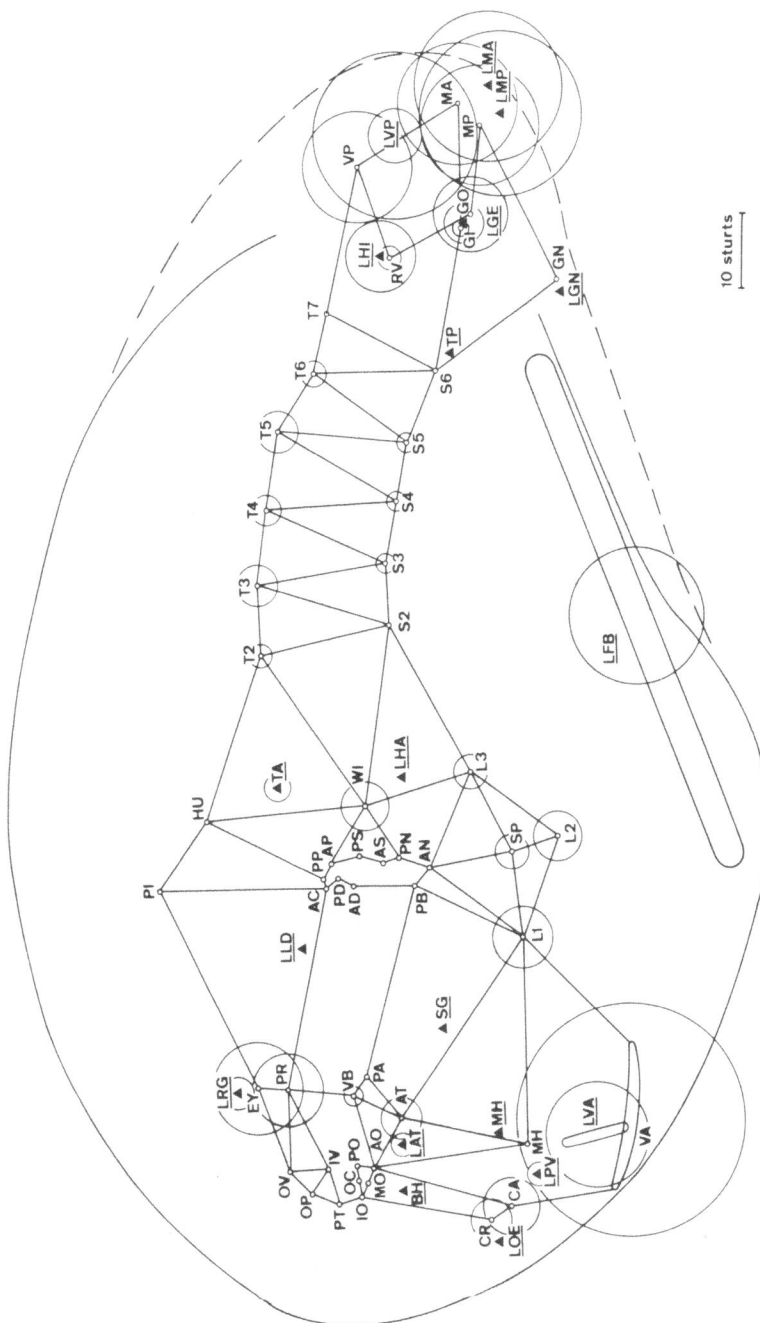

Fig. 10. Gynandromorph fate map with sites of larval and adult structures. Most of the larval sites are indicated as *shaded triangles* (▲), and all abbreviations for larval structures are *underlined*. For the construction of the embryo's midline see Section VII. Abbreviations as in Figures 6 and 8

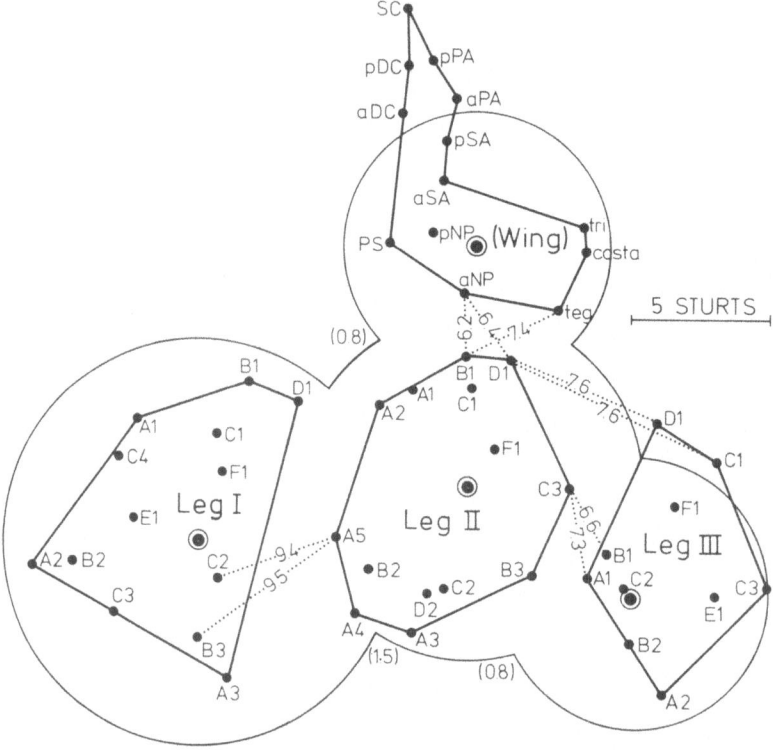

Fig. 11. Detailed gynandromorph fate map of four thoracic discs established by Wieschaus and Gehring (1976b). The diameters of the overlapping circles are calculated from the frequencies of mosaicism $(d = 2f/\pi)$. ⊙ disc centers; ● individual landmark bristles

Wieschaus and Gehring (1976b) chose the imaginal discs as an example and suggested a model in which (1) determination to be a specific disc precedes determination to be a specific bristle (for review see Nöthiger, 1972), and (2) where the intervening growth period is accompanied by some random cell mixing or variability in cell proliferation rates, i.e., any phenomenon that would elongate or contort the genotypic dividing line within a disc before final bristle determination occurs. This must lead to a great variability of dividing lines which is in fact observed (see Sec. IV). Nevertheless "fine-structure fate mapping" is possible and yields significant results (Fig. 11). Wieschaus and Gehring (1976b) found that the primordia of the four thoracic discs examined lie very close to each other and are separated by few if any nondisc cells. They further concluded that the proximity of disc primordia suggests that the different discs might share common precursor cells in the early embryo. This was confirmed by the observation that clones induced by mitotic recombination at the blastoderm stage can overlap two different discs (Wieschaus and Gehring, 1976a), or within the anterior or posterior compartments of a segment (Steiner, 1976; Lawrence and Morata, 1977; Garcia-Bellido and Ripoll, this volume).

VII. Mapping Function

In fate mapping there are some problems that are of interest from a more mathematical point of view.

It is generally observed that (1) the theoretical maximum distance of 100 sturt units is not found experimentally, (2) that map distances are not additive over large distances, and (3) that fate maps can measure up to 150 sturt units in length (e.g., Fig. 10). These findings resemble strongly the conditions of meiotic gene mapping, though fate mapping is two-dimensional. Therefore we looked at the relationship between observed distances and actual fate map distances obtained by "adding" shorter distances. This was done first with the data on which the fate map of imaginal structures (Fig. 6) are based. Plotting of observed distances against the corresponding fate map distances gave the set of points shown in Figure 12. As expected, there is no overall linear relationship. Linearity up to 10 or 15 sturts may be due (1) to the mapping procedure as such, because one principle of mapping is the preference for short distances, and (2) to only few convolutions of the dividing lines within short distances. With long distances a given observed distance corresponds to a wide range of fate map distances and vice versa. However, small and large observed distances correspond to short and long map distances, respectively.

With the general equations given in Figure 12, a straight line and curves were fitted to the set of points. The best adaptation was possible with the exponential curve.

The same procedure was done with the larval map (Fig. 8) and with the data published by Garcia-Bellido and Merriam (1969a; see Fig. 2), Gelbart (1974), and Kankel and Hall (1976; dehydrogenase system). In all cases the experimental data fitted best with an exponential curve (Fig. 13)[2]. The resemblance between these mapping functions and those of meiotic gene mapping is evident. Neglecting interference, Haldane (1919) found that the relation between recombination (p) and map distance (d) can be described by the exponential function $p = 0.5$ $(1-e^{-2d})$. In experiments with *Drosophila* (and therefore including interference), Morgan et al. (1935; cited in Perkins, 1962) found a similar curve with the asymptotic value of $p = 0.5$. None of the mapping functions of Figure 13 has an asymptotic value, because there are no theoretical considerations that would allow us to deduce a priori a mapping function for gynandromorphs except that the maximum value of observable distances should lie between 50 and 100 sturts.

The mapping functions show, as they do in meiotic gene mapping, that fate map distances are not synonymous with observed distances, but that for any given observed distance the corresponding map distance can be estimated. This may help to handle large observed distances (e.g. to map behavioral foci).

Construction of the tentative midline of the embryo by halving the distances between identical structures on the left and right sides is difficult if only cuticular structures were used (Fig. 7). However, when with these distances, corresponding map distances were estimated from the mapping function, a midline results that

---

[2] In his computer program for the construction of fate maps Flanagan (1976, 1977) used a similar mapping function, which is based on theoretical considerations.

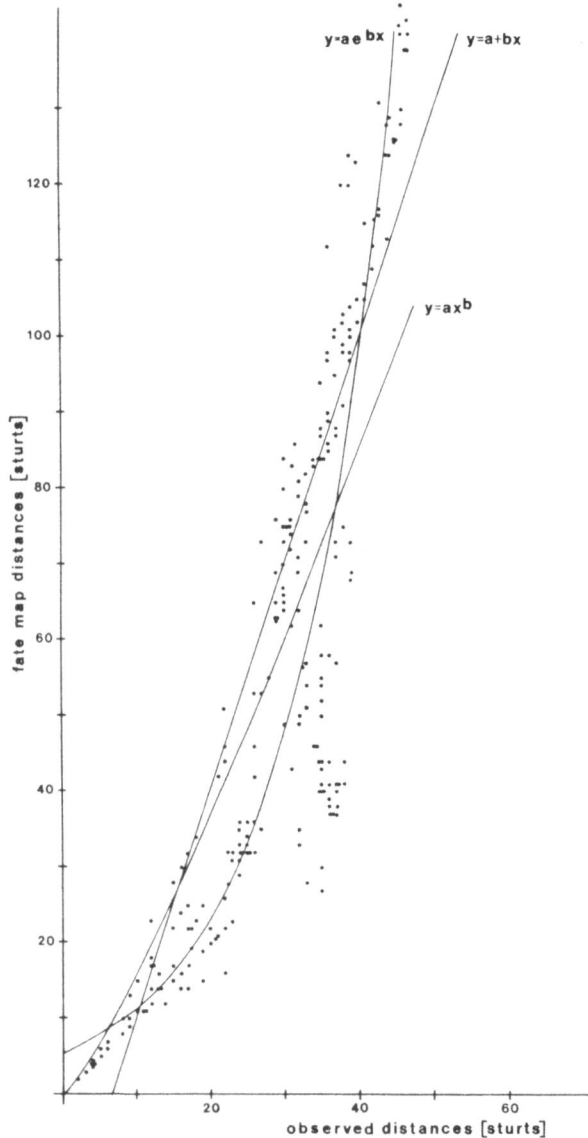

Fig. 12. Mapping function of the gynandromorph fate map of presumptive adult structures (see Fig. 6). The set of points is the result of plotting the observed distances against the corresponding fate map distances. A straight line and curves with the general equations indicated were fitted to the points. $y=a+bx$ ($a=-19.35$, $b=2.94$); $y=ae^{bx}$ ($a=5.53$; $b=0.07$); $y=ax^b$ ($a=0.94$; $b=1.22$)

includes nearly all internal organs. At the posterior pole, where no left–right distances for the internal organs are available, the presumable midline is drawn so that it includes these sites (Fig. 10).

The correspondence between the constructed midline and the real midline of the embryo can be tested in the following way (similar to Garcia-Bellido and

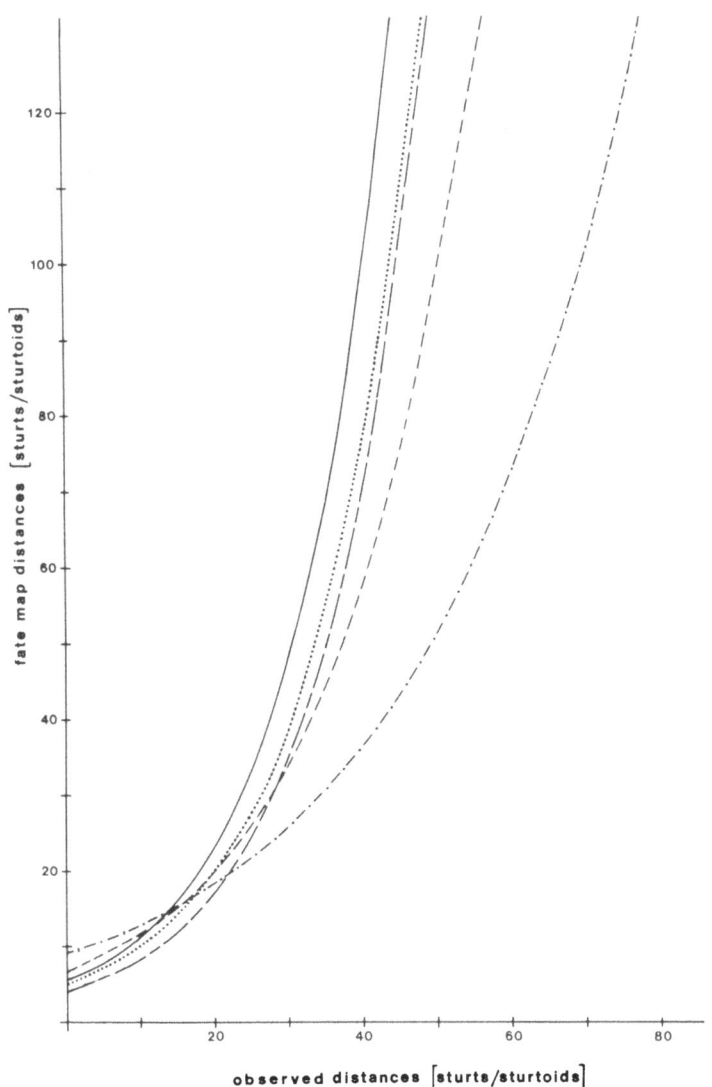

Fig. 13. Mapping functions. The procedure shown in Figure 12 was carried out for four additional gynandromorph fate maps, which all resulted in best fitting with the exponential curve $y = ae^{bx}$. ——— see Figure 12, ⋯⋯⋯ Kankel and Hall (1976; dehydrogenases system; $a = 5.14$; $b = 0.07$), — — — — — Garcia-Bellido and Merriam (1969a; see Figure 4; $a = 4.09$; $b = 0.07$), — — — — — Janning (1976; fate map of larval structures; see Figure 8; $a = 6.83$; $b = 0.05$), — · — · — Gelbart (1974; $a = 9.15$; $b = 0.03$)

Merriam, 1969a). Left–right distances differ widely along the egg axis: the smallest will be found in the anterior and posterior regions, the maximum distances in the dorsal mesothorax, i.e., notum and wing. If we assume (1) that this maximum distance corresponds to the middle of the egg and (2) that the geometry of the egg can be described in a simplified manner as a rotation ellipsoid, then the following calculations can be done.

The left–right distance in the mesothorax is observed to be 38 sturts, which are equivalent to 84 sturts on the map estimated from the mapping function (Fig. 12). Since the major axis of the egg is 500 μm and the minor axis 200 μm, half the circumference in the middle of the egg would be 100 πμm, which corresponds to 84 sturts. Therefore 1 sturt equals approximately 3.7 μm. In the anterior-posterior direction, the length on the egg's surface would be about 160 sturts (= half the circumference of the ellipse). The dimensions of the constructed midline (Fig. 10) with 165 × 100 sturts correspond well to these values. Since the fate map measures about 150 × 70 sturts (Fig. 10), this would mean that the map covers about 94% of the length and 83% of the width of the egg. This result is close to what we would expect because many of the anterior and posterior structures, such as head cuticle and -ganglia and the germ cells, respectively, are mapped, as well as dorsal and ventral organs. The positions of most of the missing structures will be in the "gaps" of the existing fate map rather than outside.

VIII. Gynandromorph Fate Maps and Embryology

Analysis of gynandromorphs leads to the construction of self-consistent fate maps in which the relative positions of progenitor cells are similar to the arrangement of corresponding structures in the adult fly. This indirect evidence for the validity of the mapping method is supported by direct embryologic observations. Poulson (1950) established a fate map of the early embryo that is based on histological data (Fig. 14). Obviously, the embryologic and gynandromorph fate maps are very similar, despite the fact that there are differences in details due to the methods used. On the other hand, comparison of the two types of fate maps may help to clarify some discrepancies in gynandromorph fate mapping. As mentioned above (Sec. VI), Kankel and Hall (1976) mapped all the alimentary structures anterior to the head (Fig. 7), whereas with the AO system the map locations were found to be anterior for anterior and posterior for posterior alimentary structures (Fig. 6), which agrees well with embryologic observations.

Poulson called his map a "tentative diagram of preblastoderm localization." This leads to the question as to which developmental stage is represented by gynandromorph fate maps. It was pointed out in Section IV that the gynandromorph mapping procedure is based upon the assumption that the position of a cell in the developmental stage, which is reflected by the map, largely decides its fate. There are several lines of evidence that this developmental stage is identical with the blastoderm.

The results from transplantation experiments have shown that cleavage *nuclei are totipotent* (Geyer-Duszýnska, 1967; Illmensee, 1968, 1970, 1972; Schubiger and Schneiderman, 1971; Zalokar, 1971, 1973; Okada et al., 1974). However, when the cleavage nuclei reach the cortical cytoplasm and later form the blastoderm it was shown by various experiments that these *cells are restricted* in their potential to form larval or adult structures. Chan and Gehring (1971) tested the developmental capacity of blastoderm cells in cell mixtures of dissociated em-

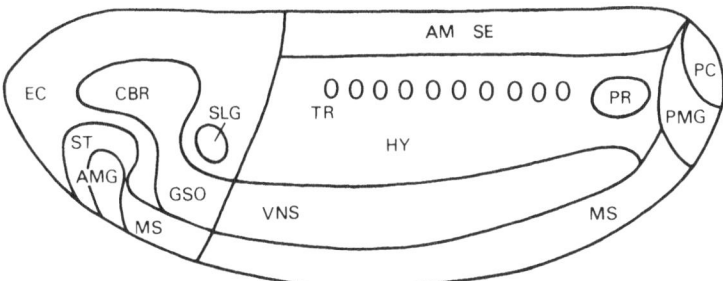

Fig. 14. Poulson's (1950) fate map based on histological observations. (Modified by E. Wieschaus, authorized by D. F. Poulson.) *AM*, amnion; *AMG*, anterior midgut rudiment; *CBR*, cerebrum; *EC*, ectoderm; *GSO*, suboesophageal ganglion; *HY*, hypoderm; *MS*, mesoderm; *PC*, pole cells; *PMG*, posterior midgut rudiment; *PR*, proctodaeum; *SE*, serosa; *SLG*, salivary gland; *ST*, stomodaeum; *TR*, trachea; *VNS*, ventral nervous system

bryos. They found that cells isolated from anterior-half embryos are determined to form head and thoracic structures, whereas cells from posterior-half embryos are determined to form thoracic and abdominal structures. In previous experiments with UV irradiation (Geigy, 1931a, b) and pricking of embryos (Howland and Child, 1935) specific defects in the adult were observed. More recently, Bownes (1975, 1976a), Lohs-Schardin and Sander (1976), and Mertens (1977) damaged defined blastoderm areas by microcautery, with a UV-laser microbeam, or by pricking, respectively. They found a strong correlation between imaginal defects and injured blastoderm sites of presumptive imaginal cells as inferred from gynandromorph fate maps.

For the formation of pole cells it has long been known that the cortical plasm participates in determinative events (e.g., Geigy, 1931a; Hathaway and Selman, 1961; Graziosi and Micali, 1974; Bownes, 1976b). This was most clearly demonstrated by transplantation experiments (Illmensee and Mahowald, 1974, 1976; Illmensee et al., 1976; Mahowald et al., 1976), which showed that posterior polar plasm when transferred to another region of the embryo is capable of inducing the formation of primordial germ cells. Experiments in which the fate of transplanted blastoderm cells was followed, showed that these cells differentiate according to origin (Illmensee et al., 1976; Illmensee, this volume). All these results indicate that blastoderm cells or groups of cells are determined to form specific structures. However, recent experiments demonstrate that at least some blastoderm cells can still give rise to cells from two kinds of imaginal discs within a body segment (Steiner, 1976; Lohs-Schardin and Sander, 1976; Wieschaus and Gehring, 1976a; Lawrence and Morata, 1977).

Direct support that the gynandromorph fate map may describe the blastoderm situation comes from mapping the primordial germ cells (Gehring et al., 1976; Nissani, 1977). Germ cells map as the most posterior element, which coincides with their known location as pole cells in the blastoderm embryo, whereas at later developmental stages the pole cells migrate anteriorly and penetrate into the embryo.

Summarizing all these experimental data we conclude that *the gynandromorph fate map reflects the blastoderm, i.e., the stage of cellularization.* This does not

mean, however, that specific sites (e.g., of bristles) in the fate map can be associated with specific blastoderm cells, since determinative events take place up to
the pupal stage. Therefore, *the fate map reflects more likely the distribution of
probabilities to form specific structures in the cortical cytoplasm of the egg.*

As shown in Section VII, the dimensions of the fate map fit well with the
dimensions of the egg. Recently, it was found by scanning electron microscopy
(Turner and Mahowald, 1976) and by using a special cytological technique (Zalokar and Erk, 1976, 1977) that the blastoderm surface is covered by about
6000 cells. Since the surface of the egg is about $260000\ \mu m^2$ or about 18 700 sturt^2
the diameter of a blastoderm cell is about 7.4 µm or 2.0 sturts[3]. This would also
mean that on the surface the length of about 160 sturts and the width of 84 sturts
correspond to about 80 and 42 cells, respectively.

A similar value of about 2 sturts per cell is found if the following gynandromorph data are used. The wing disc primordium was estimated to contain 47 cells
(Ripoll, 1972) and the diameter of this circular anlage (Wieschaus and Gehring,
1976 b) is about 15 sturts (Garcia-Bellido and Ferrús, 1975). With these figures the
diameter of a cell would be 2.2 sturts. Recently, Nissani and Lipow (1977) developed a new method for estimating the number of blastoderm cells of particular
regions using the frequencies of mosaicism in these regions. If, in their equations,
the number of 3400 blastoderm cells is replaced by 6000 cells (Turner and Mahowald, 1976; Zalokar and Erk, 1976), the number of blastoderm cells in the
mesothoracic disc anlage will be 55 and the diameter of a cell about 2 sturts.
However, discrepancies arise in the estimations of the number of primordial cells
by different methods (see Merriam, this volume). In general, gynandromorph data
yield higher estimates of numbers of primordial cells than mitotic recombination
experiments (cf, e.g., Table 1 in Nöthiger, 1972, with Table 3 in Wieschaus and
Gehring, 1976 a). As was pointed out by Wieschaus and Gehring (1976 a) gynandromorph analyses tend to overestimate, whereas mitotic recombination tends to
underestimate the number of primordial cells. However, mitotic recombination
data seem to be more likely in the proximity of the true values.

On the other hand, good estimation for the size of a primordium comes from
the frequency of mosaicism in gynandromorphs because this frequency is independent of later determinative events and/or morphogenetic processes. If we use
the data of Wieschaus and Gehring (1976 a, b) the forelegs in gynanders are
mosaic in 22%; i.e., the diameter of the circular primordium is 14 sturts. By
mitotic recombination induced at the blastoderm stage they estimated the number of primordial cells to be 8 per foreleg anlage. With these figures the diameter
of a blastoderm cell can be calculated to be 5 sturts. This value is very different
from the 2 sturts per cell calculated above: It would mean that on the blastoderm
surface the length of about 80 cells and the width of about 42 cells would correspond to 400 and 210 sturts, respectively. However, a length of 160 sturts and a
width of 84 sturts seem more likely because these values are consistent with all

---

[3] The calculation of the egg's surface S is based on the following formula for the surface
of a rotation ellipsoid with radii a and b: $S = 2\pi b \left(b + a \dfrac{\arcsin \varepsilon}{\varepsilon} \right)$; $\varepsilon = \dfrac{1}{a} \sqrt{a^2 - b^2}$.

other fate mapping data. A cell diameter of 5 sturts would imply a series of assumptions that are not yet understandable. For example, it seems most unlikely that a small anlage of eight cells (i.e., about 0.3% of half the surface) leads to a diameter of 14 sturts, which is about one-fourth of the experimentally found maximum sturt value. In general, it is believed that linearity between observed and fate map distances up to about 15 sturts is due to only a few convolutions. If, however, an area of eight cells causes a frequency of mosaicism of 22% the dividing lines must be convoluted to a great extent.

We cannot give a solution of the problem for the moment, but we suppose that a certain primordium, which leads to a certain frequency of mosaicism, must not necessarily contain only the primordial cells of the chosen structure from which the mosaic frequency was calculated, but may also contain other (e.g., larval) cells in between.

IX. Perspectives

We think that the considerations about embryology have shown that analyses of gynandromorphs can help to expand our knowledge of developmental processes even if many questions still remain open. To complete the existing fate maps additional marker systems (e.g., for musculature) are required. Most desirable would be a system that could demonstrate the distribution of genotypically different nuclei and cells in the cleavage and blastoderm stages. This would clear up for example all indirect suggestions on the extent of convolutions of the dividing lines. Here, markers other than enzymes should probably be found, e.g., methods that make use of the different chromosomal compositions in gynander cells.

The technique of fate mapping should also be applicable to mitotic recombination data (Frei and Soure-Leuthold, 1976). Since then not only fate maps of the blastoderm but of later developmental stages could be established, it seems reasonable that these maps could elucidate to some extent morphogenetic processes.

Acknowledgements. I wish to thank Drs. Rolf Nöthiger and H.J.Becker for many helpful suggestions and linguistic aid, Bernd Eickenscheidt for continuous help with the computer analysis of the gynander data, and my wife Renate for her great patience. Own work was generously supported by the Deutsche Forschungsgemeinschaft.

Note Added in Proof : Since this paper was written, the mapping function (see Sec. VII) was improved. The formula now used is

$$f(x) = a \left(e^{\frac{1}{a}x} - 1 \right) .$$

In a computer program (established by Jürgen Pfreundt) for the construction of fate maps on the basis of this mapping function, and with the data of adult gynandromorphs (Fig. 6), *a* was found to be about 30.

References

Baker, B. S.: Paternal loss *(pal)*: A meiotic mutant in *Drosophila melanogaster* causing loss of paternal chromosomes. Genetics **80**, 267—296 (1975)

Bodenstein, D.: The postembryonic development of *Drosophila*. In: Demerec, M. (Ed.): Biology of *Drosophila*, pp. 275—367. New York: Wiley 1950

Bownes, M.: Adult deficiencies and duplications of head and thoracic structures resulting from microcautery of blastoderm stage *Drosophila* embryos. J. Embryol. Exp. Morph. **34**, 33—54 (1975)

Bownes, M.: Larval and adult abdominal defects resulting from microcautery of blastoderm staged *Drosophila* embryos. J. Exp. Zool. **195**, 369—392 (1976a)

Bownes, M.: Defective development after puncturing the periplasma of nuclear multiplication stage *Drosophila* embryos. Dev. Biol. **51**, 146—151 (1976b)

Bryant, P. J., Zornetzer, M.: Mosaic analysis of lethal mutations in *Drosophila*. Genetics **75**, 623—637 (1973)

Catcheside, D. G., Lea, D. E.: Dominant lethals and chromosome breaks in ring X-chromosomes of *Drosophila melanogaster*. J. Genet. **47**, 25—40 (1945)

Chan, L.-N., Gehring, W.: Determination of blastoderm cells in *Drosophila melanogaster*. Proc. Natl. Acad. Sci. USA **68**, 2217—2221 (1971)

Falk, R., Orevi, N., Menzl, B.: A fate map of larval organs of *Drosophila* and preblastoderm determination. Nature (New Biol.) **246**, 19—20 (1973)

Flanagan, J. R.: A computer program automating construction of fate maps of *Drosophila*. Dev. Biol. **53**, 142—146 (1976)

Flanagan, J. R.: A method for fate mapping the foci of lethal and behavioral mutants in *Drosophila melanogaster*. Genetics **85**, 587—607 (1977)

Frei, H., Soure-Leuthold, A.: Development of clones within the tergum anlage of *Drosophila hydei* (Abstract). 5th Eur. Drosophila Res. Conf. (1976)

Garcia-Bellido, A., Ferrús, A.: Gynandromorph fate map of the wing-disk compartments in *Drosophila melanogaster*. Wilhelm Roux's Archives. **178**, 337—340 (1975)

Garcia-Bellido, A., Merriam, J. R.: Cell lineage of the imaginal discs in *Drosophila* gynandromorphs. J. Exp. Zool. **170**, 61—76 (1969a)

Garcia-Bellido, A., Merriam, J. R.: A preliminary morphogenetic map of the wing disc. Dros. Inf. Serv. **44**, 65—66 (1969b)

Gehring, W. J., Wieschaus, E., Holliger, M.: The use of "normal" and "transformed" gynandromorphs in mapping the primordial germ cells and the gonadal mesoderm in *Drosophila*. J. Embryol. Exp. Morph. **35**, 607—616 (1976)

Geigy, R.: Action de l'ultra-violet sur le pôle germinal dans l'oeuf des *Drosophila melanogaster* (castration et mutabilité). Rev. Suisse Zool. **38**, 187—288 (1931a)

Geigy, R.: Erzeugung rein imaginaler Defekte durch ultraviolette Eibestrahlung bei *Drosophila melanogaster*. Wilhelm Roux' Archiv **125**, 406—447 (1931b)

Gelbart, W. M.: A new mutant controlling mitotic chromosome disjunction in *Drosophila melanogaster*. Genetics **76**, 51—63 (1974)

Geyer-Duszýnska, I.: Experiments on nuclear transplantation in *Drosophila melanogaster*. Preliminary report. Rev. Suisse Zool. **74**, 614—615 (1967)

Graziosi, G., Micali, F.: Differential responses to ultraviolet irradiation of the polar cytoplasm of *Drosophila* eggs. Wilhelm Roux' Archiv **175**, 1—11 (1974)

Haldane, J. B. S.: The combination of linkage values and the calculation of distances between the loci of linked factors. J. Genet. **8**, 299—309 (1919)

Hall, J. C., Gelbart, W. M., Kankel, D. R.: Mosaic systems. In: Ashburner, M., Novitski, E. (Eds.): The Genetics and Biology of *Drosophila*, Vol. Ia, pp. 265—314. New York: Academic Press 1976

Hathaway, D. S., Selman, G. G.: Certain aspects of cell lineage and morphogenesis studied in embryos of *Drosophila melanogaster* with an ultra-violet microbeam. J. Embryol. Exp. Morph. **9**, 310—325 (1961)

Hotta, Y., Benzer, S.: Mapping of behaviour in *Drosophila* mosaics. Nature (Lond.) **240**, 527—535 (1972)

Hotta, Y., Benzer, S.: Mapping of behavior in *Drosophila* mosaics. In: Ruddle, F. H. (Ed.): Genetic Mechanisms of Development, pp. 129—167. New York: Academic Press 1973

Howland, R. B., Child, G. P.: Experimental studies on development in *Drosophila melanogaster*. I. Removal of protoplasmic materials during late cleavage and early embryonic stages. J. Exp. Zool. **70**, 415—427 (1935)

Illmensee, K.: Transplantation of embryonic nuclei into unfertilized eggs of *Drosophila melanogaster*. Nature (Lond.) **219**, 1268—1269 (1968)

Illmensee, K.: Imaginal structures after nuclear transplantation in *Drosophila melanogaster*. Naturwissenschaften **57**, 550—551 (1970)

Illmensee, K.: Developmental potencies of nuclei from cleavage, preblastoderm, and syncytial blastoderm transplanted into unfertilized eggs of *Drosophila melanogaster*. Wilhelm Roux' Archiv **170**, 267—298 (1972)

Illmensee, K., Mahowald, A. P.: Transplantation of posterior polar plasm in *Drosophila*. Induction of germ cells at the anterior pole of the egg. Proc. Natl. Acad. Sci. USA **71**, 1016—1020 (1974)

Illmensee, K., Mahowald, A. P.: The autonomous function of germ plasm in a somatic region of the *Drosophila* egg. Exp. Cell Res. **97**, 127—140 (1976)

Illmensee, K., Mahowald, A. P., Loomis, M. R.: The ontogeny of germ plasm during oogenesis in *Drosophila*. Dev. Biol. **49**, 40—65 (1976)

Janning, W.: Aldehyde oxidase as a cell marker for internal organs in *Drosophila melanogaster*. Naturwissenschaften **59**, 516—517 (1972)

Janning, W.: Distribution of aldehyde oxidase activity in imaginal disks of *Drosophila melanogaster*. Dros. Inf. Serv. **50**, 151—152 (1973)

Janning, W.: Entwicklungsgenetische Untersuchungen an Gynandern von *Drosophila melanogaster*. I. Die inneren Organe der Imago. Wilhelm Roux' Archiv **174**, 313—332 (1974a)

Janning, W.: Entwicklungsgenetische Untersuchungen an Gynandern von *Drosophila melanogaster*. II. Der morphogenetische Anlageplan. Wilhelm Roux' Archiv **174**, 349—359 (1974b)

Janning, W.: Entwicklungsgenetische Untersuchungen an Gynandern von *Drosophila melanogaster*. III. Einige Beobachtungen an larvalen Gynandern. Verh. Dtsch. Zool. Ges. **1974**, 134—138 (1975)

Janning, W.: Entwicklungsgenetische Untersuchungen an Gynandern von *Drosophila melanogaster*. IV. Vergleich der morphogenetischen Anlagepläne larvaler und imaginaler Strukturen. Wilhelm Roux's Archives **179**, 349—372 (1976)

Kankel, D. R., Hall, J. C.: Fate mapping of nervous system and other internal tissues in genetic mosaics of *Drosophila melanogaster*. Dev. Biol. **48**, 1—24 (1976)

Lawrence, P. A., Morata, G.: The early development of mesothoracic compartments in *Drosophila*. An analysis of cell lineage and fate mapping and an assessment of methods. Dev. Biol. **56**, 40—51 (1977)

Lewis, E. B., Gencarella, W.: *Claret* and nondisjunction in *Drosophila melanogaster* (Abstract). Genetics **37**, 600—601 (1952)

Lindsley, D. L., Grell, E. H.: Genetic variations of *Drosophila melanogaster*. Carnegie Inst. Wash. Publ. **627** (1968)

Lohs-Schardin, M., Sander, K.: UV-microirradiation of *Drosophila* embryos (Abstract). 5th Eur. Drosophila Res. Conf. (1976)

Mahowald, A. P., Illmensee, K., Turner, F. R.: Interspecific transplantation of polar plasm between *Drosophila* embryos. J. Cell Biol. **70**, 358—373 (1976)

Mertens, M.: Larval and imaginal defects after pricking of blastoderm stages: Experiments complementary to morphogenetic fate maps of *Drosophila*. Dros. Inf. Serv. **52**, 134—135 (1977)

Milne, Jr., C. P.: Morphogenetic fate map of prospective adult structures of the honey bee. Dev. Biol. **48**, 473—476 (1976)

Morgan, T. H., Bridges, C. B., Schultz, J.: Report of investigations on the constitution of the germinal material in relation to heredity. Carnegie Inst. Wash. Yearbook **34**, 284—291 (1935)

Nissani, M.: Cell lineage analysis of kynurenine producing organs in *Drosophila melanogaster*. Genet. Res. (Camb.) **26**, 63—72 (1975)

Nissani, M.: Cell lineage analysis of germ cells of *Drosophila melanogaster*. Nature (Lond.) **265**, 729—730 (1977)

Nissani, M., Lipow, C.: A method for estimating the number of blastoderm cells which give rise to *Drosophila* imaginal discs. Theor. Appl. Genet. **49**, 3—8 (1977)

Nöthiger, R.: The larval development of imaginal disks. In: Ursprung, H., Nöthiger, R. (Eds.): Results and Problems in Cell Differentiation, Vol. 5, pp. 1—34. Berlin-Heidelberg-New York: Springer 1972

Okada, M., Kleinman, I. A., Schneiderman, H. A.: Chimeric *Drosophila* adults produced by transplantation of nuclei into specific regions of fertilized eggs. Dev. Biol. **39**, 286—294 (1974)

Parks, H. B.: Cleavage patterns in *Drosophila* and mosaic formation. Ann. Entomol. Soc. Am. **29**, 350—392 (1936)

Pasztor, L. M.: Unstable ring-X-chromosomes derived from a tandem metacentric compound in *Drosophila melanogaster*. Genetics **68**, 245—258 (1971)

Patterson, J. T., Stone, W.: Gynandromorphs in *Drosophila melanogaster*. Univ. Tex. Publ. No. **3825**, 1—67 (1938)

Perkins, D. D.: Crossing over and interference in a multiply marked chromosome arm of *Neurospora*. Genetics **47**, 1253—1274 (1962)

Petters, R. M.: Blastoderm fate map construction from *Habrobracon juglandis* gynandromorphs (Abstract). Genetics **83**, s 57—s 58 (1976)

Petters, R. M.: A morphogenetic fate map constructed from *Habrobracon juglandis* gynandromorphs. Genetics **85**, 279—287 (1977)

Portin, P.: Analysis of the negative complementation of Abruptex alleles in gynandromorphs of *Drosophila melanogaster*. Genetics **86**, 309—319 (1977)

Poulson, D. F.: Histogenesis, organogenesis, and differentiation in the embryo of *Drosophila melanogaster* Meigen. In: Demerec, M. (Ed.): Biology of Drosophila, pp. 168—274. New York: Wiley 1950

Ripoll, P.: The embryonic organization of the imaginal wing disc of *Drosophila melanogaster*. Wilhelm Roux' Archiv **169**, 200—215 (1972)

Schubiger, M., Schneiderman, H. A.: Nuclear transplantation in *Drosophila melanogaster*. Nature (Lond.) **230**, 185—186 (1971)

Steiner, E.: Establishment of compartments in the developing leg imaginal discs of *Drosophila melanogaster*. Wilhelm Roux's Archives **180**, 9—30 (1976)

Sturtevant, A. H.: The *claret* mutant type of *Drosophila simulans*: a study of chromosome elimination and cell-lineage. Z. Wiss. Zool. **135**, 323—356 (1929)

Szabad, J., Schüpbach, T., Wieschaus, E.: Mosaic analysis of the larval hypoderm of *Drosophila melanogaster* (Abstract). 5th Eur. Drosophila Res. Conf. (1976)

Turner, F. R., Mahowald, A. P.: Scanning electron microscopy of *Drosophila* embryogenesis. 1. The structure of the egg envelopes and the formation of the cellular blastoderm. Dev. Biol. **50**, 95—108 (1976)

Wieschaus, E.: Clonal analysis of early development in *Drosophila melanogaster*. Thesis, Yale Univ. (1974)

Wieschaus, E., Gehring, W.: Clonal analysis of primordial disc cells in the early embryo of *Drosophila melanogaster*. Dev. Biol. **50**, 249—263 (1976a)

Wieschaus, E., Gehring, W.: Gynandromorph analysis of the thoracic disc primordia in *Drosophila melanogaster*. Wilhelm Roux's Archives **180**, 31—46 (1976b)

Williamson, R. L., Kaplan, W. D.: The developmental focus of the suppression of Hk^2 leg shaking by $para^{ts-1}$ in *Drosophila melanogaster* (Abstract). Genetics **83**, s 82—s 83 (1976)

Zalokar, M.: Transplantation of nuclei in *Drosophila* melanogaster. Proc. Natl. Acad. Sci. USA **68**, 1539—1541 (1971)

Zalokar, M.: Transplantation of nuclei into the polar plasm of *Drosophila* eggs. Dev. Biol. **32**, 189—193 (1973)

Zalokar, M., Erk, I.: Division and migration of nuclei during early embryogenesis of *Drosophila melanogaster*. J. Microsc. Biol. Cell. **25**, 97—106 (1976)

Zalokar, M., Erk, I.: Phase-partition fixation and staining of *Drosophila* eggs. Stain Technol. **52**, 89—95 (1977)

Mitotic Recombination and Position Effect Variegation

HANS J. BECKER

Zoologisches Institut der Universität, München, FRG

I. Introduction

Mitotic recombination and position effect variegation have widely different underlying mechanisms.

In mitotic recombination two out of four chromatids of a pair of homologous chromosomes exchange fragments in a process similar to the one in meiotic prophase. Unlike meiotic recombination, mitotic recombination is a fairly rare event. Spontaneously, one event—in any one of the four chromosome pairs of *Drosophila melanogaster*—takes place in an estimated 100–1000 cell divisions. Mitotic recombination can be induced artifically by the application of a number of chemical and physical agents that can increase its incidence about 100-fold or more. In qualitative and quantitative studies, twin mosaic spots are most commonly assayed. They arise in epidermal structures of the adult insect as the result of the genotypic change following recombination. From their features conclusions are drawn on the nature of the recombinational event or on the fate of the genetically marked cells during development.

Position effect variegation, on the other hand, is exhibited regularly in animals with chromosomal rearrangements that have euchromatic genes transposed into heterochromatin. There are two basic events, then, that lead to this phenomenon. One is the origin of the chromosomal rearrangement, spontaneous or induced. The other is the event that leads to the variegated phenotype. This latter event cannot be induced; and it is the far more puzzling of the two. The pattern of some variegated phenotypes is similar to the clonal pattern of recombinational mosaic spots, and this similarity might justify the treatment of the two phenomena within one chapter.

Both phenomena have been reviewed in detail only recently (mitotic recombination: Becker, 1976; position effect variegation: Spofford, 1976). These articles should be consulted for more details and for their more complete quotation of the literature. In this chapter the basic features of both will be described first. Then the contribution of mitotic recombination to an understanding of position effect variegation will be discussed. Most of the work reported in this chapter has been done with *Drosophila melanogaster*. Only when other species were used will the name be given in the text.

II. Mitotic Recombination

A. General Remarks

Occurrences of mitotic recombination were observed and described as early as 1925, when Bridges found mosaic spots with yellow body color and normal bristles in $+Mn/y+$ females. He thought that the $+Mn$ X chromosome had been eliminated. However, Stern in 1936, using $y+/+sn$ flies, found: "Among 15 spots 2 were yellow and not-singed, 2 others not-yellow singed, and 11 were twin spots consisting of a yellow not-singed area adjacent to a not-yellow singed." "It turned out that the overall solution was based on the unexpected existence of somatic crossing over, not on chromosomal loss" (Stern, 1971). Today we prefer to use the term mitotic recombination; it describes not only the above-mentioned spontaneous somatic crossing-over, but it includes induced as well as gonial recombinations. The results of both these latter types of recombination had been observed earlier, but their underlying mechanism first was recognized in 1936 by Friesen, by Neuhaus, and by Stern and Doan.

Cytologically mitotic recombination has not been investigated to any extent. Cooper (1948) and Kaufmann (1934) have seen chiasma-like structures in all three large chromosomes of *Drosophila melanogaster* at pro- and metaphase stages. Whether these represent the morphologic counterpart of mitotic recombination, has still to be investigated.

When chromosomes undergo mitotic recombination in a somatic cell that is heterozygous for a marker gene, this can lead to two daughter cells, one homozygous for one marker allele and the other homozygous for the other. These two cells are a twin mosaic spot. By subsequent cell divisions they can form a visible clone.

When chromosomes undergo recombination in a gonial cell of a heterozygous genotype, this can lead to two clusters of recombinant offspring, the clusters being complementary in their genotypes.

Both somatic and gonial mitotic recombination are known to occur spontaneously. The incidence of both types of events can also be increased by artificial induction. The properties of spontaneous and induced mitotic recombinations will be treated separately.

B. Spontaneous Mitotic Recombination

Most of what is known about spontaneous mitotic recombination is reported in the detailed work of Stern (1936). His results will be summarized here and they will be supplemented by later work.

In most of his experiments Stern made use of the enhancing effect of *Minutes* on mitotic recombination. The system he used was the occurrence of yellow and singed mosaic spots on the head, the thorax, and the abdomen of the adult fly. In one of the first experiments he excluded reductional segregation of centromeres during anaphase as the origin of mosaic spots. The absence of this type of segregation makes it impossible to decide whether there is recombination at the two-strand stage of the cell cycle, i.e., during the G_1 phase. There is no reason to believe that chromosomes should not be able to break and rejoin during that phase. So far, however, there is no way of discovering mitotic recombination during that phase.

The distribution of recombination events along the chromosomes proved to be quite different from the distribution of meiotic recombinations. More detailed comparisons between mitotic recombination and cytologic chromosome maps (Becker, 1974) showed that the events are distributed more or less evenly over all chromosome regions. The frequency of mosaic spots for any chromosome marker depends, therefore, largely on its cytologic distance from the centromere, regardless of the type of chromatin that occupies that chromosome section.

Genetic differences, however, have been found between strains with regard to the frequency of mosaic spots for only one marker (Weaver, 1960). The *Minutes* as modifiers of that frequency (Kaplan, 1953) have already been mentioned. Other genetic elements evoke recombination in the male germ line (Hiraizumi et al., 1973; Slatko and Hiraizumi, 1975). Heterozygous inversions affect mitotic recombination in the same way that they affect meiotic recombination; therefore, such inversions also prevent the formation of mosaic spots that result from mitotic recombination. Mitotic recombination within the inverted region appears to result in loss of cells rather than in loss of recombinant strands, at least when the recombinations are induced (Becker, 1969).

The same is true for mitotic recombination in special chromosomes, such as rings (Brown et al., 1962) and attached-X's (Schwartz, 1954; Brown and Welshons, 1955; Baker and Swatek, 1965). The recombinant cells survive as long as they receive a balanced chromosome complement. Dicentric rings seem to be lost more readily than dicentric rods. Loss of the former in ring/rod heterozygotes leads to single-X daughter cells, while dicentric rods appear to interfere with cell division and lead to cell death.

In mosaic studies, where use was made of mitotic recombination, a high ratio of single spots was often found where twin spots were expected. The nature of the single spots has been discussed extensively (Becker, 1976). There seem to be various factors involved in their origin, one prominent one being the loss or the undetectability of one twin spot partner.

Temperature has repeatedly been claimed to influence the frequency of mitotic recombination, particularly by Stern and Rentschler (1936) and by Brosseau (1957). In some cases it increased, in others it decreased the frequency; there is, so far, no evidence that it acts consistently in one direction.

Spontaneous spermatogonial (Stern and Doan, 1936; Lindsley, 1955) and oogonial mitotic recombination (Neuhaus, 1936) were inferred to have taken place in cases where recombinants appeared in clusters of identical phenotypes. In cases of spermatogonial recombination their identification is unequivocal since there is no meiotic recombination in males. The cases of oogonial mitotic recombination referred to were recombinations between X and Y chromosomes in attached-X/Y females.

It is difficult to study the nature of the spontaneous mitotic recombination event, largely because it occurs so rarely and because there are, so far, no means to study the genetic constitution of mosaic spots. Stern (1969) found mosaic spots that suggested recombination within the *white* locus. Tokunaga (1973) found, after X-ray treatment of larvae, spots that appeared to be the result of mitotic recombination within the *lozenge* locus. Similar experiments of Kelly (1974), however, led her to conclude that those intragenic recombinations were nonreciprocal.

Attempts to disclose the nature of the recombinational events have, therefore, been made in recent years using induced mitotic recombinations. This was felt to be all the more important, since induced recombinations are much more valuable as tools in fields where cells have to be marked genetically. Aside from being more numerous than spontaneous events, they can be induced at almost any time the experimenter chooses.

C. Induced Mitotic Recombination

In most experiments this type of recombination has been induced by X-rays (for detailed references see Becker, 1976). A few other experiments were made using γ-rays (Lefevre, 1947; Whittinghill, 1951), UV (Meyer, 1954; Prudhommeau and Proust, 1969; Prudhommeau, 1972) or chemical agents (Auerbach, 1945, 1946; Sobels, 1956; Whittinghill and Lewis, 1961; Alderson, 1967; Becker, 1975).

The first extensive work was done by Friesen (1936). He X-rayed heterozygous *rucuca* males before he mated them with homozygous *rucuca* females. From the last two or three in a total of five successive broods he found clusters of recombinant offspring, all members of one cluster having the same genotype, and often two clusters with complementary genotypes within the offspring of one male. These properties as well as the generally erratic appearance of recombinants led him to conclude that they resulted from induced mitotic recombinations in spermatogonia.

Friesen tried to bring the recombinant chromosomes into an homozygous condition. In most cases he succeeded, even where recombination had taken place in euchromatin. This showed that the induced recombinations had supposedly taken place at homologous loci. This might seem surprising, since X-ray-induced mitotic recombination has been shown to be due to breakage and reunion (Haendle, 1971a, b). If, on the other hand, the assumption of Lee (1975) is correct that repetitious DNA components along the chromosome play a roll in the recombination process of induced chromosome breaks, breakage at nonhomologous sites within a repetitious unit might still lead to viable recombinant strands.

In this context, it is interesting that the induced recombination events of Friesen (1936) do not seem to be as evenly distributed over the chromosome as spontaneous mitotic recombinations, but rather more concentrated around centric heterochromatic regions of the chromosome (Becker, 1974), where repetitious DNA sequences are located.

Mitotic recombination in the female germ line is most clearly shown when it is induced in females whose meiotic recombination is suppressed, such as in c(3)G females (Whittinghill, 1938). When it is induced in females with normal meiotic recombination, an increase of recombination near the centromere and a decrease in distal chromosome regions has been found consistently (Mavor, 1923; Mavor and Svenson, 1924; Muller, 1925; Whittinghill, 1951). The distal decrease is not surprising. Any mitotic recombination near the centromere leads—when the recombinant strands go to opposite poles—to homozygocity in the daughter oocytes and thereby masks any subsequent meiotic recombination.

Quite recently, Wieschaus and Szabad (1978) solved the problem of gonial recombination being masked by subsequent meiotic recombination in normal females. They used maternal effect markers, one altering the morphology of the egg (fs(1)K10) and the other the phenotype of the resultant progeny (maroon-like). They found, as Friesen (1936) had found for the male germ line, that complementary clones in the offspring of single females can differ extremely in size, indicating the existence of stem cells, which divide differentially into a cytoblast and another stem cell.

The most thorough attempt to unravel the nature of the X-ray-induced mitotic recombination event has been made by Haendle (1971a, b). She showed that there are at least two types of events involved, characterized by different dose and different dose-rate responses. One is a break type with a rapidly acting repair system that is induced by the hard portion of the 100 kV spectrum. The other is a more slowly repaired type of break that can be induced by all the investigated spectra. In addition to the two types of breakage-fusion mechanisms there exists a further component, which was termed "X-ray-induced chromosome pairing". Especially designed experiments led to the assumption that this induced pairing is a necessary prerequisite for induced mitotic recombination, and they showed that its intensity is dose dependent (Haendle, 1974).

The problem of delayed recombination after treatment or of breaks in chromosomal subunits has not yet been investigated thoroughly. Thus far, however, there is no indication that a major proportion of induced mitotic recombination is due to delayed effects of the inducing agent, even in chemically induced mitotic recombination (Auerbach, 1945, 1946; Whittinghill and Lewis, 1961; Alderson, 1967; Becker, 1975). Therefore, the mosaic spot frequency and spot sizes depend largely on the time of treatment (Becker, 1957; Garcia-Bellido and Merriam, 1971). This makes induced mitotic recombination a useful tool in studies of development.

D. Developmental Conclusions Drawn from Mitotic Recombination

In this section mitotic recombination in the eye anlage of *Drosophila melanogaster* will be discussed, since these results will be used to interpret cases of position effect variegation in Section IV.

The eye of *Drosophila* consists of roughly 18000 cells, arranged into about 800 ommatidia. Each eye develops as part of an eye-antennal disc that first becomes visible toward the end of the embryonic stage. Two discs form the whole head of the fly, except for the mouth parts. One disc forms the left half of the head, the other the right half. The cells of the eye anlage multiply during the three larval instars, and they differentiate into ommatidia during the late third larval instar and during the pupal stage.

Mitotic recombination was induced by X-ray treatment during the larval instars (Becker, 1957). The animals were heterozygous for the two eye color alleles *white (w)* and *white-coral (w^{co})*. The gene is located at the tip of the X chromosome at locus 1.5. Mitotic recombination between the markers and the centromere results, if the two recombinant strands are separated, in two homozygous cells, one w/w, the other w^{co}/w^{co}. After multiplication, they develop into two adjacent partners of a twin mosaic spot, one *(w/w)* without pigment and the other *(w^{co}/w^{co})* darker than the rest of the intermediately colored heterozygous eye *(w/w^{co})*. After treatment of young larvae spots are large but few; after treatment of older larvae spots are small but numerous. To determine the size of a mosaic spot, the number of ommatidia were counted. These counts are only approximations since a single ommatidium on the margin of a spot can contain pigment cells of both genotypes (Benzer, 1973; Hofbauer and Campos-Ortega, 1976).

In the largest mosaic spots found, a horizontal borderline going through the middle of the eye is rather conspicuous as part of a spot's margin. Strangely enough, this line does not seem to coincide with the horizontal line separating the two halves of the eye, characterized by their mirror-image symmetry of retinula cell arrangement (Benzer, 1973). In 17 of 23 mosaic spots with a size of more than 220 ommatidia, the horizontal line was the upper borderline (in the case of spots in the lower half) or the lower borderline of a spot (in the case of spots in the upper half). The other six spots stretch over both halves of the eye. For the most part, therefore, the cell lineages of the upper and the lower halves of the eye seem to be separated.

In addition to this marked central borderline there are other borderlines, especially in the lower half of the eye. The positions of these borderlines on the eye surface become clearest if tracings of the mosaic patterns of lower eye halves are superimposed. For the present purpose the eyes selected were those in which the spots reached from the horizontal midline of the eye to the margin of the lower half of the eye (Fig. 1a–i). Such a tracing is shown in Figure 1l. The outlines of mosaic spots are almost superimposed in certain regions, and the resulting pattern has been used for the scheme given in Figure 1m, in which the sections average 40 ommatidia. Spots of the size of one single section are found predominantly when larvae are treated at about the time moulting occurs between the first and the second larval instars. It was estimated, therefore, that late in the first larval instar the presumptive eye area consists of about 20 cells. Each cell gives rise to an eye section, and these sections have an average of 40 ommatidia (yielding the total adult number of 800 ommatidia). The congruence of borderlines suggests that at the end of the first larval instar each cell of the eye area seems to have been destined to form a definite section of the eye. This is not true for earlier stages. Spots two or more times larger than one of the sections schematically outlined in Figure 1m can cover any combination of sections (Table 1). Therefore, they have not yet been assigned definite places within the eye area.

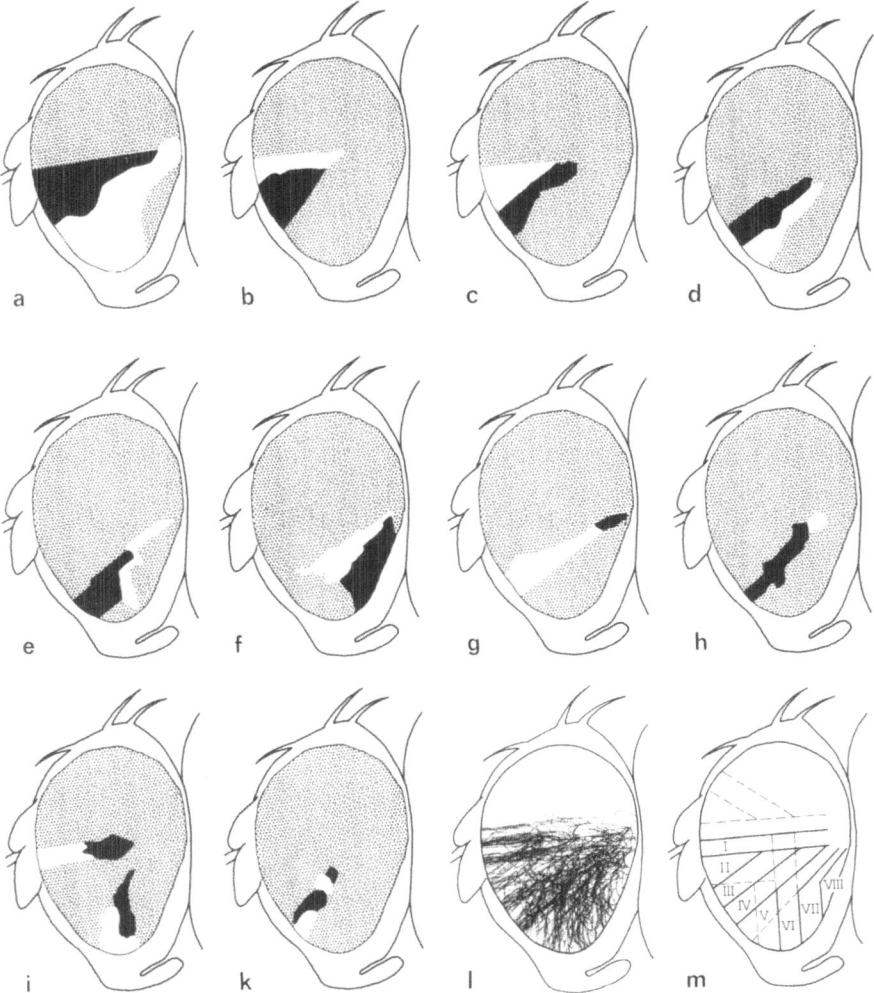

Fig. 1. (a–k) Images of w-w^{co} twin mosaic spots in the eyes of *Drosophila melanogaster* observed after irradiation of the embryos and of first and second instar larvae. (l) Tracings of mosaic spots (in the lower half of the eye) that reach from the horizontal midline to the lower eye margin. (m) Schematic representation [derived from (l)] of commonly found spot sizes, shapes, and predominant positions in the lower half of the eye *(solid lines)* and of less commonly found spot limits in the lower and upper parts of the eye *(hatched lines)*. (From Becker, 1957)

This latter point has recently been challenged. Ready et al. (1976) point out the difficulties of superimposing different mosaic spots on one eye; they claim that the accuracy of placing spots on the eye is not good enough for finding distinct clonal borders. A reinvestigation should settle the controversy. For such a study it would be advisable to use spontaneously arising clonal mosaic spots, since in X-ray-induced spots the death of cells might cause a distortion of clones.

Table 1. The randomly overlapping distribution of 52 spots the
size of two sections within sections I—VII of Figure 1m

Section						
I	II	III	IV	V	VI	VII
9		8		11		
	9		9		6	

Independent of the outcome of such a reinvestigation, it can be concluded that
at the end of the first larval instar the presumptive eye area consists of about
20 cells. Each of these cells gives rise to a clone with a distinct size and shape,
depending on its position within the eye.

The positions of twin spot partners relative to one another reveals the orientation of cell divisions in the prospective eye area. Judging from the way the bar-
and sector-shaped partners in the lower half of the eye lie next to each other
(Fig. 1 a–f), they appear to have arisen from cell divisions oriented predominantly
in a general upper-anterior to lower-posterior direction. The stretched shape of
each sector in Figure 1 m suggests that afterwards, i.e., by the second and third
larval instars, the orientation of cell divisions is roughly at right angles to the
previous one. The twin spots in Figure 1 g–k, obtained after X-irradiation of
second instar larvae, show that this is true at least for the first two or three
divisions after the change of mitotic orientation.

The latter spots demonstrate also a regularly occurring size difference between
twin partners. For 34 twin spots similar to those shown in Figure 1 g and h, the
average sizes of the proximal and distal twin partners were 14.0 and
34.5 ommatidia. Such a consistent difference has not been found in the large spots
obtained by irradiation of the first larval instar.

The end of the first larval instar, then, is distinguished by a change in the
orientation of cell divisions and, presumably, by the assignment of the prospective
eye area to a definite region within the eye anlage. Along with this, a gradient for
mitotic activity seems to be established with its center in the mid-posterior region
of the eye. Spots "induced" by irradiation at the end of the first larval instar
exhibit characteristic shapes, as shown in Figure 1 b–f and m. In the anterior
region of the eye they are predominantly bar-shaped, on the lower eye margin
they tend to be triangular, and in the lower posterior region they often have the
shape of a bent bar, as for instance the white twin partner in Figure 1 e.

The great significance of the end of the first larval instar for the development
of the eye was inferred from the experiments with genetic mosaics. Further support for this finding will be shown in the analysis of certain types of position effect
variegation.

So far, it is not known whether some morphologic change is associated with
the establishment of the eye anlage at the end of the first larval instar. It is possible
that the connection that the head anlage makes with the larval brain at this stage
is the origin of the changes.

III. Position Effect Variegation

A. The Rules

If the phenotypic effect of a gene is changed as a result of a change in its position within the chromosome set, the phenomenon is called position effect (review by Lewis, 1950). If a gene is transposed by a chromosomal rearrangement from its normal location in euchromatin into heterochromatin or vice versa, variegation of the phenotype controlled by that gene is often the result. This is called position effect variegation (reviews by Baker, 1968; Spofford, 1976). The variegated phenotype is believed to be due to the suppression of the activity of the wild-type gene in some cells but not in others.

There are variegated phenotypes known that are not based on position effect (cf. Becker, 1966). Therefore, it is necessary to name the criteria that determine whether an observed variegation is caused by position effect. These criteria are:

1. One of the breaks of a chromosomal rearrangement is in a heterochromatic region and the other is close to the gene whose phenotype is variegated.

2. The suppression of gene activity spreads from the heterochromatic break point. The gene nearest the break point is most affected; genes farther away are less affected.

3. Recombination between the affected gene and the break point changes the phenotype from variegated to wild type.

4. Reversion of the chromosomal rearrangement changes the phenotype from variegated to wild type.

5. Addition of a Y chromosome shifts the expression of the variegated phenotype toward wild type, removal of a Y chromosome toward the mutant phenotype.

6. An increase of temperature during development shifts the variegated phenotype toward wild type; a decrease toward the mutant phenotype.

7. The expression of variegation can be modulated by the parental genotype.

In the two cases that will receive the most attention in this article, the *white* locus (*w*, 1–1.5) is involved. In one case, $T(1;4)w^{258-18}$, the w^+ locus is transposed close to the heterochromatin of chromosome 4 (Demerec and Slizynska, 1937; Fig. 2). In the other case, $Dp(1;3)w^{264-58a}$, a small w^+-carrying section of chromosome 1 is inserted into the heterochromatic region at the base of chromosome 3 (Demerec, 1940; Sutton, 1940). Flies carrying mutant *w*-alleles in addition to the transposed w^+-allele, i.e., w^{258-18}/w or $w/w; w^{264-58a}$, show variegated eyes. Some sections of the eyes are normally pigmented, others are a very light shade of red (w^{258-18}/w) or white $(w/w; w^{264-58a})$. Since the eyes of w^+/w flies are normal, the transposed w^+ loci are considered partially suppressed by the adjacent heterochromatin.

In w^{258-18}, the break in chromosome 1 is just right of the *roughest* locus. In heterozygotes for *white* and *roughest*, the light areas in the eye always show the roughest phenotype, i.e., disarranged ommatidia. Not all disarranged ommatidia, however, are lightly pigmented. This phenomenon has been named "spreading effect". The suppressive action of the heterochromatin spreads either to the closest

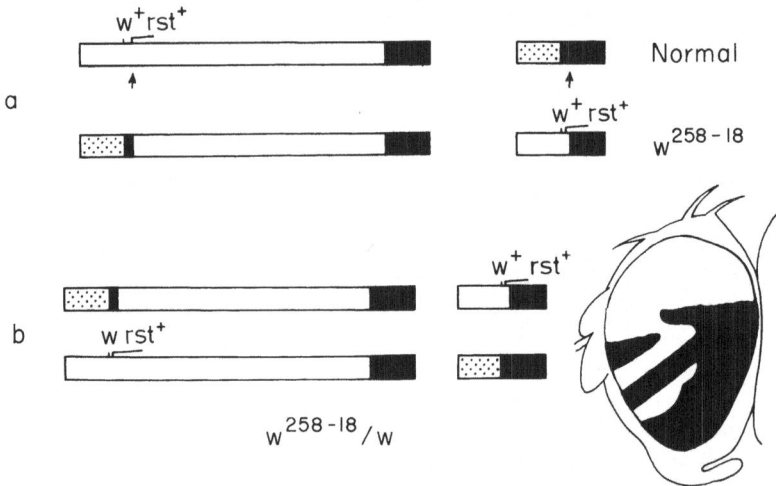

Fig. 2. (a) Origin of the translocation $T(1;4)w^{258-18}$; *left:* X chromosome; *right:* chromosome 4. *Arrows* indicate break points. Heterochromatin of X and 4 is shown in *black.* (b) Genotype and one characteristic phenotype of flies referred to in text. (From Becker, 1966)

locus, *roughest*, yielding disarranged but normally pigmented ommatidia, or its action spreads and includes the more distal *white* locus, yielding disarranged lightly pigmented ommatidia. In such flies normally arranged lightly pigmented ommatidia do not exist.

They do, however, exist in *w spl/w spl*; $w^{264-58a}$ flies (Baker, 1963). This is not in contradiction to the spreading effect, since the inserted chromosome section carrying the w^+ and the spl^+ locus, is flanked by heterochromatin on both sides.

The *white* gene is one of the few cases in which recombination between the transposed w^+ locus and the break point has restored the normal function of the locus (Judd, 1955). Reversion of position effect variegation by reversion of the chromosomal rearrangement has been claimed in two cases of *roughest* (Grüneberg, 1937; Novitski, 1961). In both cases variegation was caused by an inversion within chromosome 1, and reinversions resulted in a normal phenotype.

The modifying effects of Y chromosomes (Gowen and Gay, 1933a; Baker and Rein, 1962), of temperature (Gowen and Gay, 1933b; Chen, 1948; Becker, 1961), of the parental genotype (Noujdin, 1944; Baker and Spofford, 1959; Spofford, 1959, 1961; Hessler, 1961), and of suppressor loci (Spofford, 1962, 1967, 1969) have been known for a long time. The peculiarities of the temperature and Y-chromosome modifications will be discussed in one of the following paragraphs (IV.C).

B. Other Manifestations of Position Effect Variegation

Aside from the variegation effect on exterior phenotypes, such as the eye color discussed in the preceding paragraph, position effect variegation manifests itself at a variety of other levels.

It was the *white* locus again that was used to study variegation in internal organs such as the larval malpighian tubules (Hartmann-Goldstein, 1967). Developmental temperature shifts the phenotype in the same direction as it does in eye pigmentation, i.e., high temperature toward wild type. However, whereas *white*-dependent eye color variegation has its temperature-sensitive period in the pupal stage (Chen, 1948), the *white*-dependent pigmentation of larval malpighian tubules is sensitive to temperature during early embryonic stages (Schultz, 1956). This leads to the important conclusion that neither the gene nor the chromosomal rearrangement determines the temperature-sensitive period of a given type of variegation, but the organ in which variegation is observed and investigated. Supporting evidence for this statement will be given below.

Another system in which position effect variegation has been observed is a group of lethal inversions of the X chromosome (Baker, 1971). It could be demonstrated that the reduced viabilities of $In(1)sc^{V2}/O$ and of $In(1)sc^{8}/O$ males are caused by position effect suppression of viability genes located at the tip of the X chromosome. The reduced viability of $In(1)sc^{L8}/O$ and of $In(1)sc^{S1}/O$ males, however, is based on the suppression of genes normally located at the base of the X chromosome, most probably of the rDNA.

Recently, position effect variegation has also been described for two enzymes, for an isoamylase (Bahn, 1971), and for 6-phosphogluconate dehydrogenase (Gvozdev et al., 1973). In both cases the enzyme activity was modifiable by temperature and by addition or removal of a Y chromosome. In both cases the only isozyme suppressed was the one controlled by the transposed enzyme locus, not the one controlled by the allele on the normal chromosome. From this observation Bahn (1971) finds it most likely that variegation is due here to a block of transcription rather than translation.

This view finds support in other investigations where cytologic consequences of position effects were studied. In $T(1;4)w^{258-21}$ heterozygotes, both Rudkin (1965) and Hartmann-Goldstein (1966) describe the puffing behavior of X chromosome sections that were transposed to the heterochromatin of chromosome 4. Among 87 nuclei investigated, Hartmann-Goldstein (1966) found 46 in which section 3C11-12 was puffed in both homologs, whereas in 41 nuclei no puff was seen in the transposed 3C11-12 section. Since puffs are the sites of transcription, the two types of nuclei found represent a variegation of transcriptive activity.

In the work of Hartmann-Goldstein (1966) the puffing was found to be correlated with the euchromatic appearance of the transposed chromosome section. When 3C11-12 was heterochromatized, no puff was seen. Cytologic heterochromatization of euchromatic bands transposed to heterochromatin was described in early salivary gland chromosome studies (Prokofyeva-Belgovskaya, 1941). This heterochromatization shows the same response to temperature as do other position effect variegations. It affects only a few bands next to the break point at 25° C and affects up to about fifty bands at 14° C (Hartmann-Goldstein, 1967). The extent of cytologic heterochromatization is closely correlated to the phenotypic variegation of euchromatic markers close to the break point (Wargent and Hartmann-Goldstein, 1974; Hartmann-Goldstein and Wargent, 1975).

It is not clear yet, how heterochromatic appearance of transposed euchromatic chromosome sections and suppressed transcriptive activity are connected with the changed replication behavior in these sections. Heterochromatin is known to replicate later than euchromatin (Lima-de-Faria and Jaworska, 1968), even when it is involved in chromosomal rearrangements. In such cases, however, when certain sections of the *Drosophila* Y chromosome are transposed to other chromosomes, they do not replicate as late as those sections in the normal Y, but do replicate later than the euchromatin (Halfer et al., 1969). Another position effect involving DNA replication has been reported by Roberts (1972). The heterochromatic base of chromosome 4, which in salivary gland chromosomes is usually much thinner than the distal portion of that chromosome and often even invisible, replicates to the same width as the rest of the chromosome when transposed to the tip of chromosome 3.

The difference between the two cases should be pointed out. The former gives examples of later replication, whereas in the latter full replication is attained by a chromosome section normally subject to underreplication. This same distinction is made in the work of Ananiev and Gvozdev (1974). In their case the euchromatic tip of the X chromosome is transposed to the centric heterochromatin of that same chromosome in the duplication $Dp(1;f)R$. While in the salivary gland chromosomes most of that euchromatic material showed underreplication, one section, 1DE, was an exception. This section, however, showed late replication. Since underreplication of heterochromatin was demonstrated only for polytene chromosomes (Rudkin, 1965, 1969; Berendes and Keyl, 1967), the Russian authors conjecture that underreplication of transposed elements might also be restricted to polytene chromosomes. This assumption will be of importance in the discussion in paragraph IV.

Similar observations were made by I. J. Hartmann-Goldstein and her colleagues. They found that the replication time of transposed euchromatic segments is extended (Wargent and Hartmann-Goldstein, 1976), and also that there is a lower DNA content in transposed versus normal homologous sections (Wargent et al., 1974; Hartmann-Goldstein and Cowell, 1976). They do, however, find a difference between replication and heterochromatization. Under the conditions in which the larvae were reared (25° C), only about 14% of the cells show heterochromatization of that same chromosome section, which was found to be underreplicated. Underreplication, therefore, seems to precede cytologic heterochromatization. This is also true for extended replication (Wargent and Hartmann-Goldstein, 1976). The question has to be raised, whether the changes in replication are a variegation type of position effect. The best argument in favor of that view is furnished by the data of Hartmann-Goldstein and Cowell (1976). They show that, although all transposed sections measured were lower in DNA content than the nontransposed homologous sections, the percentage of their DNA content ranged from 80 to 100% of that of the nontransposed section. If this variability is not due to measuring errors, it could very well represent a case of variegated underreplication.

IV. Mitotic Recombination and Position Effect Variegation

A. The Clonal Character of Variegation; Clonal Initiation

In the two types of *white*-variegation described in Section III.A practically all flies show the variegated phenotype. Some areas of the eyes are normally pigmented, others lack most or all of the pigment. There are, then, at least two types of cells in the eye. They have different capabilities for synthesizing pigment although their genotypes are presumably identical. The areas that differ in color are quite similar to the clonal mosaic spots that arise after mitotic recombination, as far as their shape and their position are concerned (Becker, 1957, 1961, 1966; Baker, 1963, 1967; Janning, 1970; Fig. 2).

The normal-red and the light areas of the eyes originate, therefore, from single cells in which the ability or inability to synthesize pigment has been fixed at a certain stage of development. The process by which this is done has been called clonal initiation (Gsell, 1971).

B. Timing of Clonal Initiation

The size of the uniformly colored areas gives a rough idea of the stage of development at which the ability for pigment synthesis is fixed. For $T(1;4)w^{258-18}$ it has been determined to be the end of the first larval instar (Becker, 1961, 1966). A more precise determination of that stage has been made in the case of $Dp(1;3)w^{264-58a}$ (Janning, 1970). Janning induced mitotic recombination in the X chromosomes of larvae of the constitution $w^a lz^{50e}/w^a rb rux^2$; $Dp(1;3)w^{264-58a}$. All eyes are variegated with normal-red and white-apricot areas. The genetic constitution of cells in the induced twin mosaic spots is $w^a lz^{50e}/w^a lz^{50e}$; $Dp(1;3)w^{264-58a}$ and $w^a rb rux^2/w^a rb rux^2$; $Dp(1;3)w^{264-58a}$. The mosaic spots differ in phenotype depending on whether they are located within a normal-red or within a white-apricot area of the eye. The mosaic spots can even be variegated within themselves, e.g., one partner can lie within a normal-red area and the other partner within a white-apricot area. Twin spots of this type, however, can only be induced in cells of the eye anlage before they have been fixed for the ability or inability to form pigment. From such studies, the time of clonal initiation leading to position effect variegation in the eye has been determined. It happens between 39 and 47 h after egg deposition at 25.5° C, i.e., during the last few hours of the first larval instar.

The important point here is that in two different chromosomal rearrangements, one $(Dp(1;3)w^{264-58a})$ with breaks between 3 B 2 and 3 and between 3 D 6 and E 1, the other $(T(1;4)w^{258-18})$ with a break between 3 C 4 and 5, the time of clonal initiation is identical. This is probably true also in $T(1;4)w^{258-21}$, in which the break lies between bands 3 E 6 and 7 (Becker, 1966), i.e., farther away from the *white* locus than in the other two rearrangements. The time of clonal initiation, therefore, seems not to depend on the distance of the break from the *white* locus. It could depend either on the phenotype involved, in this case the pigmentation of the eye, or on the organ in which variegation is observed. We know that the *white*-locus-dependent variegation in the malpighian tubes has its temperature-sensitive period early in the embryo. We also know that the *roughest* variegation of $T(1;4)w^{258-18}$ and the *split* variegation of $T(1;4)w^{258-21}$ also show the pattern typical for clonal initiation at the end of the first larval instar. It is, therefore, probable that it is the eye anlage and what happens to it at the end of the first larval instar that determines the time of clonal initiation. It is likely that the time of clonal initiation coincides with the developmental processes in the eye anlage that were discussed at the end of Section II.

C. Modification of Variegation

1. Temperature

When the nature of clonal initiation is discussed, the effect of two modifiers have to be taken into consideration. The first is the way temperature modifies variegation. As is typical for all position effect variegations, with increasing temperature the variegated eyes become darker. Although the temperature-sensitive period is during the pupal stage (Chen, 1948), the darkening is based on a change of the numerical relationship between the light and the dark sections within the eye (Becker, 1961). When w^{258-18}/w flies were raised at 25° C throughout all the developmental stages, 41.0% of all 2400 eye sections investigated (300 eyes with 8 sections each) were light. However, when the larval stages of flies with the same genotype were kept at 25° C and the pupae at 19° C, the frequency of light sections among 2368 sections investigated rose to 50.9%.

The characteristic cell lineage pattern is not altered by changing the temperature during development. That is, the time of clonal initiation has not been changed; all descendants of one first larval instar anlagen cell still react as a unit. From the above given figure one has to conclude, therefore, that 9.9% of all cells (50.9 minus 41.0) in the described experiments behave differently from the others. In these cells clonal initiation seems to have influenced the *white* locus in such a way that all their progeny can synthesize pigment normally when exposed to 25° C at the time of pigment synthesis during the pupal stage. When exposed to 19° C, however, these same cells do not synthesize pigment normally.

This suggests strongly that in each cell clonal initiation determines the ability for pigment synthesis. The determination is stable to a degree that all descendants of one cell have identical abilities. However, considering the whole cell population there are not only two alternative states into which clonal initiation turns the cell, as might appear from the two alternative shades of pigmentation in the eye. There seems rather to be a gradient of different states. Each state has a temperature threshold, above which a cell's descendants form a dark section of the eye, and below which they form a light section. It looks, therefore, as though clonal initiation is as diversely variegated as the visual heterochromatization of salivary gland chromosome bands.

It has been questioned, time and again, whether the inactivation is really stabilized at the chromosomal rather than at the whole-cell level. Van Breugel (personal communication) made observations which led him to believe that clonal initiation is nothing but a fixation of a stage in development that all cells normally go through. If one would follow his arguments one could visualize that messenger is formed more rapidly in some cells than in others but that all cells produce eventually the optimal amount if nothing interferes with the production, like clonal initiation. At the time of clonal initiation the amounts of messenger already present in the cells would vary from cell to cell. Heterochromatization would prevent further transcription, so that the varying amounts of messenger would be handed down to the offspring, which would form correspondingly varied clones.

In such a case the turning-off process at clonal initiation, possibly heterochromatization of the variegating loci, could affect all homologous loci with equal strength. The variegated phenotype in the adult would represent nothing but the "variegated" amount of messenger at the time transcription is turned off, i.e., at clonal initiation. However, an explanation must still be given for the fact that in variegation usually two alternative phenotypes are observed in clones. It seems to this author that one would have to expect either two alternative types of cells at the time of clonal initiation, one type containing one amount of messenger, the other containing another. Or one would have to invoke a threshold amount of messenger below which one phenotype is produced and above which another. Although I consider this possibility not too likely, it has not been ruled out.

If clonal initiation affects the *white* locus itself, there is every reason to compare the behavior of clonal initiation with the cytologically observed behavior of transposed chromosome sections, as discussed in Section III.B. There, heterochromatization of giant chromosome bands was observed. Variegation at this level could only be scored in terms of the number of bands that were heterochromatized within a single cell. There has been no way, so far, to find and measure any differences in the degree of heterochromatization between homologous bands within one gland. However, heterochromatization seems to be preceded by underreplication or by delayed replication. If underreplication has to be excluded in the propagating type of cells composing the eye anlage, then the transposed *white* locus might undergo delayed replication and heterochromatization and that possibly to varying degrees. This change in state has been called clonal initiation. The degree of heterochromatization determines the degree of impairment of the *white*-locus function. If the pigment cells of the eyes had giant chromosomes, one might visualize a puff at the *white* locus in those cells in the pupa that synthesize pigments normally and no puff in cells without normal pigment synthesis. Whether in a group of cells a puff is formed or not depends on the degree of heterochromatization. Unheterochromatized loci form a puff, strongly heterochromatized loci form no puff. There should be intermediate states of heterochromatization, in which puff formation is dependent on the temperature during transcription. At high temperatures they can form a puff, at low temperatures they cannot.

This last paragraph has been largely conjecture, and it would be highly desirable to have some support for it. Variegation with respect to transcription has already been described (Rudkin, 1965; Hartmann-Goldstein, 1966). If, in the propagating cells of the developing salivary gland of the young embryo, there should be clonal initiation, comparable to that in the eye anlage cells of the first larval instar, the progeny of single cells within a larval salivary gland should be identical in their puffing activity in transposed chromosome sections. Such cloning studies in salivary glands are not easy but feasible.

In salivary glands, then, variegation at the level of transcription might be based on clonal initiation; in this case, one would expect to find cell clones within glands differing in puff activity. Variegation might also be found at the later stage of the fully differentiated gland, i.e., each cell might differ from every other one in its puffing activity.

This latter type of variegation at the time of transcription proper might be represented in variegated eye pigmentation by the cases of so-called salt-and-pepper variegation, e.g., in $In(1)w^{m4}$. In these cases it looks as if different pigment

cells within one ommatidium differ in their pigment content. This type of variegation, then, at the time of assumed transcription is modifiable by temperature
(Chen, 1948), but not the clonal initiation.

2. Heterochromatin

The time and mode of action of heterochromatin on position effect variegation seems to be similar to that of temperature. Variegation is suppressed toward
wild type by the addition of a heterochromatic Y chromosome and it is enhanced
toward the mutant phenotype by the removal of a Y chromosome (Gowen and
Gay, 1933a). The addition of a Y to the chromosome complement as well as the
subtraction of a Y from it also changes the numerical relationship between normal and mutant clonal sections within the eyes (Becker, 1961). Since the modifying action of Y heterochromatin had been found to be autonomous (Janning,
1970), it was possible to determine the Y sensitive stage in development (Becker
and Janning, 1977). The basis for this determination was the induction of hyper-
and hypo-Y mosaic spots at different stages in development. This was done by X-
ray inducing mitotic recombination in $w^a lz^{50e}/Y^S w^a rb \ rux^2; \ Dp(1;3)w^{264-58a}$
larvae heterozygous for a normal X chromosome and for an X carrying a Y-short
(Y^S) arm distally. This led to two daughter cells, one with two and the other with no
Y^S arms. The two-Y^S mosaic spots induced after clonal initiation had a significantly
darker phenotype than the rest of the eye, indicating that the Y heterochromatin
exerts some function after clonal initiation. The two-Y^S mosaic spots induced before
clonal initiation showed a phenotype even darker than the spots induced after
clonal initiation. The Y heterochromatin, therefore, does also act during clonal
initiation, possibly on the process of clonal initiation itself.

The surprising result of these experiments was that zero-Y^S mosaic spots were
not lighter in phenotype than the rest of the eye, neither the ones induced after nor
the ones induced before clonal initiation. This could only be interpreted by invoking
a threshold amount of some Y product that is produced in one-Y^S cells prior to
clonal initiation, i.e., in the embryo and during the first larval instar, and which is
still available for pigment synthesis during the pupal stage, even when the Y
material is taken out of the cells by the end of the first larval instar.

The fact that heterochromatin modifies variegation not only during but also
after clonal initiation and that it modifies the salt-and-pepper variegation of
$In(1)w^{m4}$ flies suggests that part of the modifying function of heterochromatin
affects the process of transcription of the *white* locus.

D. Nonrandom Distribution of Variegated Areas

One final aspect is noteworthy with regard to eye color variegation. Whereas
induced twin mosaic spots are randomly distributed over the whole surface of the
eye, the light sections in $T(1;4)w^{258-18}$ eyes show a distinct deviation from such
randomness (Becker, 1961, 1966). Light sections are more often found in the
anterior region of the eye than in the posterior. Whether in this case loci in

prospective anterior cells become more strongly heterochromatized than in prospective posterior cells, or whether the threshold for normal pigment synthesis is lower in the posterior than in the anterior region of the pupal eye, is an open question.

V. Conclusions

The study of twin mosaic spots resulting from induced mitotic recombination has given insight into the growth pattern of the primordial eye cells. Spots induced at the end of the first larval instar have characteristic sizes and shapes, depending on their position within the eye. In size they are about one twentieth of the eye. Their preferential shapes and positions, particularly in the lower half of the eye, were essentially instrumental for (1) demonstrating that variegation in the $In(1)w^{258-18}$ and the $Dp(1;3)w^{264-58a}$ eyes is clonal in nature and for (2) quantifying the different degrees of variegation, particularly when modifiers such as temperature or heterochromatin were involved. This led to the conclusion that there was clonal initiation in the eye anlagen of larvae at the end of the first instar. The time of the initiation is given by the developmental events going on in the eye anlage, and it is independent of the type of chromosomal rearrangement and of any of the modifying agents. Clonal initiation probably determines the transcriptive capabilities of the transposed loci; it suppresses them to varying degrees. Since the consequences of clonal initiation materialize only several cell generations later, i.e., during transcription in the pupa, it is difficult to visualize what happens at the chromosomal level at the time of clonal initiation proper.

It is suggested that clonal initiation is connected with heterochromatization of the genes under discussion, possibly preceded by delayed replication or possibly connected with underreplication. One could, for instance, visualize the *white* locus to be normally reiterated and, in the case of position effect variegation, to be more or less underreplicated in different cells during clonal initiation. Then each cell containing an underreplicated w^+-locus would form a clone of cells with identical properties, i.e., identical numbers of *white* gene copies. The existence of only two alternative types of areas within the variegated eyes would have to be explained by some threshold gene copy number, above which full and below which no pigment is synthesized.

Such an interpretation would explain why the state, in which clonal initiation leaves the cells, is rather stable. Its stability beyond the number of cell divisions normally taking place between initiation and transcription has been demonstrated by Gsell (1971) in serial transplantations of imaginal discs. In their seemingly strong stability the clonally initiated cells differ from determined cells, Gsell says, since determined cells regularly show transdetermination. Gsell (1971) points out another difference between determination and clonal initiation: The former affects groups of cells, the latter single cells. It is, therefore, doubtful whether clonal initiation is only a special type of determination. It is, on the other hand, rather conspicuous that clonal initiations of genes influencing eye color and eye facetation both occur at a stage at which the determination of prospective eye cells within the head anlage takes place. Clonal initiation seems to be, therefore, an indicator of normal developmental processes; however, clonal initiation itself is not yet well understood. In the future, other organs will have to be investigated in order to see whether clonal initiations take place and whether they also coincide with determinative events in their anlagen.

Acknowledgement. I am grateful to my wife, Dr. Gweneth L. Becker, for her help in preparing the manuscript.

References

Alderson, T.: Induction of genetically recombinant chromosomes in the absence of induced mutation. Nature (Lond.) **215**, 1281—1283 (1967)

Ananiev, E. V., Gvozdev, V. A.: Changed pattern of transcription and replication in polytene chromosomes of *Drosophila melanogaster* resulting from eu-heterochromatin rearrangement. Chromosoma (Berl.) **45**, 173—191 (1974)

Auerbach, C.: The problem of chromosome rearrangements in somatic cells of *Drosophila melanogaster*. Proc. R. Soc. Edin., B. **62**, 120—127 (1945)

Auerbach, C.: Chemically induced mosaicism in *Drosophila melanogaster*. Proc. R. Soc. Edin., B. **62**, 211—222 (1946)

Bahn, E.: Position-effect variegation for an isoamylase in *Drosophila melanogaster*. Hereditas **67**, 79—82 (1971)

Baker, W. K.: Genetic control of pigment differentiation in somatic cells. Am. Zool. **3**, 57—69 (1963)

Baker, W. K.: A clonal system of differential gene activity in *Drosophila*. Dev. Biol. **16**, 1—17 (1967)

Baker, W. K.: Position-effect variegation. Adv. Genet. **14**, 133—169 (1968)

Baker, W. K.: Evidence for position-effect suppression of the ribosomal RNA cistrons in *Drosophila melanogaster*. Proc. Natl. Acad. Sci. USA **68**, 2472—2476 (1971)

Baker, W. K., Rein, A.: The dichotomous action of Y chromosomes on the expression of position-effect variegation. Genetics **47**, 1399—1407 (1962)

Baker, W. K., Spofford, J. B.: Heterochromatic control of position-effect variegation in *Drosophila*. Texas Univ. Publ. **5914**, 135—154 (1959)

Baker, W. K., Swatek, J. A.: A more critical test of hypotheses of crossing over which involve sister-strand exchange. Genetics **52**, 191—202 (1965)

Becker, H. J.: Über Röntgenmosaikflecken und Defektmutationen am Auge von *Drosophila* und die Entwicklungsphysiologie des Auges. Z. Indukt. Abstamm. Vererbungsl. **88**, 333—373 (1957)

Becker, H. J.: Untersuchungen zur Wirkung des Heterochromatins auf die Genmanifestierung bei *Drosophila melanogaster*. Verhandl. Dtsch. Zool. Ges., Bonn 1960, pp. 283—291 (1961)

Becker, H. J.: Genetic and variegation mosaics in the eye of *Drosophila*. In: Moscona, A. A., Monroy, A. (Eds.): Current Topics in Developmental Biology, Vol. I, pp. 155—171. New York-London-San Francisco: Academic Press 1966

Becker, H. J.: The influence of heterochromatin, inversion-heterozygosity and somatic pairing on X-ray induced mitotic recombination in *Drosophila melanogaster*. Molec. Gen. Genet. **105**, 203—218 (1969)

Becker, H. J.: Mitotic recombination maps in *Drosophila melanogaster*. Naturwissenschaften **61**, 441—448 (1974)

Becker, H. J.: X-ray- and TEM-induced mitotic recombination in *Drosophila melanogaster*: Unequal and sister-strand recombination. Molec. Gen. Genet. **138**, 11—24 (1975)

Becker, H. J.: Mitotic recombination. In: Ashburner, M., Novitski, E. (Eds.): The Genetics and Biology of *Drosophila*, Vol. I c, pp. 1019—1087. London: Academic Press 1976

Becker, H. J., Janning, W.: Heterochromatin of the *Drosophila melanogaster* Y chromosome as modifier of position effect variegation: The time of its action. Molec. Gen. Genet. **151**, 111—114 (1977)

Benzer, S.: Genetic dissection of behavior. Sci. Am. **229**, 24—37 (1973)

Berendes, H. D., Keyl, H. G.: Distribution of DNA in heterochromatin and euchromatin of polytene nuclei of *Drosophila hydei*. Genetics **57**, 1—13 (1967)

Bridges, C. B.: Elimination of chromosomes due to a mutant (*Minute-n*) in *Drosophila melanogaster*. Proc. Natl. Acad. Sci. USA **11**, 701—706 (1925)

Brosseau, G. E.: The environmental modification of somatic crossing over in *Drosophila melanogaster* with special reference to developmental phase. J. Exp. Zool. **136**, 567—593 (1957)

Brown, S. W., Walen, K. H., Brosseau, G. E.: Somatic crossing-over and elimination of ring-X chromosomes of *Drosophila melanogaster*. Genetics **47**, 1573—1579 (1962)

Brown, S. W., Welshons, W.: Maternal aging and somatic crossing over of attached-X chromosomes. Proc. Natl. Acad. Sci. USA **41**, 209—215 (1955)

Chen, S. Y.: Action de la température sur troi mutants a panachure de *Drosophila melanogaster*: w^{258-18}, w^{m5} et z. Bull. Biol. France Belg. **82**, 114—129 (1948)

Cooper, K. W.: The evidence for long range specific attractive forces during the somatic pairing of dipteran chromosomes. J. Exp. Zool. **108**, 327—336 (1948)

Demerec, M.: Genetic behavior of euchromatic segments inserted into heterochromatin. Genetics **25**, 618—627 (1940)

Demerec, M., Slizynska, H.: Mottled *white 258-18* of *Drosophila melanogaster*. Genetics **22**, 641—649 (1937)

Friesen, H.: Spermatogoniales Crossing-over bei *Drosophila*. Z. Indukt. Abstamm. Vererbungsl. **71**, 501—526 (1936)

Garcia-Bellido, A., Merriam, J. R.: Parameters of the wing imaginal disc development of *Drosophila melanogaster*. Dev. Biol. **24**, 61—87 (1971)

Gowen, J. W., Gay, E. H.: Eversporting as a function of the Y-chromosome in *Drosophila melanogaster*. Proc. Natl. Acad. Sci. USA **19**, 122—126 (1933a)

Gowen, J. W., Gay, E. H.: Effect of temperature on eversporting eye color in *Drosophila melanogaster*. Science **77**, 312 (1933b)

Grüneberg, H.: The position effect proved by a spontaneous reinversion of the X-chromosome in *Drosophila melanogaster*. J. Genet. **34**, 169—189 (1937)

Gsell, R.: Untersuchungen zur Stabilität einer *yellow* Positionseffekt-Variegation in Imaginalscheiben-Kulturen von *Drosophila melanogaster*. Molec. Gen. Genet. **110**, 218—237 (1971)

Gvozdev, V. A., Gerasimova, T. I., Birstein, V. Ya.: Inactivation of the structural gene of 6-Phosphogluconate dehydrogenase when transferred to heterochromatin in *Drosophila melanogaster*. Genetika (USSR) **9**, 64—72 (1973)

Haendle, J.: Röntgeninduzierte mitotische Rekombination bei *Drosophila melanogaster*. I. Die Abhängigkeit von der Dosis, der Dosisrate und vom Spektrum. Molec. Gen. Genet. **113**, 114—131 (1971a)

Haendle, J.: Röntgeninduzierte mitotische Rekombination bei *Drosophila melanogaster*. II. Beweis der Existenz und Charakterisierung zweier von der Art des Spektrums abhängiger Reaktionen. Molec. Gen. Genet. **113**, 132—149 (1971b)

Haendle, J.: X-ray induced mitotic recombination in *Drosophila melanogaster*. III. Dose dependence of the "pairing" component. Molec. Gen. Genet. **128**, 233—239 (1974)

Halfer, C., Tiepolo, L., Barigozzi, C., Fraccaro, M.: Timing of DNA replication of translocated Y chromosome sections in somatic cells of *Drosophila melanogaster*. Chromosoma (Berl.) **27**, 395—408 (1969)

Hartmann-Goldstein, I. J.: Relationship of heterochromatin to puffs in a salivary gland chromosome of *Drosophila*. Naturwissenschaften **53**, 91 (1966)

Hartmann-Goldstein, I. J.: On the relationship between heterochromatization and variegation in *Drosophila*, with special reference to temperature-sensitive periods. Genet. Res. Camb. **10**, 143—159 (1967)

Hartmann-Goldstein, I. J., Cowell, J.: Position effect on the DNA content of chromosome regions in *Drosophila*. In: Jones, K., Brandham, P. E. (Eds.): Current Chromosome Research, pp. 43—50. Amsterdam: Elsevier/North Holland Biomedical Press 1976

Hartmann-Goldstein, I. J., Wargent, J. M.: Cytological observations on the interaction between two inversions responsible for position-effect variegation in *Drosophila melanogaster*. Chromosoma (Berl.) **52**, 349—362 (1975)

Hessler, A. Y.: A study of parental modification of variegated position effects. Genetics **46**, 463—484 (1961)

Hiraizumi, Y., Slatko, B., Langley, C., Nill, A.: Recombination in *Drosophila melanogaster* male. Genetics **73**, 439—444 (1973)

Hofbauer, A., Campos-Ortega, J. A.: Cell clones and pattern formation: Genetic eye mosaics in *Drosophila melanogaster*. Wilhelm Roux's Archives **179**, 275—289 (1976)

Janning, W.: Bestimmung des Heterochromatisierungsstadiums beim *white*-Positionseffekt mittels röntgeninduzierter mitotischer Rekombination in der Augenanlage von *Drosophila melanogaster*. Molec. Gen. Genet. **107**, 128—149 (1970)

Judd, B. H.: Direct proof of a variegated-type position effect at the *white* locus in *Drosophila melanogaster*. Genetics **40**, 739—744 (1955)

Kaplan, W. D.: The influence of *Minutes* upon somatic crossing over in *Drosophila melanogaster*. Genetics **38**, 630—651 (1953)

Kaufmann, B. P.: Somatic mitosis of *Drosophila melanogaster*. J. Morphol. **56**, 125—155 (1934)

Kelly, P. T.: Non-reciprocal intragenic mitotic recombination in *Drosophila melanogaster*. Genet. Res. Camb. **23**, 1—12 (1974)

Lee, C. S.: A possible role of repetitious DNA in recombinatory joining during chromosome rearrangement in *Drosophila melanogaster*. Genetics **79**, 467—470 (1975)

Lefevre, G.: The relative effectiveness of fast neutrons and γ-rays in producing somatic crossing over in *Drosophila* (Abstr.). Rec. Gen. Soc. Am. **16**, 40 (1947)

Lewis, E. B.: The phenomenon of position effect. Adv. Genet. **3**, 73—115 (1950)

Lima-de-Faria, A., Jaworska, H.: Late DNA synthesis in heterochromatin. Nature (Lond.) **217**, 138—142 (1968)

Lindsley, D. L.: Spermatogonial exchange between the X and Y chromosomes of *Drosophila melanogaster*. Genetics **40**, 24—44 (1955)

Mavor, J. W.: An effect of X-rays on the linkage of Mendelian characters in the first chromosome of *Drosophila*. Genetics **8**, 355—366 (1923)

Mavor, J. W., Svenson, H. K.: An effect of X-rays on the linkage of Mendelian characters in the second chromosome of *Drosophila melanogaster*. Genetics **9**, 70—89 (1924)

Meyer, H. U.: Crossing-over in the germ line of *Drosophila melanogaster* males following irradiation of the embryonic pole cells with ultraviolet. Genetics **39**, 982 (1954)

Muller, H. J.: The regionally differential effect of X-rays on crossing over in the autosomes of *Drosophila*. Genetics **10**, 470—507 (1925)

Neuhaus, M. J.: Crossing-over between the X- and Y-chromosomes in the female of *Drosophila melanogaster*. Z. Indukt. Abstamm. Vererbungsl. **71**, 265—275 (1936)

Noujdin, N. I.: The regularities of heterochromatin influence on mosaicism. J. Gen. Biol. (USSR) **5**, 357—388 (1944)

Novitski, E.: The regular reinversion of the *roughest* inversion. Genetics **46**, 711—717 (1961)

Prokofyeva-Belgovskaya, A. A.: Cytological properties of inert regions and their bearing on the mechanisms of mosaicism and chromosome rearrangement. Drosophila Inf. Serv. **15**, 34—35 (1941)

Prudhommeau, C.: Irradiation UV des cellules polaires de l'oeuf chez *Drosophila melanogaster*. III. Étude de la recombinaison mitotique induite chez le mâle. Mutation Res. **14**, 53—64 (1972)

Prudhommeau, C., Proust, J.: U.V. induced mitotic recombination in *Drosophila melanogaster* females and males. Abstr. 1st Eur. Drosophila Res. Conf. (1969)

Ready, D. F., Hanson, T. E., Benzer, S.: Development of the *Drosophila* retina, a neurocrystalline lattice. Dev. Biol. **53**, 217—240 (1976)

Roberts, P. A.: A possible case of position effect on DNA replication in *Drosophila melanogaster*. Genetics **72**, 607—614 (1972)

Rudkin, G. T.: The structure and function of heterochromatin. Proc. 11th Int. Congr. Genetics, The Hague, 1963, Vol. II, pp. 359—374. New York: Pergamon Press 1965

Rudkin, G. T.: Non replicating DNA in *Drosophila*. Genet. Suppl. **61**, 227—238 (1969)

Schultz, J.: The relation of the heterochromatic chromosome regions to the nucleic acids of the cell. Cold Spr. Harb. Symp. Quant. Biol. **21**, 307—328 (1956)

Schwartz, D.: Studies on the mechanism of crossing over. Genetics **39**, 692—700 (1954)

Slatko, B. E., Hiraizumi, Y.: Elements causing male crossing over in *Drosophila melanogaster*. Genetics **81**, 313—324 (1975)

Sobels, F. H.: Studies on the mutagenic action of formaldehyde in *Drosophila*. II. The production of mutations in females and the induction of crossing-over. Z. Indukt. Abstamm. Vererbungsl. **87**, 743—752 (1956)

Spofford, J. B.: Parental control of position-effect variegation. I. Parental heterochromatin and expression of the *white* locus in compound-X *Drosophila melanogaster*. Proc. Natl. Acad. Sci. USA **45**, 1003—1007 (1959)

Spofford, J. B.: Parental control of position-effect variegation. II. Effect of parent contributing *white-mottled* rearrangement in *Drosophila melanogaster*. Genetics **46**, 1151—1167 (1961)

Spofford, J. B.: Direct and parental phenotypic effects of a euchromatic variegation-suppressor locus in *Drosophila melanogaster*. Genetics **47**, 986—987 (1962)

Spofford, J. B.: Single-locus modification of position-effect variegation in *Drosophila melanogaster*. I. *White* variegation. Genetics **57**, 751—766 (1967)

Spofford, J. B.: Single-locus modification of position-effect variegation in *Drosophila melanogaster*. II. Region 3 C loci. Genetics **62**, 555—571 (1969)

Spofford, J. B.: Position-effect variegation. In: Ashburner, M., Novitski, E. (Eds.): The Genetics and Biology of *Drosophila*, Vol. I c, pp. 955—1018. London: Academic Press 1976

Stern, C.: Somatic crossing over and segregation in *Drosophila melanogaster*. Genetics **21**, 625—730 (1936)

Stern, C.: Somatic recombination within the *white* locus of *Drosophila melanogaster*. Genetics **62**, 573—581 (1969)

Stern, C.: From crossing-over to developmental genetics. Stadler Symposia **1** and **2**, 21—28 (1971)

Stern, C., Doan, D.: A cytogenetic demonstration of crossing-over between X- and Y-chromosomes in the male of *Drosophila melanogaster*. Proc. Natl. Acad. Sci. USA **22**, 649—654 (1936)

Stern, C., Rentschler, V.: The effect of temperature on the frequency of somatic crossing-over in *Drosophila melanogaster*. Proc. Natl. Acad. Sci. USA **22**, 451—453 (1936)

Sutton, E.: The structure of salivary gland chromosomes of *Drosophila melanogaster* in exchanges between euchromatin and heterochromatin. Genetics **25**, 534—540 (1940)

Tokunaga, C.: Intragenic somatic recombination in the *lozenge* cistron of *Drosophila*. Molec. Gen. Genet. **125**, 109—118 (1973)

Wargent, J. M., Hartmann-Goldstein, I. J.: Phenotypic observations on modification of position-effect variegation in *Drosophila melanogaster*. Heredity **33**, 317—326 (1974)

Wargent, J. M., Hartmann-Goldstein, I. J.: Replication behaviour and morphology of a rearranged chromosome region in *Drosophila*. In: Pearson, P. L., Lewis, K. R. (Eds.): Chromosomes Today, 1974, Vol. V, pp. 109—116. New York: Wiley 1976

Wargent, J. M., Hartmann-Goldstein, I. J., Goldstein, D. J.: Relationship between DNA content of polytene chromosome bands and heterochromatization in *Drosophila*. Nature (Lond.) **248**, 55—57 (1974)

Weaver, E. C.: Somatic crossing over and its genetic control in *Drosophila*. Genetics **45**, 345—357 (1960)

Whittinghill, M.: The induction of oogonial crossing over in *Drosophila melanogaster*. Genetics **23**, 300—306 (1938)

Whittinghill, M.: Some effects of Gamma rays on recombination and on crossing over in *Drosophila melanogaster*. Genetics **36**, 332—355 (1951)

Whittinghill, M., Lewis, B. M.: Clustered crossovers from male *Drosophila* raised on formaldehyde media. Genetics **46**, 459—462 (1961)

Wieschaus, E., Szabad, J.: The development and function of the female germline in *Drosophila melanogaster*: A cell lineage study. Dev. Biol. in press (1978)

Drosophila Chimeras and the Problem of Determination

KARL ILLMENSEE

Department of Biology, University of Geneva, Switzerland

I. Introduction

There has always been some doubt whether the fruit fly *Drosophila melanogaster* might help us to a better understanding of differentiation in higher organisms inasmuch as "it is rather unfortunate that *Drosophila*, the classical object of genetic research, should happen to have a mosaic type of development" (Needham, 1950). I do not think that it is superfluous to remind ourselves that during the past two decades, however, this gloomy notion has become irrelevant since new genetic and microsurgical approaches opened up novel ways of analyzing nucleocytoplasmic interactions during cellular diversification (reviewed by Gehring, 1976a; Schneiderman, 1976). Moreover, *Drosophila* proved to be an ideal organism for the production of mosaic individuals composed of two genetically different cell populations. By utilizing appropriate marker genes in combination with X-irradiation induced mitotic recombination (reviewed by Becker, this volume), chromosomal loss (reviewed by Janning and Wieschaus, this volume), or microinjection (discussed in this contribution), it is now possible to trace phenotypically the origin and fate of "genetically labeled" cells and their clonal descendants during the course of development.

The latter technique for making mosaics differs conceptually from the other two approaches in that cells or nuclei of a given genotype are transplanted to a *different* embryo, thereby creating a composite animal which genetically originates from at least two zygotes, whereas otherwise the genetic mosaicism results experimentally or spontaneously within the *same* individual derived from only one zygote. In order to distinguish readily between these alternative concepts, the term "chimera" has been widely accepted for microsurgically produced mosaics (reviewed by McLaren, 1976). A unique and very important feature of such a transplantation assay refers to its experimental flexibility as far as the origin of the donor material and its new (heterotopic) position in the recipient embryo are concerned. It is therefore feasible not only to analyze the developmental fate of a particular cell or nucleus, but also to reveal its prospective potential that elsewhere remains obscured under normal in situ conditions.

In the present review, I should like to focus attention on the early embryonic period of *Drosophila*, and shall discuss some of the current problems related to this phase in development during which the nuclei with their surrounding cytoplasm become segregated into cells forming a monocellular layer, at the egg periphery, the blastoderm. Are these cells already determined to participate later in differentiation of certain larval or even adult tissues and, if so, to what extent and with what kind of phenotypic complexity? Is there a regional specificity previously programmed into the egg that might be responsible for determining the future cell lineage-specific pathways? In this context, "determination" is used as an operational term to describe the *potential of a cell or its nucleus as being restricted to a particular developmental fate.* Our microinjection assay of producing chimeras provides us with the necessary in vivo system in which these questions can be analyzed at the nuclear, cellular, and cytoplasmic level.

II. Developmental Plasticity of Transplanted Nuclei

In *Drosophila*, there have recently been designed two experimental ways to test nuclei for their developmental capacity by injecting them into either unfertilized or fertilized eggs (reviewed by Illmensee, 1976). However, only with the latter approach, in which the genome of the donor coexists with that of the recipient during ontogeny, is it possible to produce chimeric animals. Thus far, embryonic nuclei up to the blastoderm stage and nuclei of cells cultured in vitro were found to be functionally integrated into several adult tissues. In order to reveal the developmental potential of transplanted nuclei in chimeric flies, two important aspects concerning the age of the recipient and the implantation site ought to be considered. First, it is necessary to inject the nuclei into the cleavage embryo at a period where cells have not yet been formed. This will allow the implanted nuclei to become segregated into cells during the regular blastoderm formation of the recipient. Second, the nuclei should be transplanted to various positions in the embryo to favor contributions to different developmental pathways (Fig. 1).

A. Nuclei of Embryonic Cells

It was previously found that anterior blastoderm nuclei, which normally would give rise to anterior structures of the fly head, became integrated into posterior structures of the abdominal fly cuticle and even differentiated into functional germ cells after transplantation to the posterior pole of genetically marked recipients (Zalokar, 1971, 1973). Similarly, anterior nuclei when transferred to a mid-lateral site of the cleavage embryo contributed to leg structures in chimeric flies (Illmensee, unpublished data). Furthermore, posterior nuclei injected into the anterior tip of the recipient participated in forming anterior structures of the adult head cuticle (Okada et al., 1974b). The same kinds of results have been obtained from interspecific transplantations combining nuclei from a num-

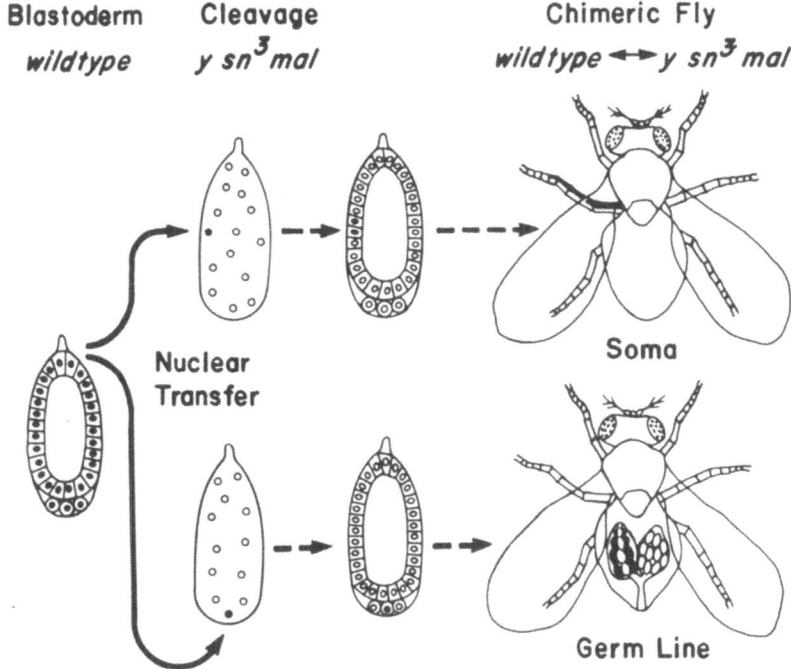

Blastoderm   **Cleavage**   **Chimeric Fly**

wildtype   *y sn³mal*   *wildtype ↔ y sn³ mal*

Nuclear
Transfer

Soma

Germ Line

Fig. 1. Nuclear totipotency at the blastoderm stage. Individual nuclei of *wildtype* blastoderm cells when taken from a given region of the embryo and injected into different sites of genetically marked *y sn³ mal* cleavage embryos will integrate in these new positions during the course of development. The resulting *wildtype* ↔ *y sn³ mal* mosaic flies show tissue contributions derived from injected nuclei in areas directly related to the former injection sites, but unrelated to the positional origin of the transplanted nuclei, thus demonstrating their developmental plasticity. The production of viable offspring clonally propagated from these regionally different blastoderm nuclei clearly reveals their totipotential state

ber of different *Drosophila* species (Santamaria, 1975). It therefore seems a valid generalization, on present evidence, to regard the embryonic nuclei prior to the cellular blastoderm as totipotent. There is no indication of developmental restrictions at the nuclear level even during gastrulation that subsequently follows the blastoderm stage. These nuclei obviously are not yet determined for a specific cell lineage and, in fact, elicit the same potential as the zygote nucleus, since viable offspring carrying the nuclear-transplant genome can be clonally propagated from a single somatic nucleus (Illmensee, 1973). Comparable results have also been obtained in amphibians where transplanted nuclei from embryonic cells promote development to the adult stage (reviewed by Gurdon, 1974).

B. Nuclei of Cells Cultured In Vitro

In recent years, succcssful attempts to establish cell cultures from *Drosophila* embryos (reviewed by Schneider, 1978) have opened up new approaches for nuclear transplantation in this genetically well-defined organism. If cell-cultured nuclei were capable of participating in differentiation of adult tissues, one could then hope to analyze in vivo the developmental and genetic potential of nuclei derived from cells selected in vitro for a particular phenotype. Thus far, only nuclei from Schneider's line 1, which frequently gave rise to larval organs, and Echalier's line 52–84, which differentiated preferentially into imaginal tissues, were able to take part in the in situ development of chimeric larvae and flies via microinjection into cleavage embryos (Illmensee, 1976). Presumably these developmental variations reflect different embryonic origins of the two cell lines and/or selection during in vitro culture, as Debec (1974) attempted to verify at the molecular level. Both kinds of nuclei participated less frequently in development, and produced much smaller patches of tissue, in comparison to contributions derived from normal blastoderm nuclei. Moreover, no progeny from the cultured-cell nuclei could be recovered, although from the total number of nuclear transplant survivors, and the frequency of somatic mosaicism obtained, one might have expected at least a few flies with mosaic gonads. Once again, it may well be that the potential of these nuclei became somewhat restricted during the long in vitro period. Further extensive transplantations with nuclei from short-term cell lines are planned in order to cycle the genome of nuclei from cultured cells through the germ line of the living organism. Nevertheless, for the time being after almost five years in culture, some nuclei still functioned normally in vivo and were able to participate in forming various kinds of somatic tissues such as gut, fat body, Malpighian tubes, and cuticle (Fig. 2). The fact that even chromosomes of the cultured-cell nuclei became polytene, as observed in cells of the Malpighian tubes, is of particular importance inasmuch as it may allow cytogenetic fine structure mapping of somatic mutations induced in vitro and analyzed in vivo.

Fig. 2a–d. Chimeric flies and larvae with tissue contributions from nuclei of in vitro cultured ▷ cells after transplantation into y sn³ mal cleavage embryos. (a) Abdominal region of a male fly whose terminal 8th to 10th segments (arrow) including genital structures all derived from a transplanted female nucleus, as judged from the vaginal plate, the anal plates, and their sex-specific bristle pattern. × 80. (b) Abdomen of a female fly whose 2nd and 3rd tergites (arrow) partly originated from a cultured-cell nucleus. × 60. In both instances, the nuclear-transplant areas show distinct deviations in size and pattern from normal cuticular pigmentation and bristle formation, cryptically expressing the *Minute* phenotype of the haplo-4 genotype of Echalier's cell line 52–84. Nevertheless, these clones are easily distinguishable from the host cuticle with *yellow (y)* pigmentation and *singed (sn³)* bristles. (c) Larval Malpighian tubes with a nuclear-transplant clone of about 20 cells (arrow). Their nuclei contain similar polytene chromosomes as the adjacent host cells. × 50. (d) Larval posterior midgut with a cluster of about 30 cells (arrow) clonally derived from a cultured-cell nucleus. × 50. As in (c) the nuclear-transplant area is histochemically visualized by the presence of aldehyde oxidase in these cells, whereas the surrounding tissue of the host does not express this enzyme due to the mutant gene *maroon-like (mal)* and, therefore, remains unstained. In both instances, the stained larval tissue derived from nuclei of Schneider's cell line 1

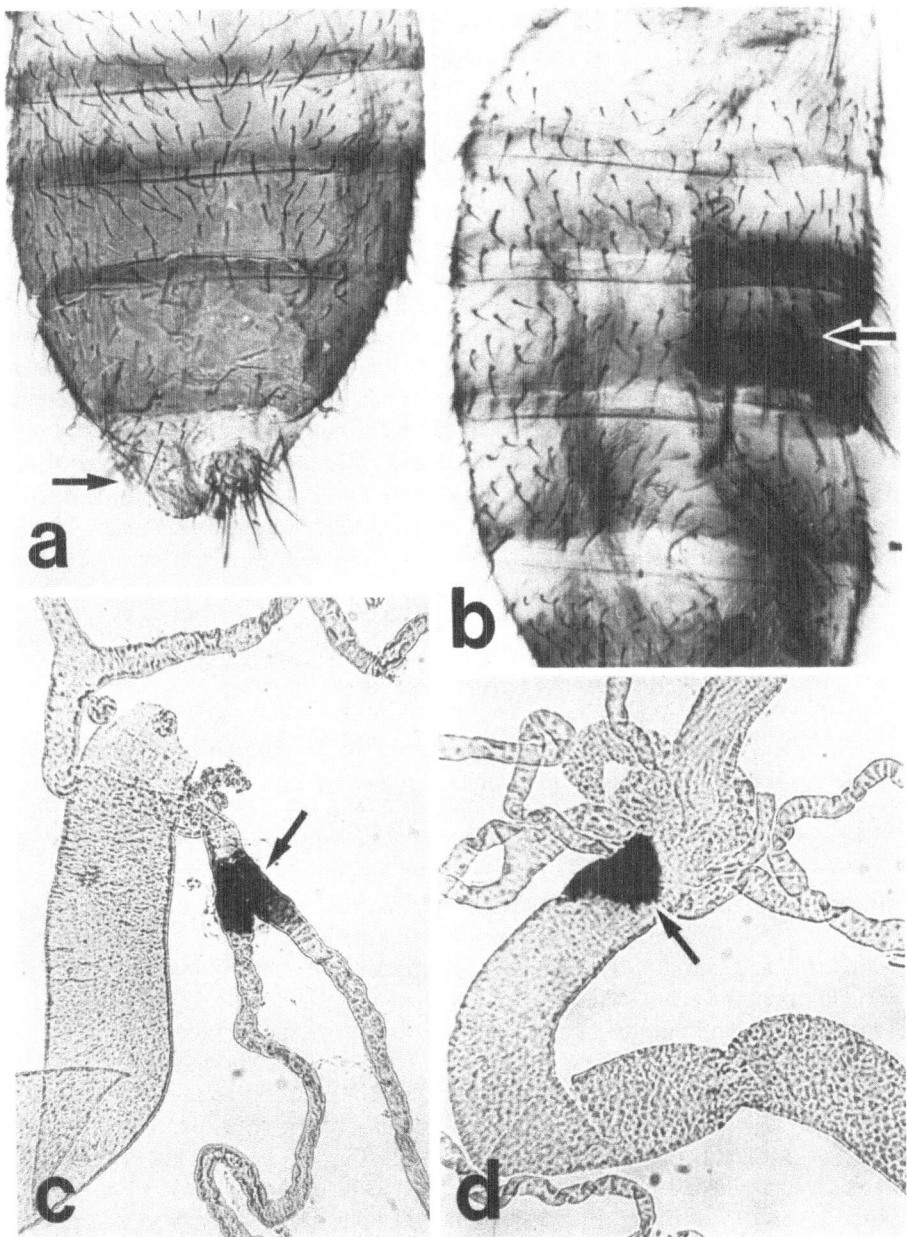

Fig. 2a–d

III. Cell Commitment and Determination
at the Blastoderm Stage

Analogous to the nuclear transfers discussed previously, we can project the same kind of problem to the cellular level and ask: do blastoderm cells show a corresponding developmental plasticity as demonstrated for their nuclei or, on the contrary, are these cells already determined for a distinct pathway?

First of all, it is necessary to uncover the developmental fate of a blastoderm cell in its normal in situ position by transplanting it to the same (homotopic) region of a genetically different blastoderm embryo. In case the cell becomes integrated, it should then phenotypically signal its clonal contribution to a particular tissue. On the other hand, by injecting a given blastoderm cell into a different (heterotopic) site of a recipient, it has to exhibit its real potential there. If the cell should again incorporate, but contribute to tissues according to the new position, we shall refer to such a cell as being pluripotent and not yet committed irreversibly to a specific fate. If the same cell, however, would rather differentiate autonomously into structures according to its original position in the blastoderm, it is considered to be determined (Fig. 3).

A. Prospective Fate in Homotopic Transfers

The first cell transfers in *Drosophila* were carried out with pole cells (Illmensee, 1973). These cells form at the posterior pole of the egg and segregate peripherally from the rest of the embryo shortly before blastoderm formation. From this peripheral position, a single pole cell can easily be picked up with a small glass capillary and subsequently be injected into the same (homotopic) region at the posterior pole of a genetically marked blastoderm recipient. What kind of developmental fate might be expected for the transplanted cell?

During gastrulation, the pole cells of the recipient and the injected pole cell invaginate into the gut, where some of them migrate to the gonads (Poulson, 1965). It was found that the injected cell (and its mitotically derived clone) became functionally integrated into the germ line of chimeric flies, and gave rise to viable offspring which expressed phenotypically the genetic markers used for the pole cell donor. By histochemical means, descendants of the transplanted cell could also be located in the larval midgut (Illmensee et al., 1976). This observation agrees well with earlier findings where specific cells of the same area (the cuprophilic cells) were missing after UV damage to the posterior region of the egg (Poulson and Waterhouse, 1960). However, since we tested histochemically for position rather than for function of the integrated cells, we could not distinguish between presumptive germ cells, which happened to be confined to the midgut during their migration, and some specialized cells of this particular gut segment.

Several conclusions may be drawn from the homotopic pole cell transfers:

1. the origin of the adult germ cells can be traced directly to the pole cells of the early embryo. This, of course, has already been demonstrated indirectly by UV irradiating the posterior pole of preblastoderm embryos which developed

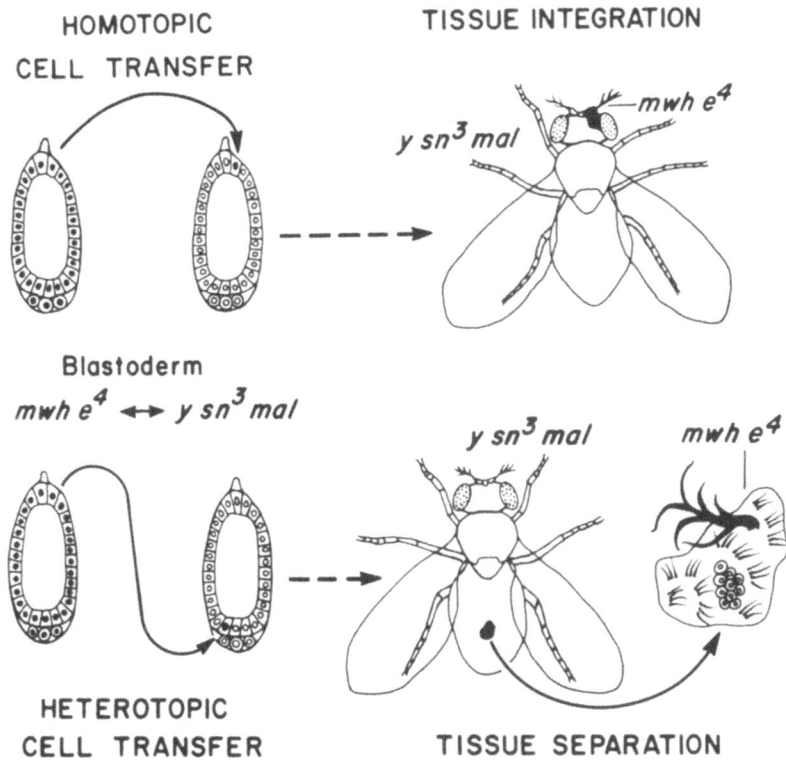

Fig. 3. Cellular determination at the blastoderm stage. Single *mwh e*^{4} blastoderm cells from a particular region of the embryo are transplanted to either the same (homotopic) or a different (heterotopic) area of genetically marked *y sn*^{3} *mal* recipients in order to follow their fate during normal development. The diagram illustrates that, for example, anterior cells injected anteriorly will integrate and contribute to anterior structures phenotypically recognized in *mwh e*^{4}↔*y sn*^{3} *mal* flies. However, anterior cells transplanted to the posterior region of blastoderm embryos will not integrate and form posterior tissues of the fly, but will separate and differentiate autonomously into anterior adult structures. This positional specificity of individual blastoderm cells with respect to location in the embryo and differentiated phenotype in the fly reflects their restricted developmental capacity

into sterile flies (Geigy, 1931). Further confirmation for localizing primordial germ cells at the posterior egg region came from gynandromorph data (Gehring et al., 1976), although this indirect method of fate-mapping the germ line led to quite some conflicting calculations (Falk et al., 1973).

2. By comparing the number of progeny clonally derived from the injected pole cell with the total offspring obtained from the chimeric host, it is possible to calculate the number of precursor cells set aside for the germ line, that is about 8 of the approximately 50 pole cells. This figure agrees well with estimations gained from histological data (Sonnenblick, 1965). However, it remains to be seen whether the actual germ-cell lineage refers to a specific group or merely happens to be a random sample of the pole cell population.

3. The primordial germ cells are already determined genetically with respect to their future sex, and will differentiate to mature eggs or sperm in chimeras only

Table 1. Microsurgical transfer of single blastoderm cells from particular regions of the *Drosophila* embryo to either the same (homotopic, e.g., a→a) or different (heterotopic, e.g., a→p) positions of blastoderm recipients

Region used in cell transfers	$mwh\ e^4$ embryos from which cells were collected	$y\ sn^3\ mal$ embryos into which cells were implanted	$y\ sn^3\ mal$ larvae and flies developed			
			Total	With $mwh\ e^4$ cells		
				Inte-grated	Sepa-rated	Total (%)
Homotopic						
a→a	163	238	119	23	2	(21.0)
p→p	79	108	72	16	2	(25.0)
v→v	86	127	80	9	—	(11.3)
l→l	33	70	53	5	—	(9.4)
Heterotopic						
a→p	203	252	139	2	15	(12.2)
p→a	151	196	85	1	6	(8.2)
v→a	52	88	35	3	—	(8.6)
v→p	91	143	93	5	2	(7.5)
l→a	40	64	29	—	2	(6.9)
l→p	46	76	51	—	3	(5.8)

Abbreviations used for regions of the blastoderm embryo selected for microinjection: a, anterior; p, posterior; v, ventral; l, lateral.

if the pole cell donor and the recipient are of the same sexual genotype. Functional germ cells do not appear in heterosexual combinations (Illmensee, 1973 and unpublished data; Van Deusen, 1976).

4. The usefulness of pole cell transfers as an in vivo assay for analyzing maternally inherited mutants has recently been shown for *deep orange* and *maroon-like*, both of them found to be cell-autonomously related to the germ line and not affecting the mesodermal components (follicle cells and ovarian sheath) of the gonads (Marsh et al., 1977; Marsh and Wieschaus, 1977).

Besides mapping the primordial germ cells at the posterior pole of the egg via homotopic transplantation, the same kind of technique has subsequently been applied to somatic cells from various regions of the blastoderm in order to uncover their developmental fate. Necessary prerequisites for this study are, of course, functional integration of the transplanted cell in the homotopic position of another embryo, and phenotypic identification of the clonal descendants by cell-autonomous genetic markers (Table 1). At present, however, our transplantation fate map cannot yet be compared in detail with the calculated map positions obtained from gynandromorph data. Nor does it allow us to refer from the extent and frequency of mosaicism in the few chimeras so far produced to a certain number of tissue-related primordial cells at the blastoderm, as previously described for the germ line.

Nevertheless, for the time being, it is already possible to trace the mitotic progeny of individual blastoderm cells through ontogeny as a true and coherent clone (Fig. 4a, e). Little or no cell mingling in *Drosophila* enables a precise cell lineage analysis which, by the way, is extremely difficult to achieve in the mouse because of considerable cell migration during mammalian development (see Gardner, this volume).

B. Determinative State in Heterotopic Transfers

After having followed a particular cell of the blastoderm embryo through tissue differentiation to the adult stage, we wanted to know more about its actual potential which could not be fully revealed in situ, but became apparent in our bioassay. The conclusions about the prospective capacity of a given cell type are primarily based on the assumption that those cells which survive and develop normally after transplantation remain undamaged by the microsurgical procedure.

What happens to a blastoderm cell after removal from its normal posterior location and subsequent transfer to the anterior pole of another blastoderm embryo? We demonstrated that such a cell (and its mitotic progeny) did not participate in development of the head cuticle, but rather became separated and differentiated autonomously into posterior adult structures (Fig. 4d). The same held true for various other heterotopic combinations in which, for example, anterior cells gave rise to antennal cuticle, and midlateral cells differentiated into leg and wing tissue after posterior injection. Pole cells still found their way either actively or passively to the gonads, and yielded functional eggs or sperm after transplantation to the midventral site of genetically marked blastoderm recipients (Illmensee and Mahowald, 1976). A quite consistent pattern emerged from our heterotopic transfers in that a blastoderm cell, unlike its nucleus, always gave rise to imaginal structures according to its origin rather than its new position in the embryo. This suggests that determination may occur as soon as cells form during blastogenesis, since the somatic and primordial germ cells are irreversibly restricted in their developmental potential. On the other hand, it looks as if some cells still retain a certain flexibility insafar as midventral cells either became integrated into fat body and Malpighian tubes after posterior implantation (Illmensee, 1976), or participated in differentiation of anterior midgut after anterior transplantation (Fig. 4c and Table 1).

Comparable results with respect to anterior versus posterior determination have been obtained from in vitro cell mixing experiments between genetically different blastoderm embryos and their subsequent in vivo culture in flies (Chan and Gehring, 1971). While some reservations should be expressed about the latter experimental design which allows neither the drawing of conclusions at the single cell level, nor the blastoderm cells to continue to develop in a normal embryonic environment, it definitely has its value in grossly defining the state of determination along the anteroposterior axis of the *Drosophila* embryo. How specific is determination at the blastoderm stage, however? Can a blastoderm cell-derived clone be confined exclusively to a particular disc (the anlage of an imaginal structure such as a leg) or does the clone overlap and extend into different disc derivatives?

Except for a few instances where clones arising from mitotic recombination covered wing and leg regions (Wieschaus and Gehring, 1976) as well as haltere and leg cuticle (Steiner, 1976), or where wing and leg tissue originated from a singly transplanted blastoderm cell (Fig. 4b), usually the clones were limited to disc-specific adult structures. Along these lines, it has indirectly been concluded from dissociation—reaggregation experiments with embryonic cells that there might even be some compartmentalization within a disc primordium as far as anterior versus posterior areas are concerned (Garcia-Bellido and Nöthiger, 1976).

In conclusion, we can say that the weight of evidence is thus pointing towards determinative events which occur already at the blastoderm stage, and seem to program intrinsically the fate of cells in the various positions of the early embryo. At present, however, we are not able to be more concrete about the specificity with which single blastoderm cells become restricted in their developmental capacity. It has recently been suggested that in a process of *step-wise determination* (Gehring, 1976b), those cells which originally are committed only to a disc primordium in the embryo become further specialized in their potential *within* a particular disc during the course of development.

IV. Cytoplasmic Localization of the Germ Line

We have previously seen that the cells and not the nuclei appear to be determined as early as during blastogenesis and, therefore, conclude that the determinative information has to reside in the cytoplasm. It has been known for some time that very early in the embryogenesis of a number of different animal species, a precocious segregation of morphogenetic potential takes place resulting in the

---

Fig. 4a–e. Larval and adult tissue contributions clonally derived from single *mwh e^4* blasto- ▷ derm cells after homotopic (a, e) and heterotopic (b–d) transplantation into *y sn^3 mal* blastoderm hosts. (a) Abdominal region of a chimeric female fly whose 6th sternite partly originated from a posterior → posterior cell transfer at the blastoderm stage. This clone (arrow) of *ebony* (*e^4*) cuticle and *multiple wing hair* (*wmh*) trichomes is clearly distinguishable from the surrounding *yellow* (*y*) cuticle and *singed* (*sn^3*) bristles of the host. × 250. (b) Leg (1) and wing (w) tissue which developed from a midlateral blastoderm cell after its transplantation to the posterior pole of a blastoderm embryo. The cuticular *mwh e^4* structures with leg-specific bracts and wing-specific trichomes did not become integrated but segregated clonally as one cluster of tissue which differentiated autonomously in the abdomen of the host fly. × 150. (c) Larval anterior midgut with a clone of darkly stained cells (*arrow*) derived from a single midventral blastoderm cell after anterior implantation and detected histochemically by testing for aldehyde oxidase activity. × 75. (d) Male genital structures which clonally arose from a posterior blastoderm cell after its anterior implantation. The two anal plates (*ap*) differentiated as separate structures without contributing to anterior tissues of the fly. × 150. (e) Eye-antennal disc of the third larval instar with a patch of stained cells (*arrow*). This aldehyde oxidase-positive clone in the antennal part of the disc, histochemically visualized by the enzyme test, is originated from an anterior→anterior cell transfer at the blastoderm stage

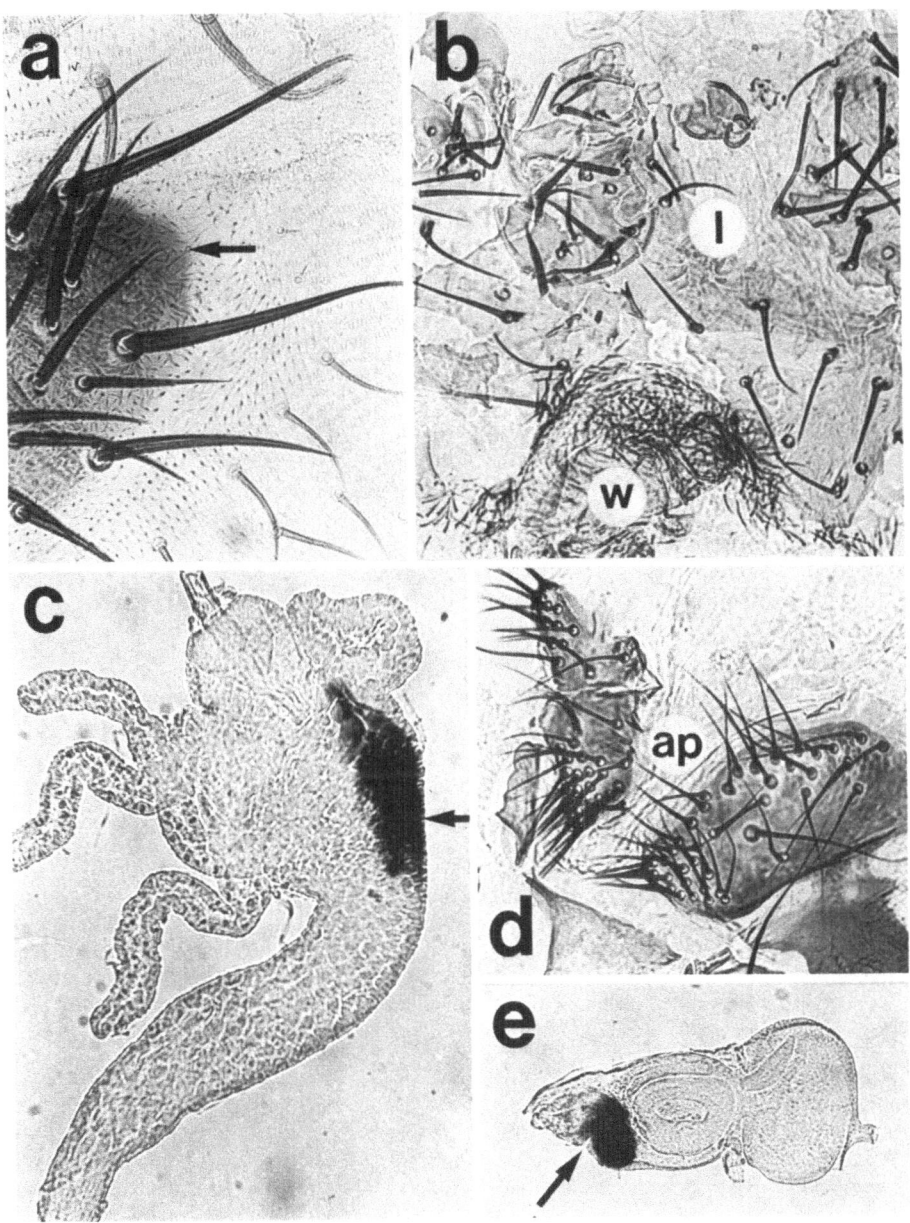

Fig. 4a–e

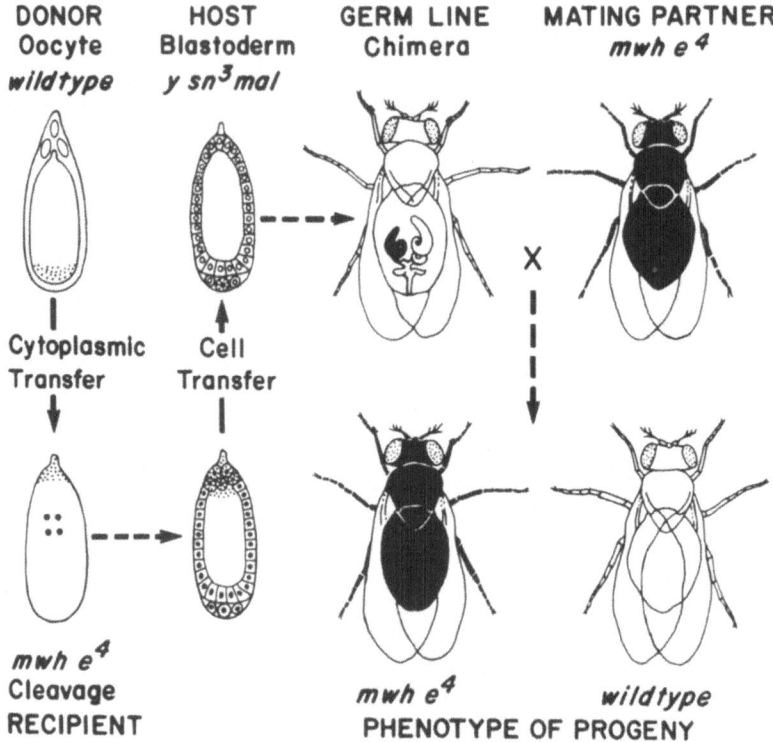

DONOR Oocyte *wildtype*	HOST Blastoderm *y sn³ mal*	GERM LINE Chimera	MATING PARTNER *mwh e⁴*

Cytoplasmic Transfer

Cell Transfer

X

mwh e⁴ Cleavage RECIPIENT

mwh e⁴   *wildtype*

PHENOTYPE OF PROGENY

Fig. 5. In vivo assay for germ cell determinants localized in a particular region of the *Drosophila* egg. Peripheral cytoplasm from the posterior pole of a *wildtype* oocyte is injected into the opposite site, the anterior tip, of a *mwh e⁴* cleavage embryo. During blastoderm formation, the injected cytoplasm segregates into several cells which are successively transferred to the posterior pole of *y sn³ mal* blastoderm embryos. There, they migrate together with the pole cells of the hosts into the gonads to establish a *mwh e⁴ sn³ mal* mosaic germ line in females as well as males. The hosts develop to flies and are mated to *mwh e⁴* partners. The appearance of *mwh e⁴* individuals among the progeny derived from some of these crosses demonstrates that the transplanted cells with injected cytoplasm function as germ cells

Fig. 6a and b. Microinjection of *posterior* polar cytoplasm from oocytes into the *anterior* ▷ region of early cleavage embryos. During blastoderm formation, the transplanted ooplasm exhibits different morphogenetic properties, depending on the developmental stage of the donor oocyte (for description of stages, see King, 1970). (a) Polar plasm from a stage-14 oocyte is separated into a cell with normal ultrastructural morphology. Characteristic subcellular organelles, the polar granules *(p)*, are clustered near the nucleus *(N)* with its nuclear body *(nb)* and are easily distinguishable from mitochondria *(m)*. The cytoplasm of mature oocytes already contains the morphogenetic potential to segregate into pole cells. The appearance of nuclear bodies, usually not present in anterior nuclei, further demonstrates the capacity of posterior ooplasm to induce normal pole cells. × 15000. (b) Transplanted polar plasm from a stage-11 oocyte and the polar granules *(p)* are found together with some nuclei *(N)* in an acellular region near the injection site. The cytoplasm of young oocytes is apparently not yet capable of forming pole cells, nor does it induce nuclear bodies. × 15000

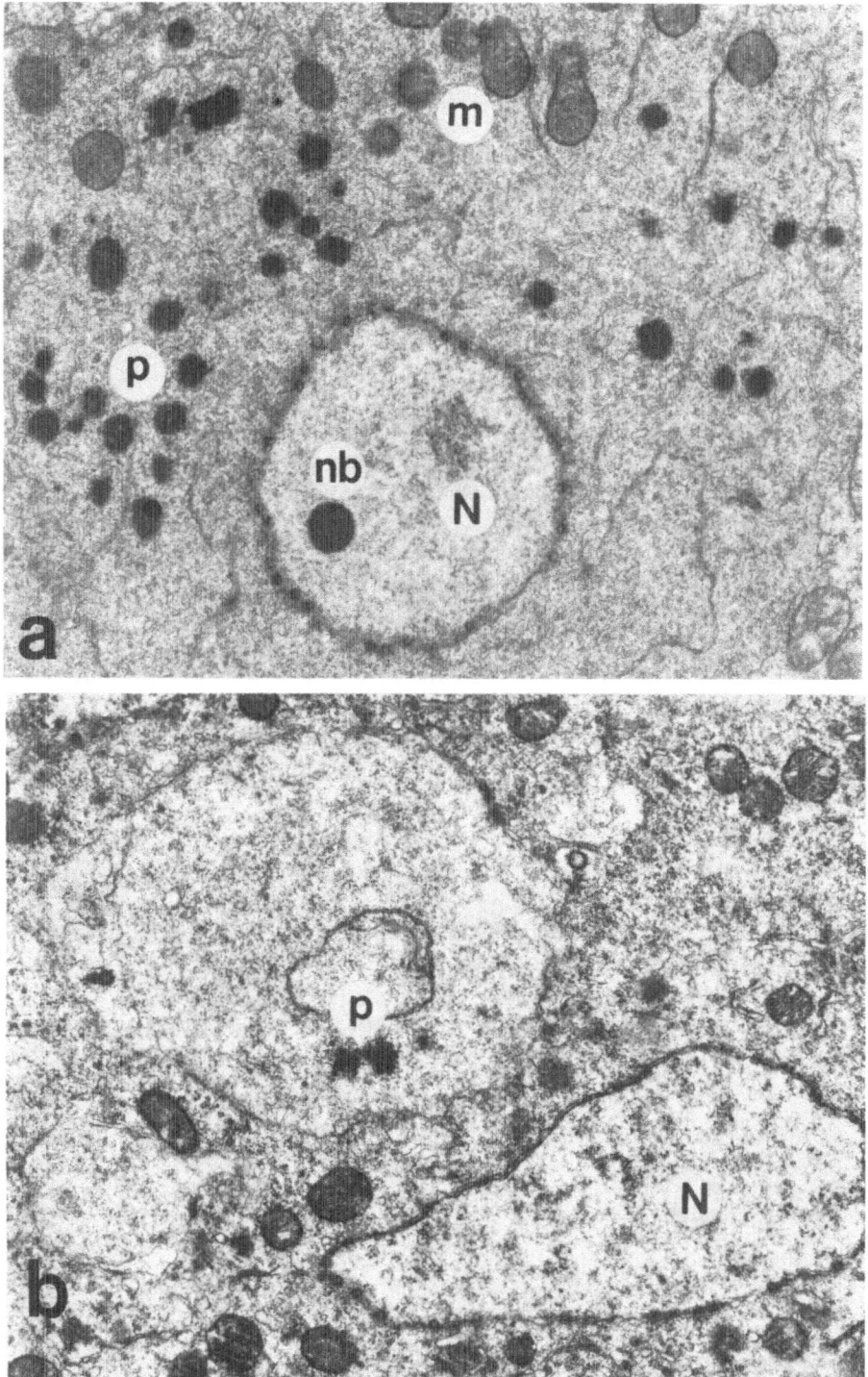

Fig. 6a and b

partitioning of localized cytoplasmic information of the egg into stem cells for future developmental pathways (reviewed by Davidson, 1976).

Probably the most conspicuous example for a cell lineage originating from a particular cytoplasmic region of the egg refers to the germ line in nematodes, amphibians, and insects (reviewed by Beams and Kessel, 1974; Eddy, 1975). The formation of germ cells can be prevented by UV irradiating the vegetal hemisphere of the *Rana* egg (Smith, 1966) or the posterior pole of the *Drosophila* egg (Okada et al., 1974a), and is restored after injection of cytoplasm collected from homotopic regions of nonirradiated eggs. In these studies, however, it has not been shown that germ cell determinants per se are already present in the egg. Demonstrating the presence of such determinants requires a bioassay for this particular cytoplasm with respect to regional autonomy, regardless of its position in the egg and functional specificity, i.e., development of germ cells that give rise to viable offspring.

Our approach to this problem in *Drosophila* actually consists of two distinct micromanipulations: (1) the transplantation of posterior polar plasm from eggs or even oocytes to heterotopic positions in cleavage embryos and (2) the transfer of cells formed at the site of cytoplasmic injection to the posterior pole of genetically different blastoderm embryos. The second operation is absolutely essential because the experimentally induced cells containing the injected cytoplasm (together with nuclei of the recipient) are not genetically distinguishable from surrounding normal cells. In the new hosts bearing several genetic markers, however, the experimental cells can be recognized phenotypically by cycling them through the germ line in chimeric flies (Fig. 5). Such a heterotopic bioassay provides a way of testing the posterior polar plasm for its autonomous inductive ability to form pole cells, and its specific determinative potential to take part in germ cell formation.

A. Regional Autonomy and Functional Specificity of Germ Plasm

In following this experimental scheme, we have shown that primordial germ cells can be induced in the anterior and ventral site of the *Drosophila* egg (Illmensee and Mahowald, 1974, 1976). Characteristic cytoplasmic constituents, the polar granules, served as a suitable morphological marker to identify the experimental cells. These conspicuous subcellular organelles are exclusively located at the pos-

---

Fig. 7a and b. Interspecific transplantation of polar cytoplasm between *Drosophila* embryos. ▷ (a) Normal pole cell of *D. immigrans* at the early gastrula stage. The polar granules *(p)* form large aggregates with rodlike substructural features that is quite different from the spherical and electron-dense fibrous structure of *D. melanogaster* polar granules. The nuclear body *(nb)* shows peripheral irregularities. (For comparison, see Fig. 6). × 18000 and *insert* × 40000. (b) Experimentally induced hybrid pole cell after transfer of posterior polar cytoplasm from a *D. immigrans* egg to the anterior region of a *D. melanogaster* cleavage embryo. The polar granules *(p)* form larger entities, do not aggregate into the same enormous cluster as in *D. immigrans*. But do, however, exhibit the typical rodlike substructure. Fibrous bodies *(fb)* near the nucleus *(N)* retain some structural characteristics of the polar granule. The nuclear body *(nb)* resembles more closely the irregular *D. immigrans* type rather than the regular *D. melanogaster* form. × 22000 and *insert* × 40000

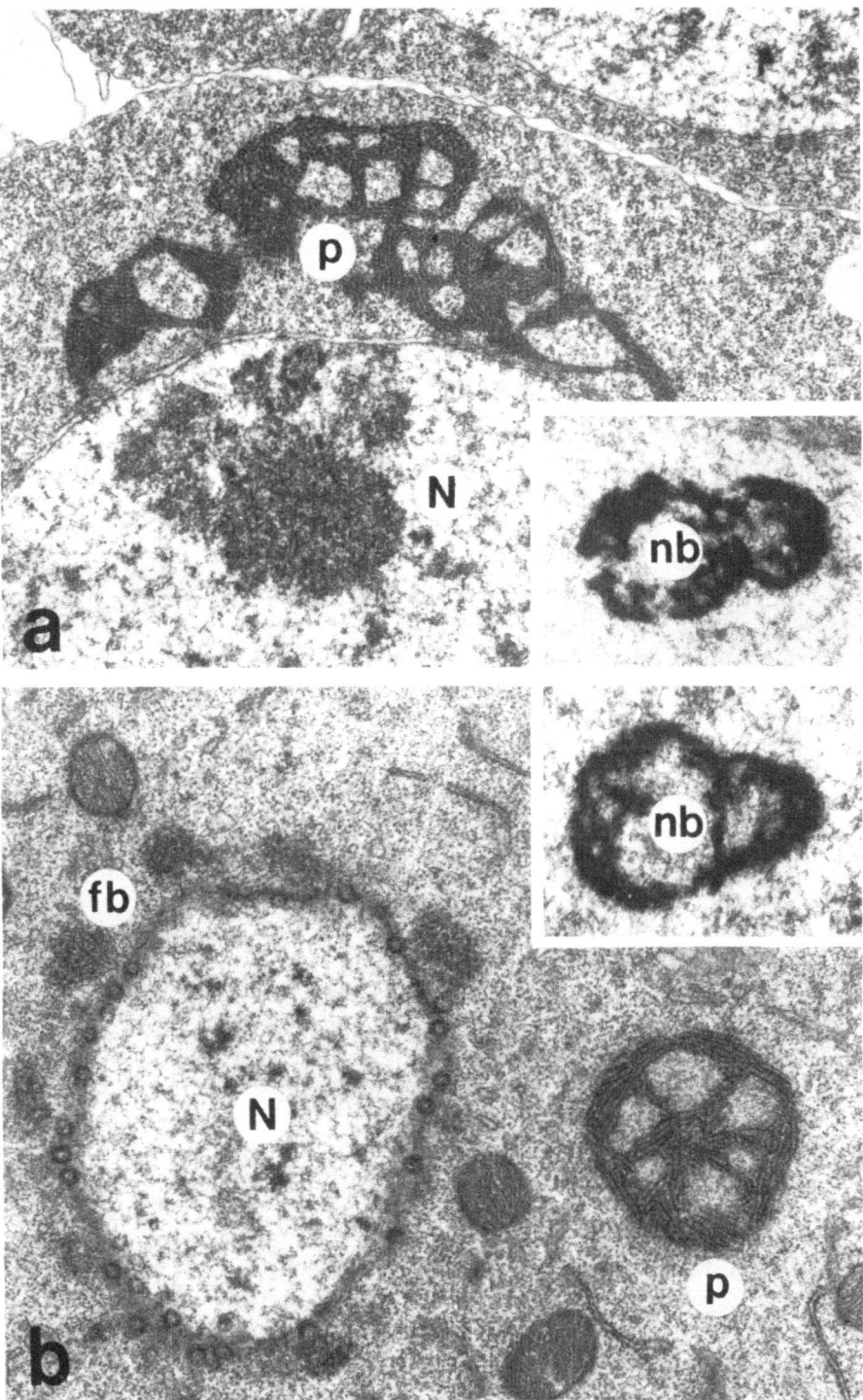

Fig. 7a and b

terior pole of the egg, where they become segregated into the pole cells (Mahowald, 1962). Ultrastructural analysis revealed that immediately after nuclei mig-
rated into the injected cytoplasm, several cells formed and separated from the
remaining embryo just as the pole cells normally bud off at the posterior tip of the
egg. Such early segregation of the induced cells reflects the morphogenetic autonomy of the posterior cytoplasm. Its functional specificity has been demonstrated
by the production of viable offspring derived from the experimentally induced
germ cells.

The presence of cytoplasmic determinants in the *Drosophila* egg poses two
closely related questions: is fertilization necessary for the polar plasm to be
functional, and if not, at what time during oogenesis does this particular cytoplasm acquire its determinative ability? In a subsequent series of experiments, we
found that the posterior cytoplasm from unfertilized eggs and late oocytes
(stages 13 and 14) was capable of inducing functional germ cells, whereas ooplasm
from earlier stages was not (Illmensee et al., 1976). Obviously, the posterior cytoplasm of oocytes younger than stage 13 has not yet attained its full morphogenetic
potential (Fig.6). A reasonable explanation for this failure may be that not all of
the constituents necessary for pole cell formation are localized at the posterior
region of immature oocytes. Since we always transplanted cytoplasm in toto, it is
not possible to find out which cytoplasmic components are responsible for the
specific developmental effect. By isolating subcellular fractions, and characterizing them biochemically and genetically, it should be possible to analyze their role
in germ cell determination.

B. The Action of Germ Plasm Across Genetic Boundaries

We now come to the question of whether or not the germ cell determinants are
species-specific in their mode of action. It is possible to test this by means of
cytoplasmic transplantation between different *Drosophila* species. Characteristic
organelles of the posterior egg cytoplasm, the polar granules, serve again as
critical marker since they exhibit structural features quite distinctive for each
species so far analyzed (Mahowald, 1968). Because of clearly recognizable differences between species in polar granule morphology, we can find out in interspecific combinations whether the transplanted pole plasm functions normally in a
hybrid cell whose nucleus is of one species, whereas its cytoplasm is derived from
a different one.

The results obtained from heterotopic transfers of *D. immigrans* egg cytoplasm
to *D. melanogaster* embryos clearly demonstrated that hybrid pole cells indeed
formed (Fig. 7). Their normal function could be disclosed in chimeric flies which
gave rise to hybrid cell-derived offspring (Mahowald et al., 1976). In conjunction
with the progeny test, it was possible to search for cytoplasmic inheritance of
polar granules by analyzing ultrastructurally the eggs and embryos originating
from the hybrid cells. A possible continuity of germ plasm to the next generation
has previously been proposed on purely morphological grounds (Mahowald,
1971). In our bioassay, however, we did not find any evidence for cytoplasmic

inheritance concerning the polar granules, since their morphology followed the nuclear genome, and not the cytoplasm.

Whatever the nature of germ cell determinants may be, the interspecific transplantations indicate that these components are to some extent conserved in evolution. A similar conclusion can be drawn from homotopic transplantations between different *Drosophila* species (Okada et al., 1974a). It remains to be shown how universally the determination of the germ line has been preserved across genetic boundares.

V. Concluding Remarks

We have seen here that our present state of knowledge on determinative events which occur during early embryogenesis of *Drosophila* does not allow us to be more precise in terms of a spatial and temporal program built into the egg. Towards an understanding of the mechanisms involved in determination, it is first necessary to establish a bioassay system in which the various entities (cell, nucleus, cytoplasm) can be analyzed independently with respect to their developmental capacity under in situ conditions. The production of *Drosophila* chimeras via microinjection into early embryos has proved to be of tremendous value in trying to study nucleocytoplasmic interactions during cellular diversification. In heterotopic *nuclear* transplantations, it has been shown that nuclei from blastoderm cells in different areas of the embryo remain totipotent. In contrast, as soon as cells form during blastogenesis, region-specific restrictions in developmental potential occur at the *cellular* level. Clearly, the determinative information has to reside in the egg cytoplasm, as revealed in heterotopic *cytoplasmic* transfers, and can even be traced to a particular region of the oocyte as far as the germ line is concerned. At present, it is not known whether a similar localization becomes established also for the somatic line of *Drosophila*. How far the various cytoplasmic determinants exist as such in the egg is a question not easy to answer. It remains a great challenge to search for these factors that are causally related to specific determinative processes, and may interact with discrete portions of the genome to control cell lineage-specific patterns of gene expression during development.

Acknowledgements. I should like to thank Dr. A.P.Mahowald for a fruitful collaboration during a very pleasant stay at Indiana University and Dr. T. Markow for careful reading of the manuscript.

References

Beams, H.W., Kessel, R.G.: The problem of germ cell determinants. Int. Rev. Cytol. **39**, 413—479 (1974)

Chan, L.-N., Gehring, W.J.: Determination of blastoderm cells in *Drosophila melanogaster.* Proc. Natl. Acad. Sci. USA **68**, 2217—2221 (1971)

Davidson, E.H.: Gene Activity in Early Development. New York: Academic Press 1976

Debec, A.: Isozymic patterns and functional states of *in vitro* cultured cell lines of *Drosophila melanogaster*. Wilhelm Roux' Archiv **174**, 1—19 (1974)

Deusen, E., Van: Sex determination in germ line chimeras of *Drosophila melanogaster*. J. Embryol. Exp. Morphol. **37**, 173—185 (1976)

Eddy, E. M.: Germ plasm and the differentiation of the germ cell line. Int. Rev. Cytol. **43**, 229—280 (1975)

Falk, R., Orevi, N., Menzel, B.: A fat map of larval organs of *Drosophila* and preblastoderm determination. Nature (New Biol.) **246**, 19—20 (1973)

Garcia-Bellido, A., Nöthiger, R.: Maintenance of determination by cells of imaginal discs of *Drosophila* after dissociation and culture *in vivo*. Wilhelm Roux' Archiv **180**, 189—206 (1976)

Gehring, W. J.: Developmental genetics of *Drosophila*. Ann. Rev. Genet. **10**, 209—252 (1976a)

Gehring, W. J.: Determination of primordial disc cells and the hypothesis of stepwise determination. In: Lawrence, P. (Ed.): Insect Development, pp. 99—108. Oxford: Blackwell 1976b

Gehring, W. J., Wieschaus, E., Holliger, M.: The use of "normal" and "transformed" gynandromorphs in mapping the primordial germ cells and the gonadal mesoderm in *Drosophila*. J. Embryol. Exp. Morphol. **35**, 607—616 (1976)

Geigy, R.: Action de l'ultra-violet sur le pôle germinal dans l'oeuf de *Drosophila melanogaster* (castration et mutabilité). Rev. Suisse Zool. **38**, 187—288 (1931)

Gurdon, J. B.: The Control of Gene Expression in Animal Development. Cambridge: Harvard University Press 1974

Illmensee, K.: The potentialities of transplanted early gastrula nuclei of *Drosophila melanogaster*. Production of their imago descendants by germ-line transplantation. Wilhelm Roux' Archiv **171**, 331—343 (1973)

Illmensee, K.: Nuclear and cytoplasmic transplantation in *Drosophila*. In: Lawrence, P. (Ed.): Insect Development, pp. 76—96. Oxford: Blackwell 1976

Illmensee, K., Mahowald, A. P.: Transplantation of posterior polar plasm in *Drosophila*. Induction of germ cells at the anterior pole of the egg. Proc. Natl. Acad. Sci. USA **71**, 1016—1020 (1974)

Illmensee, K., Mahowald, A. P.: The autonomous function of germ plasm in a somatic region of the *Drosophila egg*. Exp. Cell Res. **97**, 127—140 (1976)

Illmensee, K., Mahowald, A. P., Loomis, M. R.: The ontogeny of germ plasm during oogenesis in *Drosophila*. Dev. Biol. **49**, 40—65 (1976)

King, R. C.: Ovarian Development in *Drosophila melanogaster*. New York: Academic Press 1970

Mahowald, A. P.: Fine structure of pole cells and polar granules in *Drosophila melanogaster*. J. Exp. Zool. **151**, 201—215 (1962)

Mahowald, A. P.: Polar granules of *Drosophila*. II. Ultrastructural changes during early embryogenesis. J. Exp. Zool. **167**, 237—262 (1968)

Mahowald, A. P.: Origin and continuity of polar granules. In: Reinert, J., Ursprung, H. (Eds.): Results and Problems in Cell Differentiation, Vol. II, pp. 158—169. Berlin-Heidelberg-New York: Springer 1971

Mahowald, A. P., Illmensee, K., Turner, F. R.: Interspecific transplantation of polar plasm between *Drosophila* embryos. J. Cell Biol. **70**, 358—375 (1976)

Marsh, J. L., Wieschaus, E.: Germ-line dependence of the *maroon-like* maternal effect in *Drosophila*. Dev. Biol. **60**, 396—403 (1977)

Marsh, J. L., Van Deusen, E. B., Wieschaus, E., Gehring, W. J.: Germ line dependence of the *deep orange* maternal effect in *Drosophila*. Dev. Biol. **56**, 195—199 (1977)

McLaren, A.: Mammalian Chimaeras. Cambridge: University Press 1976

Needham, J.: Biochemistry and Morphogenesis. Cambridge: University Press 1950

Okada, M., Kleinman, I. A., Schneiderman, H. A.: Restoration of fertility in sterilized *Drosophila* eggs by transplantation of polar cytoplasm. Dev. Biol. **37**, 43—54 (1974a)

Okada, M., Kleinman, I. A., Schneiderman, H. A.: Chimeric *Drosophila* adults produced by transplantation of nuclei into specific regions of fertilized eggs. Dev. Biol. **39**, 286—294 (1974b)

Poulson, D. F.: Histogenesis, organogenesis, and differentiation in the embryo of *Drosophila melanogaster* Meigen. In: Demerec, M. (Ed.): Biology of *Drosophila*, pp. 168—274. New York: Hafner 1965

Poulson, D. F., Waterhouse, D. F.: Experimental studies on pole cells and midgut differentiation in Diptera. Aust. J. Biol. Sci. **13**, 541—567 (1960)

Santamaria, P.: Transplantation of nuclei between eggs of different species of *Drosophila*. Wilhelm Roux' Archiv **178**, 89—98 (1975)

Schneider, I., Blumenthal, A. B.: *Drosophila* cell culture. In: Wright, T. R. F., Ashburner, M. (Eds.): The Genetics and Biology of *Drosophila*, Vol. II, pp. 266—315. New York: Academic Press 1978

Schneiderman, H. A.: New ways to probe pattern formation and determination in insects. In: Lawrence, P. (Ed.): Insect Development, pp. 3—34. Oxford: Blackwell 1976

Smith, L. D.: The role of germinal plasm in the formation of primordial germ cells in *Rana pipiens*. Dev. Biol. **14**, 330—347 (1966)

Sonnenblick, B. P.: The early embryology of *Drosophila melanogaster*. In: Demerec, M. (Ed.): Biology of *Drosophila*, pp. 62—167. New York: Hafner 1965

Steiner, E.: Establishment of compartments in the developing leg imaginal discs of *Drosophila melanogaster*. Wilhelm Roux' Archiv **180**, 9—30 (1976)

Wieschaus, E., Gehring, W. J.: Clonal analysis of primordial disc cells in the early embryo of *Drosophila melanogaster*. Dev. Biol. **50**, 249—263 (1976)

Zalokar, M.: Transplantation of nuclei in *Drosophila melanogaster*. Proc. Natl. Acad. Sci. USA **68**, 1539—1541 (1971)

Zalokar, M.: Transplantation of nuclei into the polar plasm of *Drosophila* eggs. Dev. Biol. **32**, 189—193 (1973)

Estimating Primordial Cell Numbers in Drosophila Imaginal Discs and Histoblasts

JOHN R. MERRIAM

Department of Biology, University of California, Los Angeles, CA, USA

I. Introduction

A primary question in developmental biology is: What is the genetic basis by which groups of cells adopt different developmental fates? In *Drosophila*, cleavage nuclei prior to blastula stage are held together in a syncytial mass; both indirect (fate maps) and direct experiments (nuclear transplantation) show cleavage nuclei to be without developmental specificity. The suggestion has been made that earliest determination occurs at the blastula stage (Chan and Gehring, 1971); during gastrulation differentiation of cells in the amnioserosal layer and of other tissues necessary for embryonic life is clearly visible (cf. Turner and Mahowald, 1977). Other cells have been identified that are apparently set aside and only later give rise to the tissues of the adult. These sets of cells, called imaginal discs and histoblasts, are named for the structures which they will form. Thus, the mesothoracic leg, for example, is formed from the second leg imaginal disc and the prothoracic leg by the prothoracic leg disc. The critical observation for us is that adult structures, populated wholly by imaginal disc descendants, can be genetically mosaic (Sturtevant, 1929). This has been interpreted as an indication that these structures are formed from groups of cells, that the adult structures they will give rise to are polyclonal (Crick and Lawrence, 1975) in origin. Likewise, transdetermination (Hadorn, 1965) appears to be a change in the developmental fate shared by a small group of cells (Gehring, 1967).

To better understand the mechanism of determination it is of interest to know what is the initial number of cells set aside (or "restricted") to form a primordium. The main problem in estimating the initial number of determined cells is that we do not know when determination occurs or which stage of the rudiment is the extension of the determined state. Conversely, the importance of cell numbers and the time and number of cell divisions lies in the inferences about when determination of the imaginal anlagen takes place. The distinctiveness of the imaginal cells and the characteristic onsets of mitoses in the larval stages (Auerbach, 1936; Madhavan and Schneiderman, 1977) argue that the determinative restrictions occur in the embryo. With that in mind, we can attempt a resume of the polyclone number at each restriction. For a long time it was considered that imaginal discs became individually determined in early development (see review by Gehring and Nöthiger, 1973), and thus a distinction could be made between wing determination, leg determination, and so on. However, recent experiments on clonal analysis of early embryos (Steiner, 1976; Wieschaus and Gehring, 1976a; Lawrence and Morata, 1977) have shown that at blastoderm, the anterior wing and anterior second leg share a common pool of cells at a time when the anterior and posterior wing cells are already segregated. At that time the anterior and posterior leg cells are likewise segregated. Thus, at blastoderm the imaginal disc anlagen still do not exist as such. This fact also indicates that when both the wing and leg disc rudiments appear they are already subdivided into two groups of cells (those anterior and those posterior) that are differently determined. Garcia-Bellido and co-workers (1973, 1976) have described compartmentalization as a process to increasingly subdivide anlage into more and more specific developmental sectors; in their view the anterior and posterior groups of cells forming the wing and leg discs are compartments (Lawrence and Morata, 1977).

Although most of the observations on mosaic flies were reported before it was recognized that the wing and leg discs are each formed by two compartments, the results on mosaic flies can be interpreted roughly to estimate the sizes of the early compartment polyclones formed at the initial developmental restrictions. This review presents the detailed results in the literature of scoring mosaic imaginal disc anlage in gynandromorphs and by mitotic recombination induced early in development. The data are used to illustrate different means of estimating polyclone numbers for the discs and histoblasts. Scoring the frequency of mosaic discs and histoblasts in gynandromorphs is a convenient way of establishing the relative sizes between different discs but the only way of estimating the absolute number of cells in a polyclone is by measuring the minimal patch size in mosaics. In addition the recent histologic observations of Madhavan and Schneiderman (1977) are available for comparison and are in general agreement with the more indirect estimates for the discs and histoblasts. Additional observations are necessary using compartments rather than discs as units for reporting before the accuracy of the estimated polyclone numbers can be tested.

II. Minimal Mosaic Patch

The most useful technique for estimating primordial cell numbers derives from mosaic patterns that are established before the primordial cells are set aside. In *Drosophila* mosaic patterns can be formed by elimination of an X chromosome from some of the cleavage nuclei to form a gynandromorph. In the usual gynandromorph one of the daughter cleavage nuclei descended from the original XX (female) zygotic nucleus appears to have lost an X chromosome at the first division. The nucleus receiving only one X chromosome is XO, or male forming, and that remaining X chromosome may carry genetic markers (Lindsley and Grell,

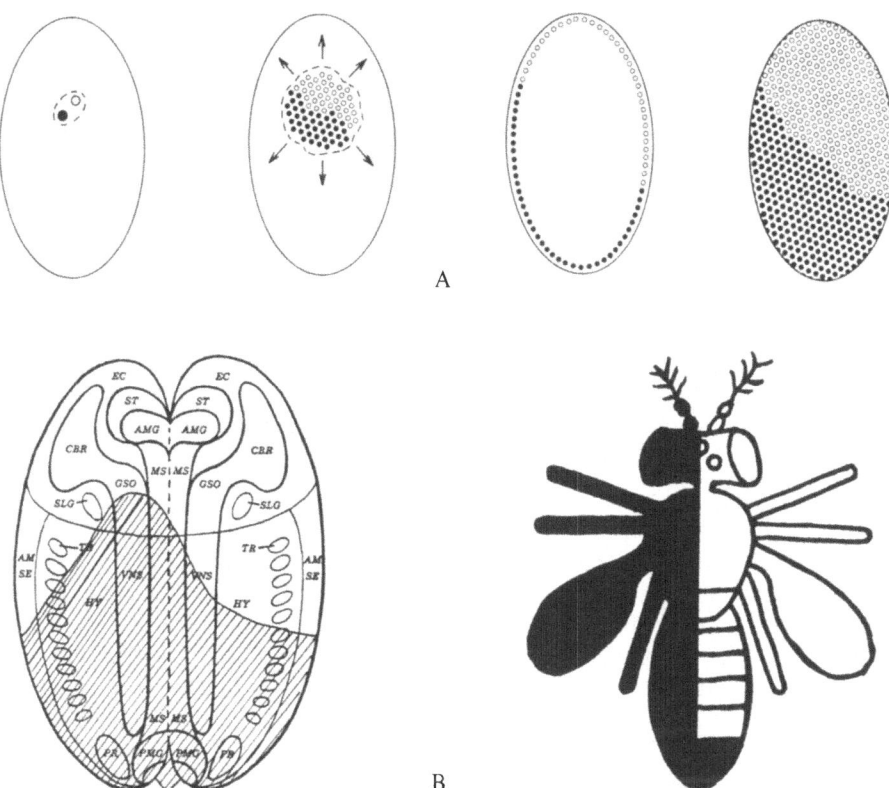

Fig. 1. (A) Schematic diagram illustrating formation of a gynandromorph. Starting with a female egg, one of the X chromosomes is lost during the first nuclear division. The nuclei divide in a syncytium and, eventually, the nuclei migrate to the surface of the egg, forming a monolayer of cells (the blastoderm). (B) The distribution of male and female cells overlays the standard fate plan (Poulson, 1950) showing the regions of the blastoderm destined to give rise to the various larval organs. The adult is a composite formed from some female and some male discs

1968) to distinguish its descendants from those of its sister XX nucleus. This is illustrated in Figure 1. Although gynandromorphs are often formed where a clean dividing line seems to separate the male half of the fly from the female half, the line of demarcation sometimes passes through the structures derived from a single disc. This is recognized as a mixed disc, which is descended from a mosaic anlage. In Figure 2 are shown several examples of mixed wing discs where cells of one sex are in a minority and form a patch with regard to the majority of cells in that disc. In these flies the patch may be either male or female; cells of the same sex as the patch are always found in other parts of the cuticle of the same gynandromorph. This ensures that the gynandromorph was formed early rather than by an elimination after the wing anlage was formed.

Gynandromorphs can be routinely produced by elimination of the unstable ring-X chromosome (Hinton, 1955) or by use of the ca^{nd} allele (Sturtevant, 1929; Lewis and Gencarella, 1952), the *mit* allele (Gelbart, 1974), or the *pal* allele (Baker,

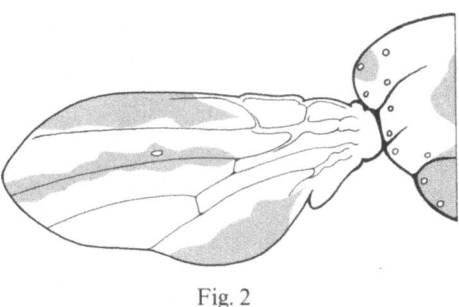
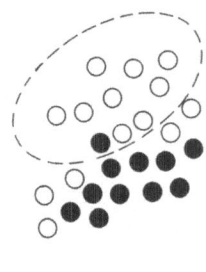

Fig. 2   Fig. 3

Fig. 2. *Hatched areas* mark examples of six independent minimal mosaic patches found by Ripoll (1972) in different regions of the notum and wing surface

Fig. 3. The distribution of differently marked cells in a gynandromorph blastoderm in relation to the origin of an imaginal anlage *(dashed outline)*

1975). Elimination of the maternal X chromosome (ca^{nd} allele) or the paternal X chromosome (*pal* allele) is usually at the first or second cleavage division, giving rise to a large proportion of gynandromorphs with one side male and the other side female. The *mit* allele conditions the loss of either X chromosome at the fourth or fifth cleavage division; consequently only an average of one-tenth of the gynandromorph is male. The chance of a disc or histoblast anlage being mixed should be reduced in a *mit* mosaic fly (one in which an elimination has occurred) although the distribution of patch sizes in mosaic disc or histoblast anlagen is unchanged. Flies with large mosaic patterns can also be produced by injection of marked nuclei into preblastoderm eggs (Zalokar, 1971; Okada et al., 1974), but this technique has not yet been employed in estimating primordial cell numbers.

The blastoderm layer of a gynandromorph is considered to be sectored into large male and female patches. The mosaic pattern, different for each fly, is established by the cellular blastoderm stage and is superimposed upon the standard developmental fate plan (Poulson, 1950). In that plan, each structure is derived from certain blastoderm cells directly or from the descendants of blastoderm cells; each region of the blastoderm ultimately gives rise to some structure(s) in the embryo, larva, or adult fly. If a given adult structure is mosaic in an individual, that fly represents a case where the boundary separating male from female cells fell within the portion of the blastoderm giving rise to the structure (Fig. 3). The mosaic adult structure may be one-half male and one-half female, or some other fraction. We assume that a half-and-half structure represents a case in which approximately half of the cells forming the anlage were male and half female. If all precursor cells go through the same number of divisions (see below), a case of one-eighth marked structure means at least eight cells are involved in forming the anlage. In theory, the minimal mosaic patch is the smallest fraction of the structure that can be marked; it represents the case where one of the cells is marked differently from the remaining cells forming the anlage. The number of cells forming the anlage is estimated as the reciprocal of the fraction of area in the structure making up the minimal mosaic patch. The application of this approach

Table 1. Number of cells estimated from minimal mosaic patch in gynandromorphs

Disc	Smallest size patch	Number of cells cited	Reference
Wing	1/8–1/16	8	Stern, 1940
Wing	1/12	12	Garcia-Bellido and Merriam, 1969
Wing	1/47	40	Ripoll, 1972
Leg	1/20	20	Bryant and Schneiderman, 1969
2nd antennal segment	1/8–1/10	9	Postlethwait and Schneiderman, 1971
Eye antenna	1/13	>13	Garcia-Bellido and Merriam, 1969
Tergite	1/8	8	Garcia-Bellido and Merriam, 1971 b

to several of the imaginal discs is summarized in Table 1; examples of specific applications to the well-studied wing disc are further discussed here.

Stern (1940) was the first to attempt to estimate the absolute number of cells entering into the mesonotal portion of the wing disc. His measurements were of the proportion of surface covered by the smaller area on 26 mosaic half mesonota.

Approximate area:	1/24	1/16	1/8	1/4	1/3	1/2
Number of cases:	1	2	11	8	1	3

He interpreted these data as indicating that at least eight or sixteen cells make up the mesonotal part of the prospective wing disc. Garcia-Bellido and Merriam (1969) scored 93 mosaic wing discs on the basis of 12 landmarks. In their hands the approximate area was given as the fraction of number of the landmarks marked out of the 12 scored as male in the mosaic disc. The numbers of cases they found are as follows:

Fraction:	1/12	2/12	3/12	4/12	5/12	6/12	7/12	8/12	9/12	10/12	11/12
Number of cases:	5	11	15	7	11	7	11	4	8	7	7

Since the wing blade was included as a single landmark in these data, the estimates are appropriately limited to the size of the mesonotal region. Like Stern's results, these data suggest 12 as the approximate number of cells making up that part of the wing disc. It must be kept in mind of course, that the twelve cells referred to cannot be distinguished by this method between being part of the cellular blastoderm stage or descendants of blastoderm cells. Nor can the possibility be distinguished that more than 12 cells form the initial mesonotal region. The estimate of Garcia-Bellido and Merriam could not be made more precise because of the limit in the number of landmarks. A more striking example comes from their analysis of the eye anlage on the basis of 13 landmarks: The smallest fractions dividing the eye anlage (1/13 male and 12/13 male) were also the most frequent mosaic classes. They concluded only that more than 13 cells form the eye anlage.

In an ideal system, the number of developmentally distinct landmarks would be large enough that the smallest patch observed would always include 2 or more landmarks. At least the ideal system would make clear the frequency distribution of the number of clones as multiples of the smallest mean patch size. In the tergites an average of 49 bristles per side are developmentally distinct in that any one can be in a male patch while adjacent bristles are female. Garcia-Bellido and Merriam (1971b) summarized the distribution of the number of cases in 66 mosaic half tergites according to the numbers of male bristles. Their results are:

Average # male bristles:	3	6*	12*	15	18*	24*	30*	33	36*	42*	45
Number of cases:	9	7	11	4	8	5	3	6	9	4	0

The *indicates those classes thought to represent multiples of the average of six bristles, corresponding to one out of eight cells initially forming each half tergite. The argument that the class of three bristles represents mitotic recombination and not the smallest patch is made in a section below. The frequencies of cases are uniformly (not binomially) distributed by numbers of bristles because the chance of any one of the eight initial cells being male is not independent of any of the other seven cells being male (cf. Fig. 1).

With better gene markers of the cuticle (f^{36} in the bristles, *mwh* in the trichomes) it is possible to recognize small patches of marked tissue and use the simple area of the clone to measure patch sizes. Ripoll (1972) scored patches of the extreme f^{36a} allele that mark wing trichomes in addition to bristles. Thus he was able to observe mosaic patches on the wing blade as well as the notum. A sample of six minimal mosaic patches he observed with the compound microscope in different gynandromorphs is shown in Figure 2. The number of adult cells included in each small patch was measured; the average calculated from the smallest patches was 1/47 of the total wing disc (excepting the pleura and post notum, which could not be scored). His distribution of patch areas was not given. Postlethwait and Schneiderman (1971) plotted the distribution of the % male tissue marked with y and f^{36a} in 60 mosaic second antennal segments. Inspection of their Figure 10 suggests about 8–10% of the segment forms the peak of the smallest male patches observed; approximately seven other peak sizes are observed, which more or less correspond to multiples of the smallest size. These data suggest 10–12 cells form the initial antennal segment. However, Postlethwait and Schneiderman arrived at essentially the same number through an alternative interpretation of the ideal frequency times size plot. On the grounds that a sample of n cells can yield no more than $n-1$ mosaic classes, they concluded that nine is the number of cells initially forming the second antennal segment. Using the smallest size has perhaps fewer problems as an indicator of initial cell number in practice, because variations in the growth or scoring of the marked cells might be expected to obscure any of the sharp size classes predicted by theory.

One potential problem with the minimal patch method that can usually be excluded is late loss of one chromosome contributing to small patches. Ring-X chromosomes in particular would be susceptible to loss through sister chromatid exchange. However, Ripoll (1972) failed to find male spots of more than one to two bristles in the mounted thoraces of nongynandromorph sisters with the unstable ring, $In(1)w^{vC}$. Also, the data of Garcia-Bellido and Merriam (1969) given above show that the minimal female patch gives the same results as the minimal male patch. If late loss were a major problem, smaller male than female patches would be expected.

Several authors (e.g., Wieschaus and Gehring, 1976a; Lawrence and Morata, 1977) have commented that the minimal patch method is likely to yield overestimates of the primordial cell numbers. If mitotic growth is not uniform over the entire disc, clones in some regions might undergo fewer mitotic divisions and, hence, underestimate the true $1/n$ fraction. Likewise, if some regions of the disc undergo programmed cell death, or are not included in external cuticle that can be marked, clones involving these regions would also underestimate the true $1/n$ fraction. For these reasons, Ripoll adopted 40 (rather than 47) for his estimate of the number of primordial wing disc cells.

Another criticism of the minimal patch method is that it usually considers the least frequent classes of mosaic anlagen. By increasing the sample size the chance increases of obtaining a deviant example (for whatever reason), resulting in a less accurate estimate of primordial cell number. While this reservation is necessarily true, and must always be kept in mind, there is no other method by which the absolute number of primordial cells can be estimated. It forms the standard by which other estimates must be checked. Methods based on the frequency of mosaic structures (Morata and Garcia-Bellido, 1976) can only refer to the relative primordial sizes at best. Nonetheless, because the minimal patch method lacks statistical efficiency in that inferences are drawn from the smallest number of cases, it is often preferable to concentrate on the frequencies with which structures are mosaic.

III. Mosaic Frequencies

It is perhaps surprising that as high as 25% of individual eye antennal or leg discs are mosaic in gynandromorphs. In *Drosophila* the gynandromorph mosaic patterns are thought to be highly contiguous (e.g., Fig. 1), unlike, for example, mosaic patterns in the mouse (Nesbitt, 1974). As a result the sample/patch ratio (Hutchinson, 1973) is a special case in *Drosophila* gynandromorphs since the patches are large and the ratio is less than one. That is, the area of the sample (the polyclone number set aside to form the anlage) is probably much less than the size of the patch of marked cells. In this sense, patch refers to the coherent growth of a single clone; a gynandromorph may be minimally formed by two patches, or more depending on mixing. With either chimeric or X inactivation mosaic mice, the sample/patch ratios for adult organs are usually greater than one (Nesbitt, 1974). The only other case of the sample/patch ratio being less than one is the development of human tumors studied in women heterozygous for the A and B electrophoretic forms of G6PD. Linder and Gartler (1965) first showed that the benign uterine leiomyoma tumors are usually characterized by one electrophoretic type (either G6PD A or B) in heterozygous women; adjacent samples of tissue even as small as 1 mm^3 are invariably mixed. Such results are taken to mean that the leiomyoma tumors are probably monoclonal, or of single-cell

Table 2. Mosaic frequencies in imaginal tissues in *Drosophila* gynandromorphs

Tissue	Part of disc	Mosaic %	Reference
Wing	Entire disc	12	Garcia-Bellido and Merriam, 1969
		12.5	Bryant, 1970
		19[a]	Ripoll, 1972
		16	Janning, 1974b
		14.7	Wieschaus and Gehring, 1976b
		20	Lawrence and Morata, 1977
		14.7	Bryant, personal communication
	Wing blade	18	Ripoll, 1972
	Anterior blade	18.4	Lawrence and Morata, 1977
	Posterior blade	8	Lawrence and Morata, 1977
	Notum	11.5	Ripoll, 1972
Leg	Entire disc	25	Bryant and Schneiderman, 1969
		18–11	Janning, 1974b
		22–15.5	Wieschaus and Gehring, 1976b
Eye antenna	Entire disc	23.8	Garcia-Bellido and Merriam, 1969
		23	Janning, 1974b
	2nd antennal	18	Postlethwait and Schneiderman, 1971
	segment	5	Janning, 1974b
Tergites	4th half segment	8	Garcia-Bellido and Merriam, 1971b
		6	Janning, 1974b
Genitalia	Entire	21	Gehring et al., 1976
	Outer	23	Janning, 1974b
	Inner	12	Janning, 1974b
Malpighian	Entire	51	Janning, 1974a
tubules	Anterior	37	Janning, 1974b
	Posterior	40	Janning, 1974b
Germ line	Entire	25	Lee et al., 1967
		17	Nissani, 1977

[a]   As cited in Nissani and Lipow, 1977.

origin. Similar studies have since been extended to other neoplasms with the result that several kinds of human tumors, both benign and malignant, are characteristically found to be monoclonal in origin (reviewed by Fialkow, 1976). Since all of the *Drosophila* anlagen studied can be observed to be polyclonal in origin, the analysis of their mosaic frequencies may shed some light on the degree of cell mixing that occurs during early development.

The frequencies for which tissues are mosaic in gynandromorphs are summarized in Table 2. It is, of course, expected that the chance of an anlage being mosaic in gynandromorphs should rise as the size of its primordial pool increases. Anlage with large starting pools should be mosaic more frequently than those with smaller starting pools. Those tissues that are estimated by the minimal patch method to start from only a few cells (8–12 for the tergites) are also observed to be infrequently mosaic (about 6–8% of individual half tergites: Garcia-Bellido and Merriam, 1971b; Janning, 1974b). The wing, estimated by minimal mosaic patches to start from about 40 cells, is observed to be mosaic in 15–20% of cases.

Some of the variation observed between studies in mosaic frequencies for any disc or histoblast may be due to observer differences; biologic differences, such as

Baker's (1975) conclusion that the X chromosome from *pal* fathers can be lost at the second or third cleavage division, must surely also be important. Viability of marked cells may also play a role: Gelbart (1974) found fewer *y* than sn^3 male patches among the mosaic daughters of *mit* females. In some reports, the mosaic frequencies are clearly low owing to the difficulty of adequately scoring a tissue for mosaics with available markers. That is probably the case with the lower frequency of wing mosaics observed by Garcia-Bellido and Merriam (1969). In other cases, mosaic tissues are overlooked when scoring with the dissecting microscope but are recognized when scoring with the compound microscope. That is particularly true for the legs and the tergites. For these reasons many failures to record mosaic structures in earlier reports (summarized in Garcia-Bellido and Merriam, 1969) are discounted and are not presented here.

Although unexplained variations exist, a comparison of the mosaic frequencies between discs or histoblasts is used to estimate the relative size of their anlagen. One metric given by Hotta and Benzer (1972, 1973) is the size of a circular ancestral pool of cells with a radius expressed in sturts. The radius is given by $r = (f/\pi) \times 100$, where f is the mosaic frequency of a tissue. This stems from the model that every mosaic boundary twice intersects an assumed circular circumference outlining the anlage; or $C = 2f \times 100$ sturts. By comparing the longest distances observed between landmarks within individual disc primordia with the diameter calculated from mosaic frequencies, Wieschaus and Gehring (1976b) argue that a circular primordium is a better description of determined wing and leg discs than are either elliptical or rod-shaped primordia.

Assuming circular primordia, the relative areas of discs can be compared from the calculated radii; the data from Table 2 on the wing blade parts and the notum, yield the following comparisons on the calculated radii and areas:

Structure	% Mosacism	Radius	Area	Reference
Anterior blade	18.4	5.9 sturts	108	Lawrence and Morata, 1977
Posterior blade	8	2.6 sturts	20	Lawrence and Morata, 1977
Wing blade	18	5.7 sturts	103	Ripoll, 1972
Notum	11.5	3.7 sturts	42	Ripoll, 1972

Lawrence and Morata (1977) estimate that the anterior wing starts from the descendants of about five times as many cells as does the posterior wing. From Ripoll's data the wing blade is seen to derive from 2.5 times as many cells as the notum. The areas calculated in this way are only relative and do not refer to absolute cell number; however, if the number of cells forming one structure and its mosaic frequency were accurately known, possibly they could be used to calibrate the primordial size of other structures. The tergites are one example: Each has a radius of about 2.6 sturts, an area of 20, and are formed from about 8–10 cells (Table 1 and below). Using these numbers for calibration suggests that roughly 50–60 primordial cells should form the wing blade (50 anterior, 10 posterior) and 20 cells should form the notum. These numbers are larger than expected, probably because they are related not to be blastoderm, but to some later stage, when the primordia are restricted. If the tergite and wing primordia are not set aside at the same time, a systematic bias would be introduced which could inflate the pool sizes. Another caveat to be noted is that discrepancies introduced by variations in mosaic frequencies between studies are increased by squaring the radii to find the relative areas. For example, recent measurements in the mosaic frequency of the wings range from 14.7% (Wieschaus and Gehring, 1976 b), to Ripoll's (1972) 19%, and 20% by Lawrence and Morata (1977). Wieschaus and Gehring (1976 b) observed a range of mosaic frequencies between the legs, from 22% in the foreleg to 15.5% in the metathoracic leg. From these frequencies the calculated radii and areas are found to be:

Structure	% Mosaicism	Radius	Area	Reference
Wing disc	14.7	4.7 sturts	69	Wieschaus and Gehring, 1976 b
Wing disc	19	6.0 sturts	115	Ripoll, 1972
Wing disc	20	6.4 sturts	128	Lawrence and Morata, 1977
Leg I	22	7.0 sturts	154	Wieschaus and Gehring, 1976 b
Leg III	15.5	5.0 sturts	78	Wieschaus and Gehring, 1976 b

Thus, depending on which mosaic frequency is used for the wing, there is a twofold difference in area. These data and those of Janning (1974 b) also show a twofold difference in area between the foreleg and the metathoracic leg; it is not possible from this approach to determine whether the difference reflects a smaller primordial pool for the metathoracic leg or that it is less easy to recognize mosaics in the metathoracic leg.

A more powerful method of relating the number of cells whose descendants populate a disc to the observed mosaicism of the disc is given by Nissani and Lipow (1977). For the sake of model building they assume the blastoderm of a first cleavage division gynandromorph represents a sphere equally bisected by a single division plane. The circumference of the division plane of the surface of the sphere describes a "great circle;" if the descendants of blastoderm cells in some small region give rise to a disc the mosaic frequency of the disc represents the probability of the great circle passing through that region on the blastoderm surface, which is related to its size. Of course this is an overestimate of primordium size since the dividing line is probably not a simple great circle. Chances are it is pushed into wiggles or even into multiple sectors (Hotta and Benzer, 1972). On the model of a great circle, however, the geometric relation of Nissani and Lipow between the area represented on the blastoderm and the mosaic frequency (P) is

$$\frac{n}{3400} = \frac{1}{2}(1 - \sqrt{1 - P^2})$$

n is the number of cells in the primordium; 3400 is the number of cells thought to populate the total blastoderm.

A comparison of the values of n with the number of primitive cells obtained from the minimal mosaic patches did suggest to Nissani and Lipow that they are largely equivalent. These numbers are reproduced below. However, certain points concerning the method must be raised before a detailed consideration of the numbers can be made here. First, more recent observations by Turner and Mahowald (1976) and by Zalokar and Erk (1976) indicate that about 6500 cells (rather than 3400) are involved in forming the blastoderm layer. That means the more recent calculated value for each primordial cell number is about twice the size calculated by Nissoni and Lipow. With the same mosaic frequency, the calculated starting size of the wing disc increases from about 31 cells to about 60 cells, for example.

Second, depending on the degree of cell intermingling before or after blastoderm formation, the observed mosaic frequency may lead to spurious overestimates of the number of primordial cells. A condition which inflates the apparent mosaic frequency is the possibility mentioned above that the blastoderm might be multiply sectored into several large patches. If this is so, the chance of any region or structure being mosaic increases and the size of the primordial pool based on mosaic frequency is overestimated. An increase in the frequency of mosaic structure is also predicted if cell intermingling occurs during gastrulation. That can be demonstrated in the following way: Consider a model of n blastoderm cells that give rise to the whole mesothoracic segment on a side. By the time cells are restricted to be wing or leg and are set apart there are $2^x n$ descendants of the original n cells, where x is the number of cell generations. Assuming the mosaic pattern is fixed at blastoderm, the probability of the sample of n cells being mosaic is also fixed. However, if cell movements intermingle the cells within the segment, any subset of the $2^x n$ descendant cells has a greater chance of being mosaic than if no cells shifted around. In that case the mosaic frequency for any adult structure would approach the mosaic frequency of the entire segment. Gehring et al. (1976) advanced this argument as an explanation of why the genital disc is so frequently mosaic. If intermingling is a factor, it should also lead to local distortions in the fate map relations between structures in the same segment. Many of the sturt distances within segments should be greater than the distance between segments. In fact, Wieschaus and Gehring (1976b) have observed what they term "map expansion," which they explain in this way. On the other hand, inspection of many gynandromorphs does not support the idea of an extensive homogenization of the cells forming a segment (Garcia-Bellido and Merriam, 1969). Stern (1940) also noted that patches are rarely inclosed by the other patch within a disc or segment.

Because there is little a priori knowledge of the degree of mixing or cell intermingling in gynandromorphs, it is instructive to see how the number of blastoderm cells calculated by Nissani and Lipow compares with the minimal mosaic patch data. The ratio should be representative of how much mixing actually occurs in gynandromorphs. Here the blastoderm cell numbers calculated by Nissani and Lipow for several discs on the basis of 3400 blastoderm cells (mosaic frequencies listed in Table 2) are compared with the number recalculated

on the basis of 6500 blastoderm cells and with the minimal mosaic patch estimates:

Structure	% Mosacism	$n/3400$	$n/6500$	Mosaic patch	Ratio
Wing blade	18	28	54	40	$\cong 1.4$
Mesonotum	11.5	11	21	17.	$\cong 1.2$
Wing disc	19	31	59	40	$\cong 1.5$
Tergite	8	5	10	8—10	$\cong 1$
Antenna	18	28	54	9	$\cong 6$
Eye disc	23.8	49	94	24(?)	$\cong 4$
Leg	22	42	80	20	$\cong 4$

The ratios listed are the blastoderm numbers calculated on the basis of 6500 cells
divided by the minimal patch estimates, which refer to the polyclone number
when the discs are set aside. In all cases the ratio is at least one or higher. Since
some mitotic divisions occur between blastoderm and restriction of the discs (see
below), the actual blastoderm cell numbers should never be more (and probably
always less) than the minimal mosaic patch number. Hence, the observation of
ratios above one is taken to mean there is probably considerable additional
mixing into multiple marked sectors in a gynandromorph. Whether additional
mixing occurs later in the formation of the eye antenna or leg discs is not clear. At
any rate, this lends support to an explanation of why as high as one out of four or
five of some discs is mosaic.

It may be possible to ascertain more directly the effect of mixing into multiple
sectors. The malpighian tubules are formed early in gastrulation from existing
cells at the origin of the hind gut (Poulson, 1950); they are frequently (51%)
mosaic in Janning's 1974a observations. He also observed that the minimal patch
size in the mosaic tubules corresponds to 3 out of the 425 cells in the tubules; that
suggests 140 as the approximate number of cells that differentiate to form the
tubules. (They undergo no further mitotic divisions.) On the basis of 6500 blastoderm cells and Nissani and Lipow's formula, the calculated blastoderm input
would be about 454 cells with 51% mosaics. Thus, additional mixing before blastoderm formation is possible. In this case, it is also possible that there is more
than one origin of the malpighian tubules, as indicated in Janning's 1974b blastoderm fate map (his Fig. 1). More than one origin of the malpighian tubules would
contribute to the high mosaic frequency observed overall for the tubules; it would
be better to consider each tubule individually since they are frequently discordant
in mosaicism. Among 345 gynandromorphs, that Janning (1974a) observed, 92
were mosaic in both the fore and hind tubules, with another 36 mosaic in the fore
tubules only and another 47 mosaic in the hind tubules only.

One example where a high mosaic frequency in a tissue probably stems from
sampling earlier in development, from a total pool size of less than 3400 cleavage
nuclei, is the germ line (Table 2). By progeny-testing attached-X-bearing females
visibly mosaic for the y^+ Y chromosome following irradiation of their y^+ Y-bearing
fathers, Lee et al. (1967) found that 18/72, or 25%, were mosaic in their gonads as
well as in the somatic cuticle. Nissani (1977) found that, out of 215 males mosaic

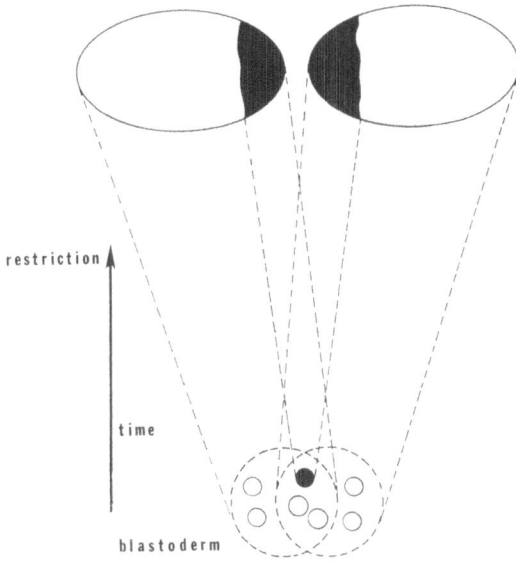

restriction

time

blastoderm

Fig. 4. Regions of overlap. Because restriction of the disc primordia takes place several cell divisions after blastoderm, the descendants of a marked cell in the blastoderm may contribute to more than one disc anlagen

for the $y^+ Y$ chromosome (from *pal* fathers), 36 (17%) gave rise to both y and y^+ sons when crossed to y females. According to Sonnenblick (1950) the germ line is established from the migration of 2–11 preblastula nuclei into the pole cell region (from histologic observations). Zalokar and Erk (1976) note that 12–18 pole cell nuclei are set aside after the eighth or ninth cleavage division. Lee et al. calculated that the average equivalent of 1.5 independently sampled nuclei are taken to form the germ line. This number was determined from the best fit to n in the relation of the chance of getting a mixed germ line equal to $1-(p^n+q^n)$ where $p=0.61$, the amount of y^+ tissue and $q=0.39$, the average amount of y tissues in the cuticle. Since the nuclei forming the germ line are probably not independently sampled, as Lee et al. recognized, the actual number of nuclei set aside to form the germ line must be somewhat larger. Consideration of the mosaic patch method in terms of Lee et al.'s results of progeny testing suggests that large numbers (ca. 20–30) of preblastoderm nuclei are involved: One female issued two y^+ sons and 50 y sons, indicating that only 1/25th of her gonads may have carried the $y^+ Y$ chromosome. Van Deusen (1976) created germ line chimerics by transplanting marked pole cells. The genotype of the donor pole cells appeared, on the average, in less than half the progeny when a germ line was mixed, so his frequency of obtaining mixed germ lines (15/55) is not instructive. In some broods as few as 36/267, or 15/93, or 4/23 progeny carried the donor genotype, indicating that some 6–8 cells most likely form the initial germ line before the separation into a symmetric pair of gonads (Gehring ct al., 1976). Assuming a determination of 6–8 nuclei at the 512-nuclei stage to form the future pole cells (germ line), the expected mosaic frequency calculated from Nissani and Lipow's formula would be about 21–25%.

Since only the pole cell nuclei appear to become determined before the blastula cellularization, the origin of every other tissue is calculated to begin from the 6500 cell blastoderm stage. However, because descendants of the same blastoderm cells may also populate a second nearby tissue, the estimates for different tissues cannot simply be added together to make a map of the blastoderm surface. The estimates of 54 blastoderm cells and 21 blastoderm cells (from Ripoll's 1972 data) listed above, whose descendants form the wing blade and the notum, are largely overlapping. Most of the ancestral blastoderm cells giving rise to the notum probably give rise to the wing blade as well and probably vice versa. In Figure 4 this situation is presented as regions of overlap. Another example of overlap represented on the blastoderm surface occurs between the anterior wing and leg and between the posterior wing and leg. This is because initially blastoderm cells are indeterminate between leg and wing (Steiner, 1976; Wieschaus and Gehring, 1976 a; Lawrence and Morata, 1977). Unfortunately the data on leg and wing mosaics are not available in such a way to calculate the overall sizes of the blastoderm pools that give rise to the anterior and posterior mesothoracic compartments. Inferences about these pool sizes are considered below after the timing of determination from mitotic recombination experiments can be taken into account.

IV. Mitotic Recombination

An alternative approach to estimating the size of the primordial pool of cells consists of marking descendants of individual cells in the pool by mitotic recombination. X-ray induction (typically with 1000 r in larvae or 500 r in embryos) of the recombinant event permits timing to specific embryonic or larval stages. In mitotic recombination, cells heterozygous for cell marker mutations undergo a partial exchange of homologous chromosome arms so that daughter cells are made homozygous for the cell markers. This process is schematically illustrated in Figure 5. By the judicious choice of cell markers both daughter cells may be marked to appear as twin spots on the adult cuticle. Since the homologous chromosomes must be in G_2 (or 4-strand stage) at recombination, the X-irradiation marks the time at which the exchange of arms occurs, but not the subsequent mitosis that segregates recombinant chromosomes to daughter cells.

The rationale of using early mitotic clones to estimate cell numbers is illustrated schematically in Figure 6. If one of the original primordial cells in a disc is induced to undergo an exchange and subsequently segregates marked daughter cells, the fraction of the adult disc area covered by one daughter clone should represent half of the fraction $1/n$ where n is the number of primordial cells. Table 3 lists the average patch sizes observed in mitotic recombinant spots induced at early stages. Where the time of X-irradiation events early in development is less certain (e.g., Garcia-Bellido and Merriam, 1971 a), the largest clones observed are used to represent the earliest stages. For other authors the average clone size induced at a given stage is used to avoid giving undue weight to variation in size

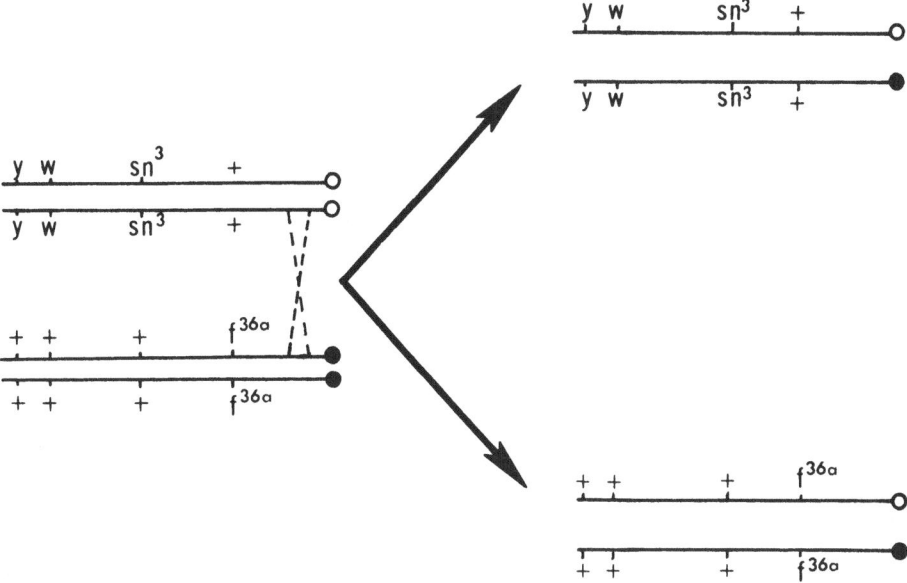

Fig. 5. Following a recombination between homologous chromosomes, segregation of the cross-over chromatids at mitosis leads to daughter cells that are homozygous for the chromosome arm distal to the cross-over. Genetic markers *y* (yellow bristles), *w* (white eyes), and *sn³* (singed bristles) are cell marker mutants on the X chromosome that permit recognition of the homozygous patches on the adult cuticle. The complementary daughter cell clones sitting side by side form a twin spot

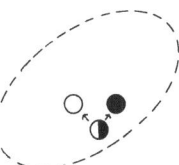

Fig. 6. Estimating the size of a polyclone by marking the descendants of a heterozygous cell assumes that the polyclone is determined *(dashed outline)* and that the cell is a member. The sum of the area of the two sister descendant cells should correspond to the minimal mosaic patch in gynandromorphs

owing to position of the clone within the disc. In addition, the data of certain authors (Bryant and Schneiderman, 1969; Bryant, 1970; Postlethwait and Schneiderman, 1971) have been corrected for the division required to segregate the marked cell.

Because of the possibility that X-irradiation also kills some of the cells in the original disc (Haynie and Bryant, 1977), there is some concern as to whether the observed size of the mitotic recombination clone is larger than half of the true fraction of $1/n$. If some cells are killed by irradiation, the surviving cells have to increase by regulative growth (or intercalary regeneration; Haynie and Bryant, 1976) the fraction of the disc each one occupies. Hence, by this deviation the

Table 3. Numbers of cells estimated from average size of mitotic recombinat patches.
E = embryo, L = larva

Disc	Average patch size	Age at irradiation	Number of cells cited	Reference
Wing	1/11	E	5[a]	Bryant, 1970
Wing	1/50	L	25	Garcia-Bellido and Merriam, 1971a
Wing	1/12	E	6	Wieschaus and Gehring, 1976a
Leg	1/20	E	10[a]	Bryant and Schneiderman, 1969
Leg	1/15	E_3	7.7	Wieschaus and Gehring, 1976a
Leg	1/36	E_{10}	18.3	Wieschaus and Gehring, 1976a
Antenna	1/7	E	3.5[a]	Postlethwait and Schneiderman, 1971
Eye antenna	1/12	E_3	6.2	Wieschaus and Gehring, 1976a
Eye antenna	1/26	E_{10}	13	Wieschaus and Gehring, 1976a
Tergite	1/17	L	8	Garcia-Bellido and Merriam, 1971b
Tergite	1/22	L	11	Guerra et al., 1973
Tergite	1/9	E_3	4.5	Wieschaus and Gehring, 1976a
Tergite	1/24	E_{10}	11.9	Wieschaus and Gehring, 1976a
Sternite	1/10	L	5	Guerra et al., 1973

[a]   Numbers corrected for absence of twin spot.

primordial cell number would be underestimated. However, Wieschaus and Gehring (1976a) found that the average clone size did not change between clones induced in embryos at 500 r and clones induced at 1000 r. The 1000-r dose would be expected to yield larger spots by killing more cells if that is a problem.

That there are clearly systematic differences between the primordial numbers obtained for the same disc from the minimal mosaic patch and from mitotic recombination can be seen by comparing the sizes listed in Tables 1 and 3. Bryant (1970) marked patches as small as 1/10 of the wing disc in gynandromorphs and as large as 1/5 (correcting for the twin spot) of the wing disc by mitotic clones induced in embryos. The clones observed by Wieschaus and Gehring (1976a) and Lawrence and Morata (1977) in the wing were also larger than the minimal mosaic patch observed by other authors.

The discrepancy between the minimal mosaic patch and the size of the mitotic clone is observed as well with structures other than the wing disc. Bryant and Schneiderman (1969) found 1/20 of the leg marked as the minimal mosaic patch, yet their mitotic clone spots were about the same without the twin spot. Wieschaus and Gehring (1976a) recovered single spots covering about 1/15 of the leg from which they estimate about 7.7 cells forming the leg disc are present at irradiation 3 h after egg laying (E_3)[1]. For the antenna, Postlethwait and Schneiderman (1971) described 1/9 as the minimal mosaic patch in the second segment. Their single mitotic clones in the antenna were also that large (1/7). Those single mitotic clones of Wieschaus and Gehring (1976a) induced at E_3 covered about 1/12 of the entire eye antenna disc; they estimate 6.2 cells forming the disc are present at E_3.

---

[1]   The terms E_3, E_{10}, or E_{12}, etc., refer to the age of the embryo in hours after egg laying and correspond to the approximate stage when irradiated.

The number of progenitor cells present in the embryo for a structure, as measured by mitotic recombination, is usually less than the number of primordial cells set aside to form the anlage, as measured by the minimal mosaic patch. The explanation for the systematic discrepancy probably lies in having induced mitotic clones before restriction of the imaginal anlagen takes place. Evidence that this is the case comes from observations that the induced clones extend over more tissue than the single imaginal disc area. Wieschaus and Gehring (1976a), Steiner (1976), and Lawrence and Morata (1977) recovered large clones that spread over both wing and leg tissues. That is, the wing and leg primordial cells could not yet have been separated into different discs at the time of irradiation in the early embryo. Postlethwait and Schneiderman (1971), Wieschaus and Gehring (1976a), and W. Baker (personal communication) observed clones in the head that extended through both antennal and eye tissues. Thus, the antenna shares cells with and is not distinctly separated from the rest of the eye imaginal disc during embryogenesis. In first instar larvae the antenna cells are reported to be more densely staining and can be counted apart from the eye cells (Madhavan and Schneiderman, 1977); this can be interpreted to mean that the two anlagen are developmentally distinct and do not share cells by that stage. However, Lawrence and Morata (personal communication) find that M^+ clones induced in 48-h *minute* larvae frequently cross between the head and antenna. In the situation illustrated in Figure 7, the antenna polyclone may include one or more cells marked by the mitotic recombination induced at E_3 (or later). The larger the fraction of the antenna made up of marked cells, the greater was the number of primordial cells in the polyclone that were descendants of the recombinant cell. It is ironic that the average clone size within the restricted tissue may approximate $1/n$, or the same size observed in the minimal mosaic patch, *without the necessity of taking twin spots into account.*

The idea that separation of antennal cells from eye cells does not occur until quite late in development may explain a possible paradox existing between the small number of primordial cells (9: Table 1) with the high mosaic frequency (18%: Table 2). That is to repeat the model developed above: The eye antennal primordium is initially large with a high mosaic frequency. If cell mixing occurs throughout development within groups of cells not determined apart from each other, the later the time in development of the determinative restriction of the cell number, the more the mosaic frequency of the subset approaches the mosaic frequency of the parent population. The mosaic frequency of the eye antenna disc is about 25%; Postlethwait and Schneiderman's (1971) mosaic frequency of the 2nd antennal segment, taken by them as representative of the entire antenna, is 18%, arguing for almost random sampling within the parent disc.

Assuming primordial cell numbers to refer to the polyclone number at the time of restriction of the individual anlage, the use of mitotic recombination to estimate the absolute primordial cell number is valid only when the time of polyclone formation is known, in order to satisfy the assumptions explicit in Figure 6. The tergites are an example of this, having good agreement between estimates formed by the minimal mosaic patch and mitotic recombination. Garcia-Bellido and Merriam (1971b) found the minimal mosaic patch to be about 1/8 of the tergite, assuming 6 bristles to mark the minimal patch. They and Guerra et

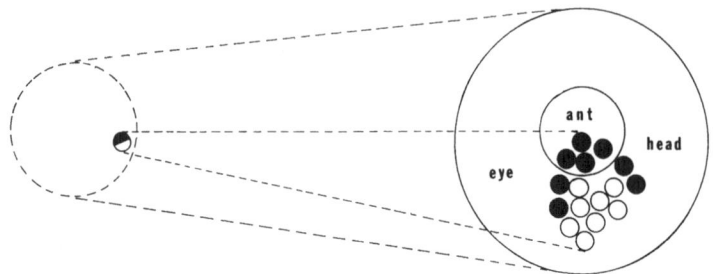

Fig. 7. Mitotic clones induced in the embryo may mark structures in the adult antenna, eye, and other head cuticle because those anlagen are not set apart from each other until late in development

al. (1973) found the size of single mitotic clones to be about one-half of that (ca. 3 bristles or 1/17–1/22 of the hemitergite) expected without scoring twin spots. These data are considered reliable; in the larvae the histoblast cells forming the adult tergites are mitotically quiescent, since they can be marked at any time to give the same size patch (Garcia-Bellido and Merriam, 1971 b; Guerra et al., 1973; see also Madhavan and Schneiderman, 1977). The data of Wieschaus and Gehring (1976a) on the tergites are also consistent with this view. They found 1/9 marked in single mitotic clones induced at E_3 and 1/24 when induced at E_{10}. Their results suggest the tergite anlagen are restricted sometime after the first cell division.

V. Histologic Observations

Imaginal anlagen cells cannot be distinguished from the larval ectoderm in the embryo and as a result cannot be directly counted. In freshly hatched larvae, however, Madhavan and Schneiderman (1977) have been able to distinguish and count the number of cells grouped into the imaginal discs. Through serial sections of larvae from several stages they were able to trace back from the mature identifiable discs to the earliest buds in the freshly hatched larvae. By 4 h after hatching the anlagen of the eye-antennal, wing, haltere, and genital discs are already formed into sac-like structures that are separated from each other. The three pairs of leg disc anlagen appear only as ventral thickenings on the epidermal wall. At this time there is considerable uniformity in the size of nuclei of cells of the various imaginal disc anlage. They are spherical and measure 1.5–2 μm in diameter (2.5–3 μm just before dividing) in contrast to those of the neighboring larval tissues such as the epidermis, fat body, or salivary gland, which measure about 3–5 μm in diameter.

The numbers of cells counted in each imaginal anlagen are listed in Table 4. The number of cells counted for each disc was extraordinarily constant from animal to animal; on four determinations they counted 37.5 ± 0.5 cells in the wing disc 4 h after hatching and 37.5 ± 1.1 cells in the wing disc 12–14 h after hatching. No wing cells were seen in division at those times. At 18–20 h after hatching, they counted 47.5 ± 7.0 cells in the wing disc; an average of 16% of these cells were

Table 4. Data from Madhavan and Schneiderman (1977) on the number of anlagen cells counted in the early first instar larvae and on the onset of mitotic activity after egg hatch

Tissue	Numbers of cells counted in early lst instar imaginal disc anlagen	% of cells blocked in division			
		12–14 h	15–17 h	18–20 h	28–30 h
Antenna only	35	0	0	0	8
Eye—head only	42	6	13	16	18
Wing	38	0	18	16	17
Haltere	20	0	0	13	16
Leg	36–45	0	0	0	10
Genital	64	0	0	0	19
Tergite	19–21	0	0	0	0
Sternite	12–13	0	0	0	0

blocked in division by colchicine treatment. Because of their compact nuclei, little cytoplasm, and close packing, the cells in the antenna region stain darkly compared to those of the eye region of the disc and they can be counted separately. For the antenna region, they counted 30.9 ± 6.0 cells 2–4 h after hatching, 33.3 ± 4.4 12–14 h after hatching, 37.2 ± 3.2 cells 15–17 h after hatching, and 35.6 ± 2.6 cells 18–20 h after hatching. The counts of the eye region were slightly larger, and they showed cells in division more quickly after hatching. The mesothoracic and metathoracic leg discs were somethat larger, at about 42 cells and 24 cells, respectively.

These data suggest, but do not prove, that each wing disc is formed from an exact number of cells (37 or 38) rather than from a variable number. Likewise, the haltere disc is always formed from 20 cells, which, at least, is clearly different from the number forming the wing disc. By extension, the number of cells forming each disc may be thought to be exact and characteristic. Another characteristic parameter for each disc is the time of onset of mitotic division within the disc anlage. In Table 4 are listed the percent of cells blocked in division by colchicine treatment. The earliest time of cell division in a disc is given by the earliest age for which blocked figures can be seen. These times are different and specific for each disc. Except for the antenna region there is an anterior-posterior gradient with cells in the anterior structures dividing earlier.

The numbers of cells counted by Madhavan and Schneiderman are in reasonable agreement with the indirect estimates obtained from the minimum mosaic patches. This suggests that the 1st larval instar rudiments are determined. Since the imaginal disc rudiments were also observed by Madhavan and Schneiderman to give rise to noncuticular tissue such as the trachea, nerves, peripodial membrane, and possibly the adepithial-muscle cells, it is clear that the numbers listed in Table 4 represent the maximum number of cells that can populate any disc anlage. Going down the list of discs in Table 4, there are more cells in the antenna and eye discs than may have been expected from the analysis of mosaics. Because of the lack of landmarks, the minimal mosaic patch of the eye-antennal disc has never been adequately established. The wing disc provided the closest match: the 38 cells they count compares to Ripoll's (1972) estimate of 40. The size of the prothoracic leg was 36 cells counted in the prothoracic leg discs and 45 cells counted in the metathoracic leg disc; this compares to 20 for each leg from minimal mosaic patches. Possibly sensory or adepithial cells increase the size of the leg discs. The genital disc has never been measured in terms of the minimal mosaic patch. It is frequently mosaic (Janning, 1974b; Gehring et al., 1976), which is in agreement with is large size from Madhavan and Schneiderman's cell counts. Cell counts in the tergites and sternites are each twice the size estimated from minimal mosaic patches. However, the cell counts of each hemitergite represent the sum of two nests of histoblast cells, one anterior nest and one posterior nest. A likely hypothesis is perhaps that only one nest makes the cuticle with bristles; the other nest of histoblasts may give rise entirely to that part of the tergite that does not contain bristles and therefore cannot be scored in mosaics. A similar hypothesis is advanced for the origin of the sternites. Thus, in all cases, the numbers of cells counted by Madhavan and Schneiderman are either close to or slightly more than the numbers obtained from mosaics. This agreement with the direct observations provides powerful support for the indirect methods of estimating polyclone number in the primitive discs.

VI. Time of Early Cell Divisions

Madhavan and Schneiderman (1977) used the same technique of serial sectioning of embryos and larvae to follow the pattern of mitotic activity in the presumptive imaginal regions of the ectoderm. Embryos at carefully timed stages were injected with colchicine to inhibit mitoses and were fixed for sectioning one-half hour later. Colchicine-inhibited mitoses are readily visable as chromatic droplets and can be counted. Several successive stages were examined:

    1.5 h (nuclear multiplication, highly synchronous)
    3 h (cellular blastoderm)
  4–4.5 h (gastrulation)
    8 h (beginning of segmentation)
  12 h (end of dorsal closure)
  20 h (just prior to hatching)

A pause in mitotic activity is normally found with the formation of the cellular blastoderm layer; the pause lasted until division started again at 4–4.5 h. There were few divisions in any ectodermal tissues after 8 h and none at all in the period between 12 and 20 h. There is an onset of mitotic activity in the imaginal anlagen during larval life. The time of onset is characteristic for each disc and usually comes in the mid-to-late first instar. The abdominal histoblasts did not divide until after puparium formation.

It seems likely that the first time of inducing mitotic recombination corresponds to the mitotic division observed in the 4–4.5 h period. Wieschaus and Gehring (1976a) found a window at E_3 that permitted survival of the embryos irradiated at that time. Irradiation during the cleavage stages invariably killed the embryos, from which they concluded that clones induced at E_3 did not represent divisions in the precellular blastoderm. The time of induced mitotic recombina-

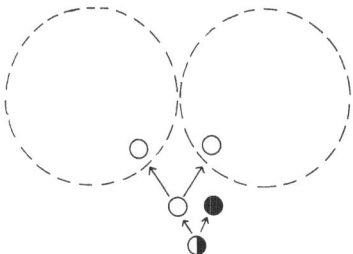

Fig. 8. Illustration of the minimum number of cell divisions needed to segregate descendants of a marked cell to two separate primordia (e.g., wing and leg)

tion does not necessarily indicate the time of the actual division since the cells have to be in the G_2 stage only. The sizes of clones obtained from embryos irradiated during another window at E_{10} were about half those obtained at E_3; this agrees with an average single round of mitosis for all presumptive imaginal cells between E_3 and E_{10}.

Inducing mitotic recombination at E_{10} means only that another mitotic division occurs sometime later. For the tergites it would seem that the next division is much delayed (after puparium formation) since the size of the clone induced at E_{10} is the same as that of clones induced during any of the larval stages. For other tissues, however, the time of the next division is not so clear. On the one hand, the observations by Steiner (1976), by Wieschaus and Gehring (1976a), and by Lawrence and Morata (1977) that the same induced clone can mark both wing and leg discs means at least two mitoses have to separate formation of the cellular blastoderm layer from the restriction of wing from leg. This situation is schematically illustrated in Figure 8. Since the wing and leg anlagen are visibly separated in freshly hatched larvae (Madhavan and Schneiderman, 1977), it is assumed both divisions take place in the embryo. In the dipteran *Dacus* there is histologic evidence (Anderson, 1963) that the mesothoracic leg and wing are part of the same primordium in early development, becoming physically separated only during larval growth. Along the same lines the reports cited of marking head, eye, and antenna tissue within single clones indicate that several cell divisions take place before the visible distinction of the antennal and eye anlagen. On the other hand, the number of blocked mitoses scored by Madhavan and Schneiderman (1977) would not seem to support more than a single round of mitoses, and almost none after E_{10}. This situation should be regarded as a potential conflict in the data, which awaits resolution.

VII. Time of Early Developmental Restrictions

It is clear from the above account that estimates of primordial cell number depend on knowing when determination occurs to set aside the specific anlage. Fortunately, enough information is available to draw some conclusions about the

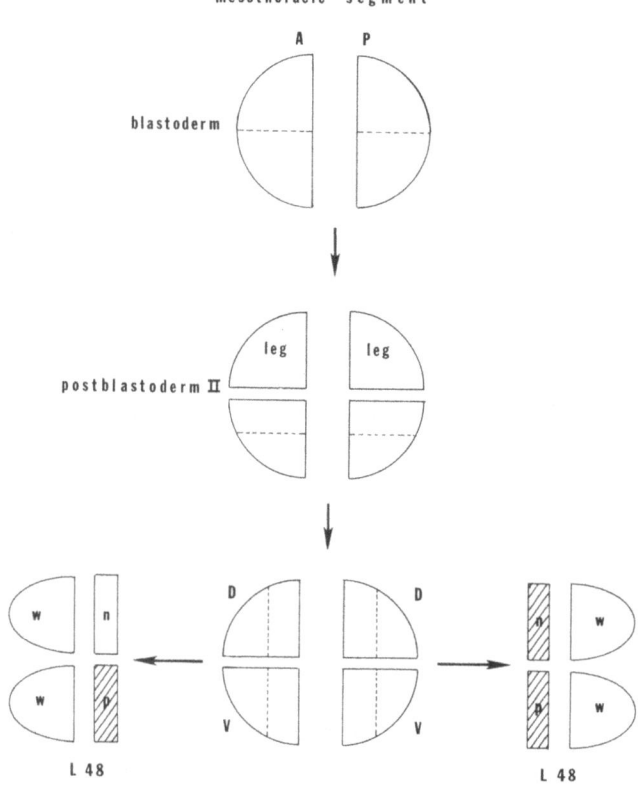

Fig. 9. Restriction of the mesothoracic-wing compartments during development. At blastoderm two compartments (*A* and *P*) exist; each is later split (postblastoderm II) between leg
and wing primordia. The wing primordia *(lower quadrants)* are later each split into dorsal,
ventral, wing, and notal sectors. The wing anlage at 48-h larval consists of eight compartments; three of these *(shaded)* cannot be scored for mosaics in the adult

time of the early developmental restrictions. Although the data on mosaics are
reported in terms of discs rather than compartments, some inferences concerning
polyclone number can be attempted at this time. The mesothoracic segment has
been studied most thoroughly and can be used for illustration. The sequence of
restriction of progressive polyclones (compartments) is presented in Figure 9. The
evidence suggests that the segment may initially (at blastoderm) consist of two
polyclones (compartments), one anterior and one posterior. After two or more cell
divisions, an additional restriction occurs to separate the wing and leg polyclones
(compartments). A form of terminology introduced by Lawrence and Morata
(1977) is useful here to specify the early stages. "Blastoderm" cells are those
formed by cellularization of the blastoderm layer; "postblastoderm I" are those
cells present after one average round of mitotic division. "Postblastoderm II" are

those cells present after two average rounds of mitotic division. From the earlier discussion it is assumed the postblastoderm II stage correlates to about E_{12}; it is the earliest that restriction of wing from leg can occur. It should be stated explicitly that this analysis is necessarily restricted to the cuticle; it may well be that in the embryonic period the cuticle shares ancestral cells with other internal tissues that cannot be scored with the genetic markers available. The fact that clones do not cross the second leg–wing border at E_{10} does not mean that the wing disc and the leg disc are segregated as such. They may be parts of more generally dorsal and ventral primordia that include other anlagen apart from cuticle. At some later time in larval development additional restrictions separate dorsal from ventral wing and proximal (notum) from distal (wing blade) compartments. Thus, by the second larval instar, each wing disc should consist of eight polyclones. Three of these, the anterior and posterior pleura and the post notum, are not marked or scored in mosaic flies.

The wing discs, observed by Madhavan and Schneiderman (1977) in the early first instar to each contain 40 cells, include both the combined anterior and posterior wing polyclones. Each leg disc contains a similar number of cells. Anterior and posterior polyclones cannot be distinguished directly, but the indirect evidence of mosaic frequencies suggests that many more cells are anterior than posterior. From Madhavan and Schneiderman (1977) the number of cells at E_{12} (postblastoderm II) forming these polyclones can be set at 40; this is in good agreement with Ripoll's (1972) observations of the minimal mosaic patch in gynandromorphs. The critical data on the number of blastoderm cells forming the anterior and posterior polyclones are lacking, but certain approximations can be made. First, 40 cells populating the initial wing disc and 40 cells populating the initial second leg disc are together the descendants (after two mitotic divisions) of 20 cells on the blastoderm surface. Second, at least two and probably three blastoderm cells form the original posterior mesothoracic segment because the posterior segment is mosaic in gynandromorphs. If three blastoderm cells form the posterior polyclone, some 12 cells are present at postblastoderm II stage for the restriction of wing from leg (6 wing and 6 leg). The remaining 17 blastoderm cells form the anterior polyclone. At postblastoderm II they have increased to 68 cells; this allows the restriction of approximately 34 anterior wing cells from 34 anterior leg cells and preserves the approximately five fivefold difference between the number of cells in the anterior and posterior polyclones. While these cell numbers are speculative at present, they can be tested further or adjusted by the method of minimal mosaic patches applied to compartments rather than to discs. Thus, assuming a second mitotic division after E_{10}, the estimated number of cells for each disc polyclone is in agreement with a postgastrulation time of determination of the separate imaginal discs. If a second round of division in the embryo cannot be supported by direct observation, the conclusion shifts to a time of developmental restriction during the larval instars.

Acknowledgement. Drs. Gines Morata and Thomas Kornberg and Mr. Dan Kass kindly helped in the preparation of the manuscript. The author's research is supported by grant GM 17096 from the USPHS.

Note Added in Proof. Baker (1978, Genetics **88**, 743—754) suggests the descendents of 6 blastoderm cells form the dorsal compartment of the eye antennal disc and the descendents of another 4 blastoderm cells form the ventral compartment. Since cells actually forming the compartments are likely not restricted until a later post blastoderm stage, the number of cells involved at restriction is doubtless larger than a total of 10.

Baker interprets the distribution of sturt distances separating landmarks to estimate the number of blastoderm cells whose descendents include the population of a compartment. This review, on the other hand, concentrates largely on techniques for estimating the initial sizes of the restricted polyclones themselves. A comparison of the numbers of cells in the mesothoracic segment (discussed here in Sec. VII) with numbers generated by Baker's method may help answer such questions as whether individual blastoderm cells contribute descendents to both the larval and the adult hypodermis, and whether all the blastoderm cells divide to form the stages called here post blastoderm I and post blastoderm II. The model proposed in Section VII is robust enough that the estimates of polyclone sizes depend on the number of cells actually set aside and not on their prior mitotic activity.

References

Anderson, D. T.: The embryology of *Dacus tryoni*. II. Development of imaginal discs in the embryo. J. Embryol. Exp. Morph. **11**, 339—351 (1963)

Auerbach, C.: The development of the legs, wings, and halteres in wild type and some mutant strains of *Drosophila melanogaster*. Trans. R. Soc. Edin. **57**, 787—816 (1936)

Baker, B.: Paternal Loss *(Pal)*: A meiotic mutant in *Drosophila melanogaster* causing loss of paternal chromosomes. Genetics **80**, 267—296 (1975)

Bryant, P.: Cell lineage relationships in the imaginal wing disc of *Drosophila melanogaster*. Dev. Biol. **22**, 389—411 (1970)

Bryant, P., Schneiderman, H.: Cell lineage, growth and determination in the imaginal discs of *Drosophila melanogaster*. Dev. Biol. **20**, 263—290 (1969)

Chan, L., Gehring, W.: Determination of blastoderm cells in *Drosophila melanogaster*. Proc. Natl. Acad. Sci. USA **68**, 2217—2221 (1971)

Crick, F. H. C., Lawrence, P. A.: Compartments and polyclones in insect development. Science **189**, 340—347 (1975)

Fialkow, P. J.: Clonal Origin of Human Tumors. Biochemica et Biophysica Acta, Cancer Reviews **458**, 283 (1976)

García-Bellido, A., Merriam, J. R.: Cell lineage of the imaginal discs in *Drosophila* gynandromorphs. J. Exp. Zool. **170**, 61—76 (1969)

García-Bellido, A., Merriam, J. R.: Parameters of the wing imaginal disc development of *Drosophila melanogaster*. Dev. Biol. **24**, 61—87 (1971 a)

García-Bellido, A., Merriam, J. R.: Clonal parameters of tergite development in *Drosophila*. Dev. Biol. **26**, 264—276 (1971 b)

García-Bellido, A., Ripoll, P., Morata, G.: Developmental compartmentalisation in the wing disc of *Drosophila*. Nature (New Biol.) **245**, 251—253 (1973)

García-Bellido, A., Ripoll, P., Morata, G.: Developmental segregations in the dorsal mesothoracic disc of *Drosophila*. Dev. Biol. **48**, 132—147 (1976)

Gehring, W.: Clonal analysis of determination dynamics in cultures of imaginal discs in *Drosophila melanogaster*. Dev. Biol. **16**, 438—456 (1967)

Gehring, W., Nöthiger, R.: The imaginal discs of *Drosophila*. In: Waddington, C. H., Counce-Nicklas, S. (Eds.): Developmental Systems: Insects, Vol. 2. New York: Academic Press 1973

Gehring, W., Wieschaus, E., Holliger, M.: The use of "normal" and "transformed" gynandromorphs in mapping the primordial germ cells and the gonadal mesoderm in *Drosophila*. J. Embryol Exp. Morph. **35**, 607—616 (1976)

Gelbart, W.: A new mutant controlling mitotic chromosome disjunction in *Drosophila melanogaster*. Genetics **76**, 51—63 (1974)

Guerra, M., Postlethwait, J. H., Schneiderman, H. A.: The development of the imaginal abdomen of *Drosophila melanogaster*. Dev. Biol. **32**, 361—372 (1973)

Hadorn, E.: Problems of determination and transdetermination. Brookhaven Symp. Biol. **18**, 148—161 (1965)

Haynie, J. L., Bryant, P. J.: Intercalary regeneration in imaginal wing disc of *Drosophila melanogaster*. Nature (Lond.) **259**, 659—662 (1976)

Haynie, J. L., Bryant, P. J.: The effects of X-rays on the proliferation dynamics of cells in the imaginal wing disk of *Drosophila melanogaster*. Wilhelm Roux's Archives **183**, 85—100 (1977)

Hinton, C. W.: The behavior of an unstable ring chromosome of *Drosophila melanogaster*. Genetics **40**, 951—961 (1955)

Hotta, Y., Benzer, S.: Mapping of behaviour in *Drosophila* mosaics. Nature (Lond.) **249**, 527—535 (1972)

Hotta, Y., Benzer, S.: Mapping of behaviour in *Drosophila* mosaics. In: Ruddle, F. (Ed.): Genetic Mechanisms of Development. New York: Academic Press 1973

Hutchinson, H. T.: A model for estimating the extent of variegation in mosaic tissues. J. Theor. Biol. **38**, 61—79 (1973)

Janning, W.: Entwicklungsgenetische Untersuchungen an Gynandern von *Drosophila melanogaster*. I. Die inneren Organe der Imago. Wilhelm Roux' Archiv **174**, 313—332 (1974a)

Janning, W.: Entwicklungsgenetische Untersuchungen an Gynandern von *Drosophila melanogaster*. II. Der morphogenetische Anlageplan. Wilhelm Roux' Archiv **174**, 349—359 (1974b)

Lawrence, P. A., Morata, G.: The early development of mesothoracic compartments in *Drosophila*. Dev. Biol. **56**, 40—51 (1977)

Lee, W. R., Kirby, C. J., Debney, C. W.: The relation of germ line mosaicism to somatic mosaicism in *Drosophila*. Genetics **55**, 619—634 (1967)

Lewis, E. B., Gencarella, W.: Claret and nondisjunction in *Drosophila melanogaster*. Genetics **37**, 600—601 (abstr.) (1952)

Linder, D., Gartler, S. M.: Glucose-6-phosphate dehydrogenase mosaicism: utilization as a cell marker in the study of Leiomyomas. Science **150**, 67—69 (1965)

Lindsley, D. L., Grell, E. H.: Genetic variations of *Drosophila melanogaster*. Carnegie Inst. Wash. Publ. No. 627 (1968)

Madhavan, M. M., Schneiderman, H. A.: Histological analysis of the dynamics of growth of imaginal discs and histoblast nests during the larval development of *Drosophila melanogaster*. Wilhelm Roux's Archives **83**, 269—305 (1977)

Morata, G., Garcia-Bellido, A.: Developmental analysis of some mutants of the bithorax system of *Drosophila*. Wilhelm Roux's Archives **179**, 125—143 (1976)

Nesbitt, M. N.: Chimeras vs. X inactivation mosaics: significance of differences in pigment distribution. Dev. Biol. **38**, 202—207 (1974)

Nissani, M.: Cell lineage analysis of germ cells of *Drosophila melanogaster*. Nature (Lond.) **265**, 729—731 (1977)

Nissani, M., Lipow, C.: A method for estimating the number of blastoderm cells which give rise to *Drosophila* imaginal discs. Theor. Appl. Genet. **49**, 3—8 (1977)

Okada, M., Kleinman, I. A., Schneiderman, H. A.: Chimeric *Drosophila* adults produced by transplantation of nuclei into specific regions of fertilized eggs. Dev. Biol. **39**, 286—294 (1974)

Postlethwait, J. H., Schneiderman, H. A.: A clonal analysis of development in *Drosophila melanogaster*: morphogenesis, determination, and growth in the wild-type antenna. Dev. Biol. **24**, 477—519 (1971)

Poulson, D. F.: Histogenesis, organogenesis, and differentiation in the embryo of *Drosophila melanogaster* Meigen. In: Demerec, M. (Ed.): Biology of *Drosophila*. New York: Hafner 1950

Ripoll, P.: The embryonic organization of the imaginal wing disc of *Drosophila melanogaster*. Wilhelm Roux' Archiv **169**, 200—215 (1972)

Sonnenblick, B. P.: The early embryology of *Drosophila melanogaster*. In: Demerec, M. (Ed.):
Biology of *Drosophila*. New York: Hafner (1950)

Steiner, E.: Establishment of compartments in the developing leg imaginal discs of *Drosophila
melanogaster*. Wilhelm Roux's Archives **180**, 9—30 (1976)

Stern, C.: The prospective significance of imaginal discs in *Drosophila*. J. Morphol. **67**, 107—
122 (1940)

Sturtevant, A. H.: The claret mutant type of *Drosophila simulans:* a study of chromosome
elimination and of cell-lineage. Z. Wiss. Zool. **35**, 323—356 (1929)

Turner, F. R., Mahowald, A. P.: Scanning electron microscopy of *Drosophila* embryos. I. The
structure of the egg envelope and the formation of the cellular blastoderm. Dev. Biol. **50**,
95—108 (1976)

Turner, F. R., Mahowald, A. P.: Scanning Electron Microscopy of *Drosophila melanogaster*.
Embryogenesis II. Gastrulation and Segmentation. Dev. Biol. **57**, 403—416 (1977)

Van Deusen, E.: Sex determination in germ line chimeras of *Drosophila melanogaster*. J. Embryol. Exp. Morphol. **37**, 173—185 (1976)

Wieschaus, E., Gehring, W.: Clonal analysis of primordial disc cells in the early embryo of
Drosophila melanogaster. Dev. Biol. **50**, 249—265 (1976a)

Wieschaus, E., Gehring, W.: Gynandromorph analysis of the thoracic disc primordia in *Drosophila melanogaster*. Wilhelm Roux's Archives **189**, 31—46 (1976b)

Zalokar, M.: Transplantation of nuclei in *Drosophila melanogaster*. Proc. Natl. Acad. Sci.
USA **68**, 1539—1541 (1971)

Zalokar, M., Erk, I.: Division and migration of nuclei during early embryogenesis of *Drosophila melanogaster*. J. Microsc. Biol. Cell **25**, 97—106 (1976)

Cell Lineage Relationships in the Drosophila Embryo

ERIC WIESCHAUS

European Molecular Biology Laboratory (EMBL), Heidelberg, FRG

I. Introduction

Embryonic development proceeds by subdivision of the embryo into groups of cells, each destined to form particular differentiated structures. A classical example of such cell groups is the imaginal discs, clusters of cells in the larva determined to form specific regions of the adult. Because the discs cannot be identified in early stages during embryogenesis, their origin and early development can best be studied using cell lineage techniques such as *mitotic recombination*. By following the contribution of individual cells to particular adult structures, it is possible to estimate the size of the different embryonic primordia and the time during development when rigid distinctions between different primordia arise.

The goal of this chapter is to review what is known about the cell lineage relationships in the *Drosophila* embryo. The most detailed data concern the precursor cells for the imaginal discs, in particular those of the thorax. To judge the extent to which the conclusions drawn from the disc work are representative of other regions of the embryo, I will also present some of the available cell lineage data for the abdominal histoblasts, the larval hypoderm, and other embryonic primordia. In the final section, the results of mitotic recombination will be discussed in relation to those obtained from other types of mosaics, particularly gynandromorphs.

II. Mitotic Recombination

Mitotic recombination provides us with a technique that permits the genetic marking of the progeny of individual cells at chosen times in development. Its use in the analysis of imaginal disc development depends on the availability of mutations that do not affect normal development of the disc, but alter the final color of the cuticle or the morphology of bristles in the adult. Because most of these

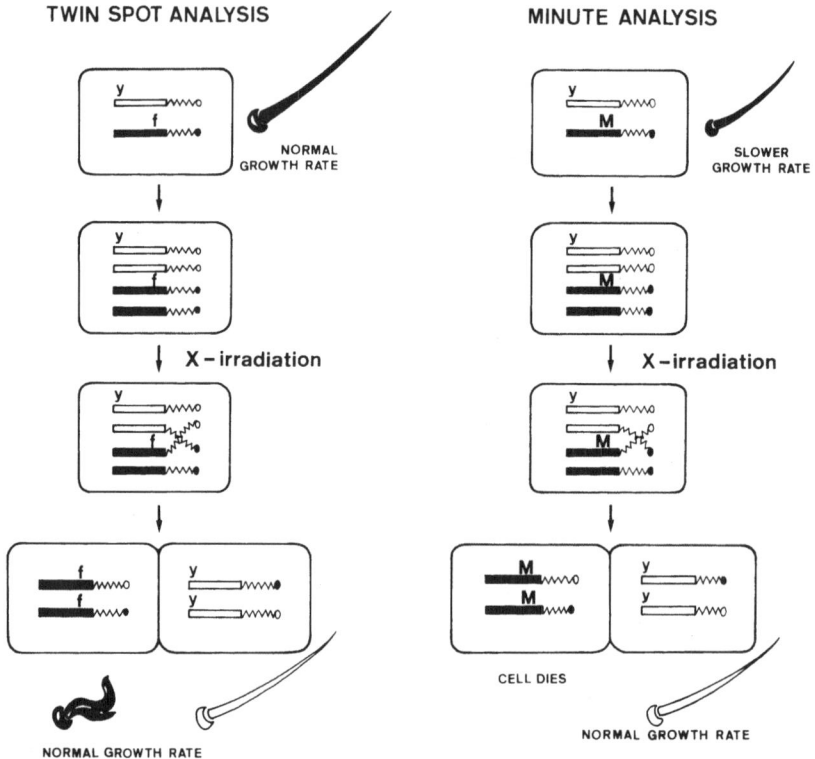

Fig. 1. Schematic representation of mitotic recombination as used in twin spot and *Minute* analyses. In both types of experiments, mitotic recombination results in the two daughter cells each being homozygous for one homolog and thus for the recessive markers it carries. In a twin spot analysis, the genotype of the cell does not affect its growth rate and the two marked patches found in the adult are representative of the normal contribution of a single cell at the time of the irradiation. The heterozygous cells in a *Minute* analysis have a slower rate of proliferation. Mitotic recombination results in a marked cell that has lost the *M* mutation and grows at a faster rate than the surrounding cells

marker mutations are recessive, heterozygotes are phenotypically wild type. If such a heterozygous individual is X-irradiated during its development, a recombination can occur in single cells such that the two daughter cells produced after mitosis are each homozygous for one homolog and thus for the recessive markers it carries (Stern, 1936; Becker, 1957; Fig. 1). The progeny of these two cells can be recognized in the adult as twin spots (i.e., two adjacent patches of differently colored adult cuticle). Where these clones arise in the embryo cannot be controlled. In embryos treated with 1000 r, mitotic recombination occurs randomly in about five cells per 1000 (Wieschaus and Gehring, 1976a). The low frequency of clonal induction makes two simultaneous inductions in the same region of the embryo very unlikely and thus almost all marked patches found in the adult can be regarded as single clones, i.e, the product of single recombined cells.

An elegant variation of this procedure for producing larger clones has been developed by Morata and Ripoll (1975) and involves the use of *Minute* mutations.

Minutes are recessive lethals that, in a heterozygous state, result in retardation of growth and cell proliferation (Dunn and Coyne, 1937; Brehme, 1939). If mitotic recombination is performed in flies heterozygous for a marker mutation and a *Minute* in *trans* configuration, the marked clone no longer carries the *Minute* mutation (Fig. 1 b) and divides at a faster rate than the surrounding *Minute* heterozygous cells. This increase in growth rate does not change the final shape or size of the resultant structure, although the relative contribution of the marked cell is much greater than its heterozygous sisters. Embryos heterozygous for *Minute* hatch at the same time as their wild-type siblings and most of the delay associated with *Minute* heterozygosity is restricted to larval development (Steiner, 1976). The use of *Minutes* in studies of embryonic cell lineage offers the practical advantage that, due to their faster larval growth rates, the clones found in the adult are larger than those produced in non-*Minute* cell lineage analysis. The larger size makes the clones easier to detect and also serves to emphasize the edges of the primordium in which the clone was induced, since regardless of where it initially arises, the faster growth eventually brings the clone to the edge.

If the irradiations are done at a time in development when the distinction between two adjacent primordia is not yet rigidly defined, a marked cell might still contribute to both. The frequency of such clonal overlap increases with the size of the clone at the time when the border between the two primordia is established. Under certain conditions a $Minute^+$ clone in a heterozygous background is more likely to overlap two primordia than a normally growing twin spot (Garcia-Bellido et al., 1973). However, this only applies when there is sufficient time for cell proliferation to make the M^+ clone large before the restriction arises. During embryogenesis, where the disc cells divide only once or twice, the faster growth of the M^+ cells is counterbalanced by the fact that in a *Minute* analysis, the marked clone arises from only one of the original two daughter cells. Thus, in the embryo the same or more clonal overlap would be found when normal twin spots are produced.

III. Number of Disc Primordial Cells and Their Proliferation During Embryogenesis

Fertilization in *Drosophila* is followed by thirteen nuclear cleavages, initially within the central yolky cytoplasm of the egg and then at the surface (Sonnenblick, 1950). The blastoderm that is formed thereafter consists of approximately 6000 cells (Turner and Mahowald, 1976; Zalokar and Erk, 1976), most of which contribute to larval organs (Poulson, 1950). The imaginal discs can only be identified 15 h later, toward the end of embryogenesis after gastrulation and subsequent cell movements have shaped the blastoderm into an almost fully differentiated larva (Auerbach, 1936; Laugé, 1967; Madhavan and Schneiderman, 1977). Despite the fact that they cannot be recognized as such, the number of precursor cells for each disc can be estimated from the relative size and frequency of clones induced using mitotic recombination at the blastoderm stage (Bryant and Schnei-

Table 1. The frequency of clones induced at the blastoderm stage in the different imaginal primordia. Clone frequencies for the legs, wings, and haltere (Steiner, 1976) are drawn from examination of 1404 flies, those for the male and female genital discs (K. Dübendorfer, 1977) from 1886 flies. The clone frequencies in the eye antennal disc and in the abdominal tergites (Wieschaus, unpublished) are based on 620 of the same flies used by Steiner and Dübendorfer. Flies were of the genotype $y(f)$; $mwh\,jv/Dp(1;3)sc^{j4}$, $y^{+}M(3)\,i^{55}\,l^{55}$, irradiated at 3.0 ± 0.5 h with 1000 r and examined in prepared specimens. The data for the abdomen have been corrected for the frequency of spontaneous clones (1.7) found in nonirradiated controls. The low frequency in the female genital disc and haltere may in part be due to the difficulties in detecting clones in structures that lack bristles. The number of cells observed in each disc in a first instar larvae are taken from actual counts made by Madhavan and Schneiderman(1977). It is not known whether all these cells contribute to the cuticle structures scored in a clonal analysis. The male and female genital discs were not distinguished from each other, and it is not certain to which disc the figure in parenthesis applies

	Eye antennal	Humerus	Wing	Haltere	Leg I	Leg II	Leg III	Abdominal tergite II–V	Male genital disc	Female genital disc
Clone frequency (per 100 discs)	3.8	1.1	2.3	0.7	2.4	2.6	2.6	1.7	3.6	1.8
Number of cells in first instar primordium (Madhavan and Schneiderman, 1977)	71	—	37	20	36	42	45	20	(64)	(64)

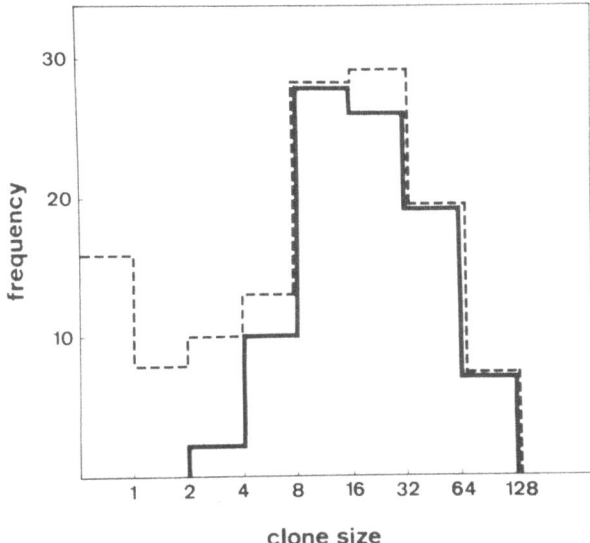

Fig.2. The size of clones induced at the blastoderm stage in the foreleg primordium. The 94 clones were obtained in a twin spot analysis where the marked cells divide at the same rate as their nonrecombined heterozygous sisters. The average size represents approximately one-ninth of the surface and indicates that about nine blastoderm cells contribute to the foreleg. The size distribution is based on examination of 1154 flies irradiated at 3.0 ± 0.5 and has been corrected for the frequency of spontaneous clones found in 378 nonirradiated controls. The clone size is measured as the number of marked bristles; the total number of bristles per leg is 234. *Dotted line* gives the initial distribution prior to correcting for the spontaneous clones

derman, 1969; Bryant, 1970; Postlethwait and Schneiderman, 1971; Wieschaus and Gehring, 1976a). When blastoderms heterozygous for recessive markers and a *Minute* mutation were irradiated with a dosage of 1000 r, about 2–3 clones were found per one hundred discs examined in the adult (Steiner, 1976; K. Dübendorfer, 1977; Wieschaus, unpubl. observations; Table 1). Since the individual blastoderm cells are not likely to vary greatly in the probability with which they undergo recombination, the similar frequency of clonal induction in the different discs indicates that they all arise from roughly the same number of blastoderm cells. Judging from the small differences that are observed, the largest would be the eye antennal disc, the smallest the haltere. Although the frequency of clones induced at the blastoderm stage does not of itself allow calculation of absolute cell numbers, the relative frequencies in different discs corresponds well to the number of cells actually counted in each disc when it is first identified in a young larva (Madhavan and Schneiderman, 1977; Table 1). The relative sizes of the different primordia are therefore maintained throughout embryogenesis and differences in the extent of cell proliferation must be small.

Estimates of the absolute number of cells in each primordium at the blastoderm stage can be made from the fraction of the adult surface occupied by the average clone, as long as the genetic markers used do not introduce differences in growth rates between the clone and its surrounding cells. Figure 2 shows the clone sizes induced at the blastoderm stage in the foreleg primordium (Wieschaus, unpublished). The size of the individual clones is extremely variable and covers a range of six cell divisions from three to 120 bristles. Since the same range is found between the two daughter clones marked in the same twin spot, the size variation is not due to differences in the number of foreleg cells in different embryos. Instead it seems that the contributions of the individual cells within the same primordium are highly variable. The clone size clusters around a mean of 26 bristles, or about one-ninth of the surface of the foreleg, implying about nine foreleg precursor cells at the time of irradiation. Extrapolation from the relative frequencies of clonal induction given in Table 1 would indicate that about 14 blastoderm cells contribute to each eye antennal disc and 2–3 to the haltere.

Disc determination in *Drosophila* is thought to proceed in a stepwise fashion throughout the entire course of development (Garcia-Bellido, 1975; Gehring, 1976 and in press). It is not known whether each step in this process must be associated with cell division. The smallest clone sizes obtainable after blastoderm irradiations might indicate the minimum number of mitoses compatible with the total programing sequence between blastoderm and metamorphosis. Such an analysis would be difficult since the size of small clones cannot be very accurately measured with the markers normally used in a twin spot analysis. Moreover the clone actually observed in the adult cuticle might only represent a fraction of the real clone, the remainder having died or formed noncuticle structures. Although it is therefore difficult to determine the minimum number of cell divisions, the *average* extent of cell proliferation occurring during embryogenesis can be estimated from the decrease in clone size, when clones are induced progressively later during embryonic development. By 10 h the average clone size is half that of clones induced at the blastoderm stage (Wieschaus and Gehring, 1976a). This would indicate that the average cell divides once during this 7-h interval. Since the clones induced at 10 h are generally not larger than those obtained from irradiations made during the first larval instar (Becker, 1957; Bryant and Schneiderman, 1969; Bryant, 1970; Garcia-Bellido and Merriam, 1971a, b; Postlethwait and Schneiderman, 1971; Guerra et al., 1973; Wieschaus and Gehring, 1976a), we can conclude that cell proliferation during the second half of embryogenesis is not extensive. Thus, on the average, a disc cell does not undergo more than one to two cell divisions during embryogenesis, although given the variability in clone sizes we cannot be certain whether this applies equally to all the cells within a given primordium.

A more detailed analysis of cell proliferation is made difficult by the sensitivity of certain embryonic stages to the irradiation used to induce mitotic recombination. Even when the survival is relatively high, as it is at the blastoderm stage (Fritz-Niggli, 1955; Wieschaus and Gehring, 1976a; Würgler and Ulrich, 1976), the size of the clones probably reflects to some extent at least the regeneration following X-ray-induced cell death. The damage following irradiation can be estimated by varying the X-ray dosage (Schweizer, 1972). Fewer cells would be killed at lower doses and the clone size would more accurately reflect the number of cells normally contributing to each disc. Detailed studies of this kind on larvae indicate that about 40–60% of the irradiated disc cells are killed by a dosage of 1000 r (Haynie and Bryant, 1977). This lethality can probably not be extrapolated to the other cells in the larvae, most of which have ceased to divide and have become polytene (Bodenstein, 1950). The animal would thus be able to survive the irradiation and the disc would have several days to regulate for the killed cells before metamorphosis. A 40–60% lethality would probably have a more drastic effect on embryos, given that, at least for early stages, it would probably not be limited to a relatively small population within the embryo. Thus, in stages at which postirradiation survival is high, the cell lethality might be lower than that found for larvae. The rather limited dosage data available for the blastoderm (Wieschaus and Gehring, 1976a) suggest that this might be the case, although a much more extensive study, ideally involving several different embryonic stages, would be necessary before the question can be finally answered.

IV. Disc Determination During Early Embryogenesis

Six pairs of discs contribute to the adult thorax (Zalokar, 1943, 1947; Bodenstein, 1950). At the base of the brain attached to the ventral larval hypoderm are the discs for the first and second legs. The discs which will form the third leg, the wing, and the haltere are grouped together and attached to the dorsal tracheae. The tiny dorsal prothoracic discs, which give rise to the humerus, are attached to the anterior end of the tracheae near the opening of the larval spiracle. The right and left foreleg discs lie fused to each other on the ventral midline. With this exception the right and left sides of the fly develop separately and fuse only at metamorphosis. This characteristic arrangement is found in the larvae at the time when it hatches from the egg (Auerbach, 1936), and is maintained until metamorphosis. Given the distinct physical separation of each disc from all others, it is not surprising that clones induced during larval development are restricted to single discs.

Gynandromorph studies (Garcia-Bellido and Merriam, 1969; Hotta and Benzer, 1973; see also Janning, this volume) indicate that, with the exception of the dorsal prothorax, the embryonic precursors for these discs lie as adjacent groups of cells on the right and left sides of the blastoderm (Wieschaus, 1974; Garcia-Bellido and Ferrus, 1975; Wieschaus and Gehring, 1976b). Adjacent thoracic discs are generally the same genotype and very few XX/XO borders cut between any two discs such that one is strictly male, the other strictly female. The shortest distance measured between adjacent discs is 6–9 sturts (for a definition of *sturt*, see Sec. IX). Since this value is smaller than distances frequently measured within discs between structures where clones do overlap, the absence of clonal overlap between particular pairs of discs can be evaluated in terms of developmental rather than spacial restrictions.

When induced at the blastoderm stage, about 10% of the clones in the wing and the second leg overlap both structures (Steiner, 1976; Wieschaus and Gehring, 1976a; Lawrence and Morata, 1977; Fig. 3). This indicates that the individual blastoderm cells are not yet assigned to specific groups corresponding to the discs observed in the larvae. The wing and second leg are both derivatives of the mesothoracic segment. A similar overlap is also found between the haltere and

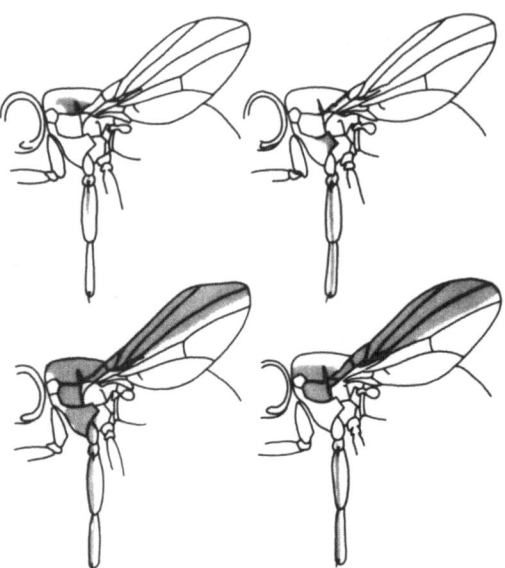

Fig. 3. Clones overlapping the wing and second leg induced at the blastoderm stage. *Shaded areas* in the upper two flies represent clones found in a twin spot analysis (Wieschaus and Gehring, 1976); *shaded areas* in the lower two flies, M^+ clones found in a *Minute* analysis (Steiner, 1976). The larger size of the M^+ clones is due to the retarded growth of the surrounding heterozygous cells

third leg, that is, between the homologous discs in the metathoracic segment (Steiner, 1976), and between the right and left foreleg (Steiner, 1976), the primordia for which are brought together by the invagination of the ventral mesoderm at gastrulation and remain adjacent throughout larval life. Within a given segment, overlap between discs seems possible as long as the primordia lie adjacent to each other in the blastoderm or are brought together shortly thereafter. No clones, however, are found that overlap between derivatives of two different thoracic segments (Steiner, 1976; Wieschaus and Gehring, 1976a; Lawrence and Morata, 1977), even though from gynandromorph data we know that the average distances measured between the different thoracic primordia are all of the same magnitude.

When irradiations are made during early larval development, clones are restricted to either anterior or posterior regions within a disc (Garcia-Bellido et al., 1973). This subdivision is regarded as the first step in a series of compartmentalizations whereby different regions within the disc are gradually assigned specific fates in the adult (Crick and Lawrence, 1975; Garcia-Bellido, 1975). Since blastoderm clones, even those which overlap between discs, also respect this anterior–posterior compartment border (Steiner, 1976; Lawrence and Morata, 1977), we can conclude that it arises prior to the establishment of the individual discs and perhaps simultaneously with the restriction of clones to single segments.

The earliest cell lineage restrictions present in the thoracic region of the embryo can be described as a series of parallel borders that divide the blastoderm cells into groups corresponding to larval segments, each segment in turn subdi-

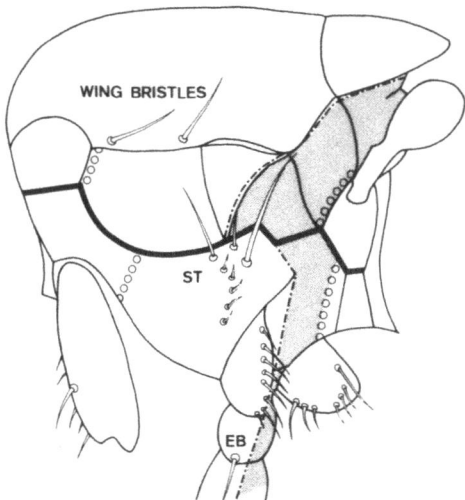

Fig. 4. The position of the segment borders (o–o) and the anterior–posterior compartment border (· – ·) in the adult thorax. At the junction of wing and haltere with the three legs, the borders are discontinuous: A particular border in the dorsal discs always meets the line of fusion anterior to the respective border in the ventral discs. The posterior compartment of the wing and second leg have been *shaded*. ST, sternopleural bristles; EB, edge bristles. (Adapted from Steiner, 1976)

vided into an anterior and posterior compartment. The borders between prospective segments might be quite sharp and continuous around the circumference of the embryo. Very little structural indication of these borders can be found in the adult thorax and their exact position can best be defined by cell lineage experiments. Here the borders are still remarkably straight except at the juncture of the wing and haltere with the three legs. The two dorsal discs seem to have been slid forward during development (Fig. 4), resulting in a discontinuity in the borders as they run from dorsal to ventral. That this discontinuity does not exist in the embryo is indicated in gynandromorph data (Wieschaus and Gehring, 1976 b), where bristles lying near the anterior–posterior border in the leg (the EB in Fig. 4) map closer to the homologous region of the wing than do leg bristles (ST, Fig. 4) which are adjacent to the wing in the adult. This shift forward also explains the peculiar disrupted appearance often found in blastoderm clones that overlap the two discs (Steiner, 1976; Wieschaus and Gehring, 1976 a).

The establishment of segmentation is presumably followed at some later point during embryogenesis by the establishment of the individual imaginal discs. In the mesothorax, this must be accompanied by the physical separation of the wing and 2nd leg primordia since in a mature embryo the 2nd leg disc is embedded in the ventral larval hypoderm, the wing disc attached to the dorsal tracheae (Auerbach, 1936). In *Drosophila*, the discs have not been identified in earlier stages and consequently we do not know when this separation arises. In a related dipteran species, *Dacus tryoni*, a common primordium for the two discs can be identified at the head invagination stage (Anderson, 1963), which occurs in *Drosophila* 10 h after fertilization. During the next hours of embryogenesis, the two primordia gradually pull apart and assume their characteristic positions in the larva. Mitotic recombination performed in *Drosophila* during these stages (7–10 h) produces no clones that overlap between the 2nd leg and wing (Wieschaus and Gehring, 1976a), implying that by this time, the two groups of cells have at least undergone a cell lineage separation. When UV microbeam irradiation is used to destroy small groups of cells on the ventral-lateral surface of a 10-h embryo, the defects observed in the adult are restricted to single discs, in contrast to the multiple defects frequently obtained when the irradiations are made at the blastoderm stage (Lohs-Schardin, personal communication). If the damage produced by the irradiation is restricted to the surface of the embryo, the fact that wing defects could be produced suggests that the wing, although perhaps already separate from the leg, has not yet assumed its larval position in the interior of the embryo.

V. Development of the Head and the Genitalia: Some Evolutionary Considerations

Segmentation is a property common to all insects whereas the particular spectrum of imaginal discs characteristic of *Drosophila* would only be found in orders that show a complete metamorphosis. From cell lineage analysis, we have concluded that establishment of segmentation is one of the first steps in embryogenesis and that this occurs prior to the time when the cells are grouped into specific discs. These results support a modified version of Haeckel's Law, whereby the earlier a developmental event occurs, the more likely it is to be shared by distantly related groups. The peculiarities of a particular species are more likely to result from modification of later developmental steps. In addition to the presence of imaginal discs, the evolution of the *Diptera* is characterized by the fusion of certain primitive segments into single structures. In *Drosophila*, this is most apparent in the head and in the genitalia where single discs give rise to structures associated with more than one segment. Cell lineage analysis allow us to determine whether primitive segmentation is established or maintained during development where it has no morphological meaning. This is probably not the case for the head but seems to hold for the genital disc.

In addition to the structures for which it is named, the eye antennal disc gives rise to most of the head cuticle, the lacinia, and the maxillary palpus (Gehring, 1966). The remainder of the head is formed by the labial (Wildermuth, 1968) and clypeolabral discs (Gehring and Seippel, 1967), neither of which has been subjected to clonal analysis. Although the segmental homologies are somewhat controversial (Ferris, 1950), the array of structures formed by the eye antennal disc would represent two to three segments in primitive insects, from the third (antennal ocular) to the fifth (maxillary) segment. No indication of this primitive segmentation is observed in the embryonic cell lineage, and clones can extend from the antenna across the eye into the palpus (Postlethwait and Schneiderman, 1971; Wieschaus and Gehring, 1976a). Nor is any clonal restriction observed that might correspond to an anterior–posterior compartmentalization, the first subdivision

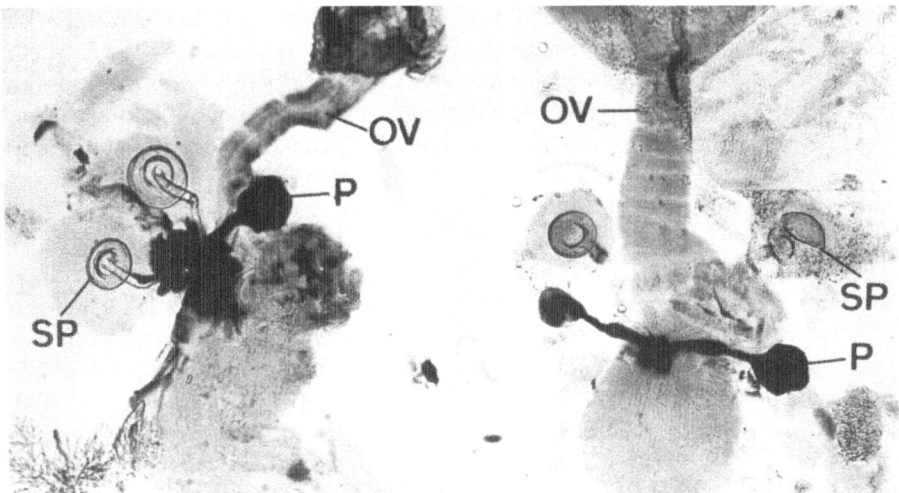

Fig. 5. Internal mosaic in the genital structures of a triplo-X diplo-X mosaic stained for aldehyde oxidase activity. The parovaria frequently differ in genotype from the other internal structures formed by the genital disc, suggesting that they arise from a separate primordium in the embryo. Triplo-X regions (R(1)w^{c}/XX, y v f mal) are *darkly stained; unstained areas* are diplo-X (XX, y v f mal). *SP*, spermatheka; *P*, parovaria; *OV*, oviduct (Schüpach, unpublished)

within the eye antennal disc becoming apparent only during early larval development (Baker, 1977). The failure to detect any embryonic subdivision is all the more puzzling in that the head region of the embryo occupies about one-third of the entire blastoderm (Poulson, 1950), and judging from the frequency of clonal induction (Table 1) and the frequency of mosaicism in gynandromorphs (Garcia-Bellido and Merriam, 1969), the eye antennal disc is at least as large as any of the thoracic primordia.

The segmental homologies are somewhat clearer in the female genital disc, where ample data exist for comparison with related *Diptera* (A. Dübendorfer, 1970, 1971; Emmert, 1972a, b). In *Drosophila*, the genital disc forms the anal plates and external genitalia as well as the internal ducts and secretory organs of the reproductive system (Ursprung, 1959). In *Musca* and *Phormia* these structures are formed by two sets of discs, the paired *Nebengenitalscheiben* associated with the eighth larval segment, which form genital structures, and a single medially located disc that forms the analia and the parovaria and is associated with the terminal segment, consisting of fused segments nine through eleven (A. Dübendorfer, 1970, 1971). Dübendorfer suggested that the single genital disc in *Drosophila* represents the fusion of these two primordia, an idea that has recently received elegant support from cell lineage analysis (K. Dübendorfer, 1977). Clones induced at the blastoderm stage extend from the analia into the parovaria and from the external genitalia into the internal reproductive ducts. No clones, however, are found to overlap analia and genitalia or in internal structures between the parovaria and the other organs of the reproductive system. Thus, the segmental restriction characteristic of blastoderm clones is maintained between the genital and anal segments even after their fusion into a single disc. That the parovaria arise from a different primordium than the genitalia is also indicated by the frequent genotypic borders between the two structures found in triplo-X diplo-X mosaics made from ring-X loss in meta females (Schüpbach, personal communication; Fig. 5). Such flies have mosaic patterns similar to gynandromorphs, but are simpler to analyze since they differentiate only female tissue.

A clonal restriction is also found between analia and genitalia in male *Drosophila*. Here, however, the segmental homologies are less clear since in all higher *Dipterans*, males possess a single genital disc associated with the fused terminal segments nine to eleven. Based on these evolutionary homologies and an analysis of mosaic genitalia in gynandromorphs, Nöthiger et al. (1977) have proposed that female genitalia, male genitalia, and analia would all be represented by separate primordia associated with the 8th, 9th, and 10th and/or 11th segments, respectively. In *Drosophila*, all three primordia are grouped together in a single genital disc. The clonal restriction observed in the blastoderm stage might therefore be explained in terms of their different segmental origins.

VI. The Adult Abdomen and Its Relationship to the Larval Hypoderm

In an adult *Drosophila* each abdominal segment is composed of a dorsal tergite and a ventral sternite. The right and left sides are fused continuously along the midline, but gynandromorph analysis indicates that they arise from separate groups of cells on opposite sides of the embryo (Garcia-Bellido and Merriam, 1969; Hotta and Benzer, 1973). Judging from the frequency and size of clones induced at the blastoderm stage, each half tergite is represented by slightly fewer blastoderm cells than that characteristic of imaginal discs (Garcia-Bellido and Merriam, 1971b; Guerra et al., 1973; Wieschaus and Gehring, 1976a). Similar to the cell lineage data from the thorax, blastoderm clones do not extend between adjacent segments (Guerra et al., 1973; Wieschaus and Gehring, 1976a; Lawrence et al., 1978). A clonal restriction is also observed between the prospective abdominal segments at the blastoderm stage in *Oncopeltus* (Lawrence, 1973). Although it is tempting to homologize the tergite and sternite in each abdominal segment to the dorsal and ventral disc present in each thoracic segment, this may not be justified from cell lineage criteria. The distance in gynandromorphs measured between tergites and sternites is much greater than that between wing and leg (20 sturts compared to six sturts) (Garcia-Bellido and Merriam, 1969; Wieschaus and Gehring, 1976). In contrast to those in the thorax, clones induced at the blastoderm stage do not overlap between dorsal and ventral primordia and within each tergite or sternite the patterns of the clones are random, offering no evidence for a subdivision into anterior and posterior compartments (Lawrence et al., 1978).

The precursors for the adult abdominal structures can be identified as clusters of diploid cells embedded in the larval hypoderm adjacent to the attachment sites

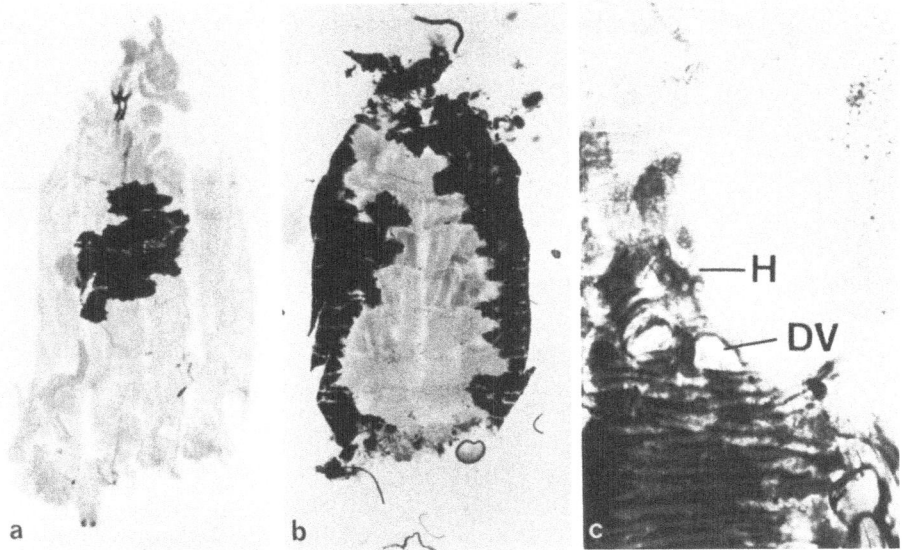

Fig. 6a–c. Hypoderm of larval gynandromorphs stained for aldehyde oxidase activity. The dark-staining areas are female, $R(1)w^{vC}/y\,v\,f\,mal$), the nonstaining areas male $(y\,v\,f\,mal)$. (a and b) Whole-mount preparations showing the typical appearance of the mosaic border and the continuity of the single male patch. (c) A mosaic ventral histoblast and the surrounding mosaic larval hypoderm. *H*, histoblast; *DV*, attachment sites of the dorsal ventral muscles. (From Szabad, Schüpbach and Wieschaus, in preparation)

for the dorsal-ventral muscles. Each dorsal histoblast is subdivided into an anterior and posterior group of cells. Given the random pattern of clones, it is doubtful that each group forms a specific part of the tergite, and it is even possible that only one of the two groups gives rise to the region of the adult tergite covered with bristles.

The cell lineage relationship between the histoblast and the adjacent larval hypoderm can be studied in larval gynandromorphs (Szabad, Schüpbach and Wieschaus, in preparation) (Fig. 6) in which the male regions are identified histochemically using the lack of aldehyde oxidase activity associated with *mal* (Janning, 1976). The distance between histoblasts of adjacent segments is about 10 sturts, similar to the value obtained between adjacent tergites and sternites in the adult. Ten sturts was also the value measured between other homologous structures in the larval hypoderm of adjacent segments. In cases where a histoblast is mosaic, the XX/XO border runs continuously between the future imaginal cells and the surrounding larval hypoderm (Fig. 6c). These results indicate that the islands of future adult cells differentiate out of the same sheet of cells that gives rise to the larval hypoderm. We do not know when or how during embryonic development this separation occurs. Future studies using mitotic recombination at carefully defined intervals might even identify a period when the larval and adult primordia still share common precursor cells.

The high frequency of mosaicism in the hypoderm of larval gynandromorphs indicates that the precursor cells are drawn from a very large area on the surface of the embryos. Mosaic hypoderms usually contain one large continuous patch of male tissue of an average size of one-half the total hypoderm surface (Fig. 6). The mosaic borders are more jagged than those found in the adult cuticle and show little tendency to run along the dorsal midline or to follow other morphological divisions. Thus, the larval hypoderm would represent a large *continuous* area of the blastoderm, not formed by the fusion of isolated primordia. There is a slight tendency of the mosaic border to coincide over short distances with the borders between segments. This might reflect some cell lineage restriction arising during embryogenesis; it might also be produced by minor infolding, cell death, or sliding between adjacent segments.

VII. Cell Lineage of the Internal Structures

That the establishment of segmentation plays an important role in the development of the ectoderm is not surprising, given that segmental morphology is the most striking feature of the insect integument. Very little indication of segmental organization is observed in the internal organs of mesodermal and endodermal origin. The few cases where a repeating segmental pattern is observed, in the larval musculature for example, might be due to a secondary interaction with the overlying segmented ectoderm. Such an interaction has been postulated on the basis of damage experiments in other insect embryos (Bock, 1941; Haget, 1953; McCrady, 1966). We do not know to what extent sharply defined clonal restrictions play a role in the development of internal structures. It is possible that segmentation is a developmental event restricted to ectodermal cells and that, given the inevitable lag time between irradiation and the formation of a clone capable of overlap (one or two cell divisions depending on whether twin spots are produced), segmentation may only be established in the overlying ectoderm after the initial invaginations have taken place.

Most cell lineage studies in *Drosophila* have been restricted to adult epidermal structures and the bristle and cuticle markers used to identify the clones in the adult are of little use in the study of internal organs. A histochemical procedure might be used to identify clones for enzyme null mutants, but such procedures become tedious if the clones are small or rare. To simplify the detection of clones induced in the embryonic nervous system, Deak (1977) has used the paralytic effect associated with the temperature-sensitive mutation *shibire* (Grigliatti et al., 1973). Flies heterozygous for this mutation irradiated at the blastoderm stage show two kinds of clones, representative of the separate origin of the motor and sensory nervous systems in *Drosophila*. About 1.4% of the adults show a leg paralysis that is temperature sensitive, the frequency depending on the temperature at which the flies are raised. This paralysis, thought from gynandromorph criteria to be dependent on the genotype of the motor neurons innervating the individual legs (Hall et al., 1973; Deak, 1976), is generally limited to single appendages, although clones are occasionally found that extend between the right and left legs of the same segment. No such overlaps are found between the legs of different segments.

Clones for *shibire* in the sensory nervous system (Deak, 1977) can be detected by the absence at high temperatures of the cleaning response normally elicited by stroking bristles in particular regions of the fly (Ghysen, personal communication). The sensory neurons are known from morphological studies (Lees and Waddington, 1942) to share a common cell lineage with the adult epidermis until metamorphosis. In accordance with this finding, clones in the sensory nervous system are found in discs that also possessed cuticle clones. This contrasts with the leg paralysis due to motor neurons found only in legs otherwise unmarked.

VIII. Cell Lineage of the Germ Line

The precursors for the germ line can be identified as a cluster of about 40 cells at the posterior pole of the blastoderm (Sonnenblick, 1950). These "pole cells" are carried into the interior of the embryo by the posterior midgut invagination and form together with the surrounding mesoderm a right and left gonad, each containing about 5–10 germ cells (Sonnenblick, 1950). The pole cells that do not contribute to the germ line are thought to be lost or to give rise to the cuprophilic cells of the midgut (Poulson, 1950).

Pole cell transplants to embryos of the opposite sex do not develop into functional gametes (Van Deusen, 1977). Differences between the genotypes of the germ line and that of the somatic tissues of the gonad probably explain why many gonads in gynandromorphs are agametic. When this agamety was used to determine the germ line genotype in gynandromorphs, the germ cell precursors could be shown to lie posterior to the precursors for all adult cuticle structures, in a position analogous to where the pole cells are observed in the embryo (Gehring et al., 1976). This mapping position has been recently confirmed using mosaics in which the chromosome loss (i.e., a marked Y chromosome) did not introduce a difference in sex of the two cell types (Nissani, 1977; T. Schüpbach, personal communication). Purely female mosaics can also be constructed of diplo-X and triplo-X tissue (Schüpbach et al., 1978). Triplo-X females are poorly viable and invariably sterile but triplo-X germ cells give rise to functional gametes when associated with diploid somatic tissue. The genotype of the germ line could be determined from the progeny of mosaic females or from histochemical staining of the gonad itself. The posterior map position of the germ line was once again confirmed. The right and left gonads were never totally different in genotype, indicating a single primordium on the posterior midline. The total germ line was mosaic in 22% of the cases and within such mosaic ovaries the individual ovarioles were often composed of both diplo and triplo-X cysts. Thus, although the number of germ cells is probably not greater than that contributing to the foreleg, single ovarioles frequently receive contributions from more than one pole cell.

The number of pole cells contributing to each female gonad is not greater than five, judging from the relative size of clones induced at the blastoderm stage (Wieschaus and Szabad, in press). During the remainder of embryogenesis, these cells divide once or twice. The high frequency of twin spots induced at the blastoderm stage indicates that both daughters of an irradiated pole cell contribute to the germ line. Thus, it is unlikely that the germ line shares common precursor cells with the midgut at this time, although both structures may still draw precursors from the same group of forty pole cells.

IX. The Relationship Between Mitotic Recombination and Gynandromorph Data: Some Theoretical Considerations

The ease with which mosaics can be produced in *Drosophila* is one of its major advantages in developmental studies. In addition to mitotic recombination, mosaic embryos can also be produced by chromosome loss, in particular by the loss of an X chromosome during early cleavage (Sturtevant, 1929; Hinton, 1955). These "gynandromorphs" contain large continuous regions of male and female tissue. Their use in the construction of embryonic fate maps (see chapters by Janning and Hall, this volume) rests on a suggestion by Sturtevant (1929) that the farther removed the primordia for two structures lie at the blastoderm stage, the more frequently they will be of different genotypes. Two primordia are said to be one "sturt" apart when the adult structures they form are of different genotypes in 1% of gynandromorphs. These frequencies are then used as distances to position the primordia relative to each other in two-dimensional anlage plans (Garcia-Bellido and Merriam, 1969; Hotta and Benzer, 1973). Although the positions of the primordia in these maps corresponds well to those deduced from histological observations (Poulson, 1950), even an approximate conversion of sturt distances into actual cells is difficult. Direct extrapolation of the distances onto the blastoderm yields values of one to two sturts per cell, depending on whether the theoretical maximum sturt value (100 sturts) is related to the long axis of the embryo (100 cells) or to the short axis (40 cells), and on the extent to which convolutions in the mosaic border are taken into consideration (see Janning for a more detailed examination of this approach). A more exact conversion over small regions of the embryo can be made by comparing gynandromorph measurements with the behavior of clones induced at the blastoderm stage. From the clone sizes presented in Figure 2, it was concluded that the foreleg arises from about nine blastoderm cells. In gynandromorph studies (Wieschaus, unpublished), the frequency of mosaicism (26%) and the greatest intradisc distances indicate that these foreleg precursor cells are drawn from an area of the blastoderm about 16.6 sturts in diameter. If these cells are all adjacent to each other, then the area of a single cell is $1/9\pi(8.3)^2$ and its theoretical diameter 5.5 sturts.

A value of 5.5 sturts per cell is much larger than the values obtained by superposition of map distances onto the blastoderm surface. This might be explained if over short distances in the embryo, the mosaic border is more convoluted than is commonly thought. Such convolutions might arise by mixing of the nuclei during their cleavage and migration into the cortex, or by occasional chromosome loss during later cleavage divisions. The more convoluted the mosaic border is at the blastoderm stage, the more likely it is to cut into a small primordium of ten cells, thus explaining the relatively high frequency of mosaicism found in the foreleg and the large interdisc distances (= primordial diameter) used in the calculation. The effect of this convolution on sturt distances between single cells can be easily visualized in an embryo where the plane of the first cleavage division, and thus the mosaic border, falls midway between the short and long axes of the egg. If no nuclear mixing and no later chromosome loss occurs, the mosaic border can be represented by an ellipse arching around the side of the embryo (Fig. 7). The length of this border is about 70 cells on one side and will approximate the average border length in a large population of gynandromorphs with variable initial cleavage planes. When the surface of this embryo is drawn in two dimensions such that each of the 6000 blastoderm cells is represented by a single hexagon (Fig. 7), the total number of mosaic cell interfaces (260) represents 1.4% of the total 18000 cell interfaces in the blastoderm. This means that the probability of the mosaic border falling between any two adjacent cells is also 1.4%. As convolutions are introduced in this border, the number of mosaic interfaces increases as does therefore the probability of the border falling between adjacent cells. In this way it is possible to predict the average appearance of mosaic blastoderms with sturt distances between adjacent cells of 5.5. The borders in such hypothetical embryos are somewhat more convoluted than the XX/ XO borders found in the adult or, for example, in the larval hypoderm (Fig. 6). However, it must be considered that the hypoderm precursor cells like those for the adult represent only a fraction of the blastoderm surface.

A figure of 5.5 sturts, like any sturt distance is actually an average of cell lineage relationships, which, as we have seen from clonal analysis, are extremely variable. In a fraction (f_o) of blastoderms, two structures A and B may be formed by the same blastoderm cell, in another fraction (f_1) by two adjacent cells, and in other fractions $(f_2$ to $f_n)$ by nonadjacent precursors n cell diameters apart. The contributions of each fraction to the final sturt value can be expressed using the following formula:

Sturt distance $_{AB}=f_o(o)+f_1(p_B)+f_2(2 \cdot p_B) \cdots +f_n(n \cdot p_B)$ where p_B is the distance between adjacent blastoderm cells and where the distance between precursors n cell diameters apart is assumed to equal $n \cdot p_B$. Since the distance between two adjacent cells equals the sum of their radii (and thus their diameter), we can substitute 5.5 sturts for p_B in the formula.

This has certain simple consequences for the interpretation of gynandromorph data. A sturt value of less than 5.5 can only be obtained if in a certain fraction of the embryos, A and B exist as common precursors. Only structures that map greater than 5.5 sturts can possibly exist as separate cell lineages in the blastoderm and as long as the distance is less than 11 sturts the two precursors must at least in some cases be represented as adjacent cells. This condition of less than 11 sturts is met for all clonal restrictions found at the blastoderm stage in the thorax, where the distance between primordia measures 7–9 sturts. On the other hand, given the variability in clone sizes and cell lineage relationships, distances greater than the average value of 5.5 sturts do not exclude the possibility of clonal overlap and in the foreleg, for example, blastoderm clones can be double that size (Fig. 8). Indeed, when *Minute* mutations are used to slow down the growth rate of

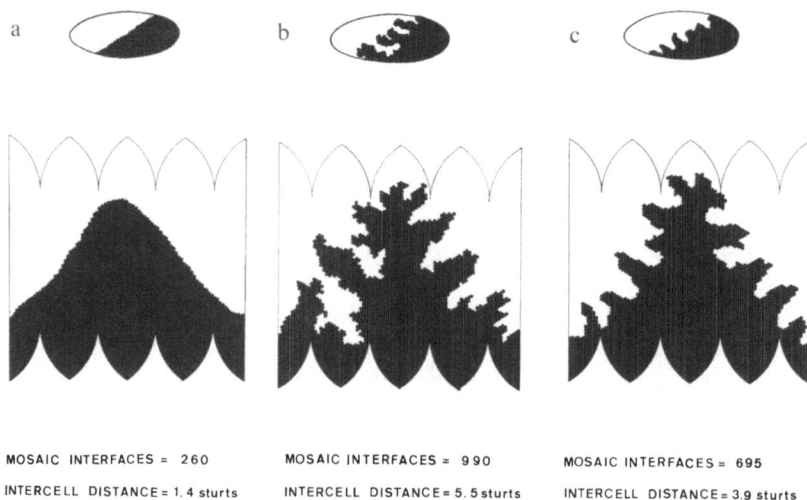

MOSAIC INTERFACES = 260   MOSAIC INTERFACES = 990   MOSAIC INTERFACES = 695

INTERCELL DISTANCE = 1.4 sturts   INTERCELL DISTANCE = 5.5 sturts   INTERCELL DISTANCE = 3.9 sturts

Fig. 7a–c. The effect of a convoluted mosaic border on sturt distances between blastoderm cells in gynandromorphs. The probability of the mosaic border falling between any two adjacent blastoderm cells equals the number of cell interfaces that separate cells of different genotypes, divided by the total number of cell interfaces. There are 6000 cells in a blastoderm and thus approximately 18 000 cell interfaces. If, as an average starting situation, a mosaic border is chosen that runs midway between the long and short axes of the egg (a), the minimum number of mosaic interfaces, counted after drawing the mosaic pattern onto hexagon paper, is about 260 (Wieschaus, unpublished). This represents 1.4% of the total blastoderm interfaces. If such a border were representative of the average border at the blastoderm stage, then the distance between adjacent blastoderm cells would be 1.4 sturts. The value of 5.5 sturts, however, is obtained from a comparison of gynandromorph data with clones induced in the foreleg primordium using mitotic recombination (see text). Such a value requires about 990 mosaic interfaces and thus a much more convoluted border (b). 5.5 sturts will be an overestimate if the number of disc precursor cells calculated in the text is artificially reduced by X-ray damage associated with the production of clones. With a 50% lethality (c), the true number of precursor cells would be double, the calculated intercell distance 3.9, and the average mosaic pattern somewhat less convoluted (c)

the surrounding cells, clones can extend between the right and left foreleg, a distance of more than 30 sturts (Steiner, 1976).

A similar formula might be proposed to relate the same sturt distance to the cell lineage relationships at later stages in development. Due to the smaller clone sizes, the p value for later stages will be smaller and the fraction of embryos in which A and B are formed from separate cells will be larger. The same applications of the formula can be made for later stages as were made for the blastoderm, as long as within the area studied, the mosaic pattern is random, that is, as long as the sturt distance between any two adjacent cells is the same. When this condition is met, two structures can only be represented by separate cell lineages in stages where the average clone size in sturts, i.e., the p value, is less than the actual sturt distance measured between the structures in the adult. The first postembryonic compartmentalization in the wing is one separating it into dorsal and ventral regions (Garcia-Bellido et al., 1973). The average distance measured across this border in the adult is 1.6 sturts (Lawrence and Morata, 1977), implying that the dorsal-ventral restrictions could only arise when the p value in the wing primordium (i.e., the average clone size) had decreased from 5.5 to 1.6. Since the diameter

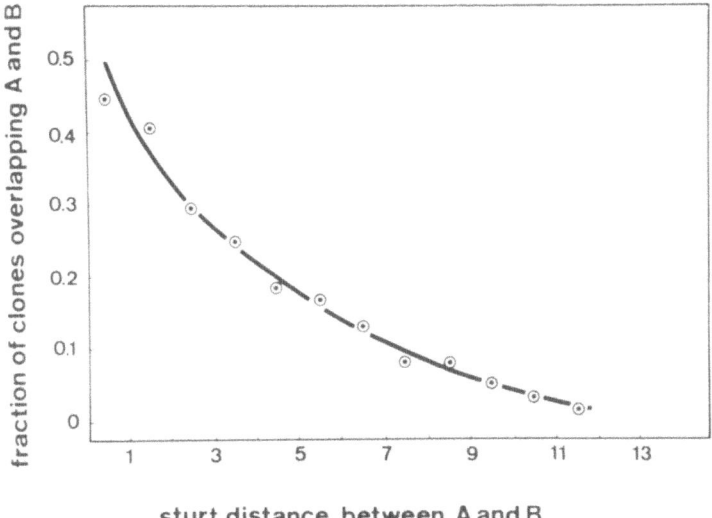

sturt distance between A and B

Fig. 8. Clonal overlap at the blastoderm stage as a function of the distance measured between two structures using gynandromorphs. The number of clones overlapping a given pair of bristles A and B is expressed as a fraction of the total clones involving either. Each point (⊙) represents the average of approximately 40 pairs of bristles in the anterior or posterior compartment of the foreleg. The sturt distances between each pair of bristles was determined from examination of 425 gynandromorphs. The clones were induced in a twin spot analysis (Wieschaus and Gehring, 1976; Wieschaus, unpublished) involving 1154 adults irradiated at the blastoderm stage

of the clone decreases by $\sqrt{2}$ with each cell division, such a conversion would require at least four rounds of divisions after the blastoderm stage. This places the dorsal-ventral compartmentalization sometime during midlarval development, an observation consistent with the findings of clonal analysis performed on larvae (Garcia-Bellido et al., 1973).

The patterns of mosaicism in gynandromorphs can be regarded as the sum of individual blastoderm clones. Both types of mosaics reflect all developmental events occurring between the time of mosaic induction and final differentiation. Since earlier cell lineage restrictions have a more dramatic effect on the mosaic pattern in the adult, it is possible to arrange these restrictions in a temporal sequence, according to the decreasing sturt distances measured between the different primordia in adult gynandromorphs. To relate this sequence to particular stages in development, we have calculated the p value for the blastoderm stage by expressing the average size of foreleg clones in sturts. Since these clone sizes might be artificially increased by X-ray damage, the value for p obtained for the blastoderm stage may be too large. The effect of this cell death is probably small, since even if one assumes a lethality of 30%, the number of precursor cells rises only to about 13 and the estimate for p drops to 4.6. A 50% lethality decreases the p value to 3.8. In any case, the sequence and intervals between the establishment of the different cell lineage restrictions are independent of X-ray-induced cell death, since it was determined from gynandromorphs where presumably no such death occurs. A distorted value for p will only shift the sequence forward or backward during development.

Acknowledgements. I would like to thank Ilan Deak, Andreas Dübendorfer, Kurt Dübendorfer, Felix Epper, Margit Lohs-Schardin, Rolf Nöthiger, Christiane Nüsslein, Trudi Schüpbach, Emil Steiner, and Janos Szabad for permission to discuss their unpublished results and for many helpful suggestions on the manuscript. My own research contributions to this review were supported by a grant of the Swiss National Science Foundation to Rolf Nöthiger (3.741–0.76).

References

Anderson, D. T.: The Embryology of *Dacus tryoni*. 2. Development of imaginal discs in the embryo. J. Embryol. Exp. Morph. **11**, 339—351 (1963)

Auerbach, C.: The development of the legs, wings and halteres in wild type and some mutant strains of *Drosophila melanogaster*. Trans. R. Soc. Edinburgh **58**, 787—815 (1936)

Baker, W. K.: A clonal analysis reveals early developmental restrictions in the Drosophila head. Dev. Biol. **62**, 447—463 (1977)

Becker, H. J.: Über Röntgenmosaikflecken und Defektmutationen am Auge von *Drosophila* und die Entwicklungsphysiologie des Auges. Z. Indukt. Abstamm.- u. Vererb.-L. **88**, 333—373 (1957)

Bock, E.: Wechselbeziehungen zwischen den Keimblättern bei der Organbildung von *Chrysopa perla*. Arch. Entwicklungsmech. Organ. **141**, 159—247 (1941)

Bodenstein, D.: The postembryonic development of *Drosophila*. In: Demerec, M. (Ed.): Biology of *Drosophila*, pp. 275—367. New York: Wiley 1950

Brehme, K. S.: A study of the effect on development of "Minute" mutations in *Drosophila melanogaster*. Genetics **24**, 131—161 (1939)

Bryant, P. J.: Cell lineage relationships in the imaginal wing disc of *Drosophila melanogaster*. Dev. Biol. **22**, 389—411 (1970)

Bryant, P. J., Schneiderman, H. A.: Cell lineage, growth and determination in the imaginal leg disc of *Drosophila melanogaster*. Dev. Biol. **20**, 263—290 (1969)

Crick, F. H. C., Lawrence, P. A.: Compartments and polyclones in insect development. Science **189**, 340—347 (1975)

Deak, I. I.: Use of Drosophila mutants to investigate the effect of disuse on the maintenance of muscle. J. Insect Physiol. **22**, 1159—1165 (1976)

Deak, I. I.: Detection of clones in the nervous system of *Drosophila melanogaster*. Wilhelm Roux's Archives **183**, 165—170 (1977)

Deusen, E. B., Van: Sex determination in germline chimeras of *Drosophila melanogaster*. J. Embryol. Exp. Morph. **47**, 173—185 (1977)

Dübendorfer, A.: Entwicklungsleistungen transplantierter Genital- und Analanlagen von *Musca domestica* und *Phormia regina*. Experientia (Basel) **26**, 1158—1160 (1970)

Dübendorfer, A.: Untersuchungen zum Anlageplan und Determinationszustand der weiblichen Genital- und Analprimordien von *Musca domestica*. Wilhelm Roux' Arch. **168**, 142—168 (1971)

Dübendorfer, K.: Die Entwicklung der männlichen und weiblichen Genital-Imaginalscheiben von *Drosophila melanogaster*. Eine klonale Analyse. Ph.D. Thesis, Univ. of Zurich, Switzerland (1977)

Dunn, L. C., Coyne, J.: The effects of the Minute mutations of *Drosophila melanogaster* on developmental rate. Hereditas **23**, 70—90 (1937)

Emmert, W.: Entwicklungsleistungen abdominaler Imaginalscheiben von *Calliphora erythrocephala*. Experimentelle Untersuchungen zur Morphologie des Abdomens. Wilhelm Roux' Arch. **169**, 87—133 (1972a)

Emmert, W.: Experimente zur Bestimmung des Anlageplans der männlichen und der weiblichen Genital-Imaginalscheibe von *Calliphora*. Wilhelm Roux' Arch. **171**, 109—120 (1972b)

Ferris, G. F.: External morphology of the adult. In: Demerec, M. (Ed.): Biology of *Drosophila*, pp. 368—419. New York: Wiley 1950

Fritz-Niggli, H.: Vergleichende Analyse der Strahlenschädigung von *Drosophila*-Eiern mit 180 keV und 31 MeV. Fortschr. Röntgenstr. **83**, 178—200 (1955)

Garcia-Bellido,A.: Genetic control of wing disc development in *Drosophila*. In: Cell Patterning. Ciba Foundation Symposium **29**, 161—183 (1975)

Garcia-Bellido,A., Ferrus,A.: Gynandromorph fate map of the wing disc compartments in
Drosophila melanogaster. Wilhelm Roux's Archives **178**, 337—340 (1975)

Garcia-Bellido,A., Merriam,J.R.: Cell lineage of the imaginal discs in *Drosophila* gynandromorphs. J. Exp. Zool. **170**, 61—76 (1969)

Garcia-Bellido,A., Merriam,J.R.: Parameters of the wing imaginal disc development of *Drosophila melanogaster*. Dev. Biol. **24**, 61—87 (1971a)

Garcia-Bellido,A., Merriam,J.R.: Clonal parameters of tergite development in *Drosophila*.
Dev. Biol. **26**, 264—276 (1971b)

Garcia-Bellido,A., Ripoll,P., Morata,G.: Developmental compartmentalization of the wing
disc of *Drosophila*. Nature (New Biol.) **245**, 251—253 (1973)

Gehring,W.J.: Übertragung und Änderung der Determinationsqualitäten in Antennenschei-
benkulturen von *Drosophila melanogaster*. J. Embryol. Exp. Morph. **15**, 77—111 (1966)

Gehring,W.J.: Determination of primordial disc cells and the hypothesis of step wise determination. In: Lawrence,P.A. (Ed.): Insect Development. Oxford: Blackwell 1976

Gehring,W.J.: Imaginal discs: Determination. In: Novitski,E., Wright,T. (Eds.): The
Genetics and Biology of *Drosophila*, Vol. 2c. New York: Academic Press (in press)

Gehring,W.,J., Seippel,S.: Die Imaginalzellen des Clypeo-Labrums und die Bildung des
Rüssels von *Drosophila melanogaster*. Rev. Suisse Zool. **74**, 589—596 (1967)

Gehring,W.,J., Wieschaus,E., Holliger,M.: The use of "normal" and "transformed" gynandromorphs in mapping the primordial germ cells and the gonadal mesoderm in *Drosophila*. J. Embryol. Exp. Morph. **35**, 607—616 (1976)

Ghysen,A.: personal communication

Grigliatti,T.A., Hall,L., Rosenbluth,R., Suzuki,D.T.: Temperature-sensitive mutations in
Drosophila melanogaster. XIV. A selection of immobile adults. Molec. Gen. Genet. **120**,
107—114 (1973)

Guerra,M., Postlethwait,J.H., Schneiderman,H.A.: The development of the imaginal abdomen of *Drosophila melanogaster*. Dev. Biol. **32**, 361—372 (1973)

Haget,A.: Analyse expérimentale des facteurs de la morphogenèse embryonnaire chez le
Coléoptère *Leptinotarsa*. Bull. Biol. France Belg. **87**, 123—217 (1953)

Hall,L., Junter,A.K., Suzuki,D.T.: Analysis of *Drosophila* behaviorial mutants by internal
genetic mosaics. Genetics **74**, S105 (1973)

Haynie,J.L., Bryant,P.J.: The effects of X-rays on the proliferation dynamics of cells in the
imaginal wing disc of *Drosophila melanogaster*. Wilhelm Roux's Archives **183**, 85—100
(1977)

Hinton,C.W.: The behaviour of an unstable ring-chromosome of *Drosophila melanogaster*.
Genetics **40**, 951—961 (1955)

Hotta,Y., Benzer,S.: Mapping of behavior in *Drosophila* mosaics. In: Ruddle,F. (Ed.): Genetic Mechanisms of Development, pp. 129—167. New York: Academic Press 1973

Janning,W.: Entwicklungsgenetische Untersuchungen an Gynandern von *Drosophila melanogaster*. IV. Vergleich der morphogenetischen Anlagepläne larvaler und imaginaler Strukturen. Wilhelm Roux's Archives **179**, 349—372 (1976)

Laug,G.: Origine et croissance du disque génital de *Drosophila melanogaster Meig*. C.R.
Hebd. Séanc. Acad. Sci. Paris **265**, 814—817 (1967)

Lawrence,P.A.: A clonal analysis of segment development in *Oncopeltus* (Hemiptera). J.
Embryol. Exp. Morph. **30**, 681—699 (1973)

Lawrence,P.A., Green,S.M., Johnston,P.: Compartmentalisation and growth of the *Drosophila* abdomen. J. Embryol. Exp. Morph. **43**, 233—245 (1978)

Lawrence,P.A., Morata,G.: Early development of mesothoracic compartment in *Drosophila*.
An analysis of cell lineage, fate mapping and an assessment of methods. Dev. Biol. **56**, 40—
51 (1977)

Lees,A.D., Waddington,C.H.: The development of the bristles in normal and some mutant
types of *Drosophila melanogaster*. Proc. R. Soc. **B131**, 87—110 (1942)

Madhavan,M., Schneiderman,H.A.: Histological analysis of the dynamics of growth of imaginal discs and histoblast nests during the larval development of *Drosophila melanogaster*.
Wilhelm Roux's Archives **183**, 269—305 (1977)

McCrady, E.: In vivo culture of *Drosophila melanogaster* embryos containing the Notch deficiencies Df(1)N 8 and Df(1)N 264-40. J. Exp. Zool. **161**, 37—52 (1966)

Morata, G., Ripoll, P.: Minutes: Mutants of *Drosophila* autonomously affecting cell division rate. Dev. Biol. **42**, 211—221 (1975)

Nöthiger, R., Dübendorfer, A., Epper, F.: Gynandromorphs reveal two separate primordia for male and female genitalia in *Drosophila melanogaster*. Wilhelm Roux's Archives **181**, 367—373 (1977)

Nissani, M.: Cell lineage analysis of germ cells of *Drosophila melanogaster*. Nature (Lond.) **265**, 729—731 (1977)

Postlethwait, J. H., Schneiderman, H. A.: A clonal analysis of development in *Drosophila melanogaster*: Morphogenesis, determination, and growth in the wild-type antenna. Dev. Biol. **24**, 477—519 (1971)

Poulson, D. F.: Histogenesis, organogenesis, and differentiation in the embryo of *Drosophila melanogaster*, Meigen. In: Demerec, M. (Ed.): Biology of Drosophila, pp. 168—274. New York: Wiley 1950

Roseland, C., Schneiderman, H. A.: Origin and regulation of the presumptive adult abdominal segments in *Drosophila melanogaster*. (in preparation)

Schüpbach, T., Wieschaus, E., Nöthiger, R.: A study of the female germline in mosaics of *Drosophila*. Wilhelm Roux's Archives **184**, 41—59 (1978)

Schweizer, P.: Wirkung von Röntgenstrahlen auf die Entwicklung der männlichen Genitalprimordien von *Drosophila melanogaster* und Untersuchung von Erholungsvorgängen durch Zellklon-Analyse. Biophysik **8**, 158—188 (1972)

Sonnenblick, B. P.: The early embryology of *Drosophila melanogaster*. In: Demerec, M. (Ed.): Biology of Drosophila, pp. 62—117. New York: Wiley 1950

Steiner, E.: Establishment of compartments in the developing leg discs of *Drosophila melanogaster*. Wilhelm Roux's Archives **180**, 9—30 (1976)

Stern, C.: Somatic crossing over and segregation in *Drosophila melanogaster*. Genetics **21**, 625—730 (1936)

Sturtevant, A. H.: The claret mutant type of *Drosophila simulans: A study of chromosome elimination and of cell-lineage*. Z. Wiss. Zool. **135**, 323—356 (1929)

Szabad, J., Schüpbach, T., Wieschaus, E.: The cell lineage relation and embryonic development of the larval hypoderm (in prep.)

Turner, F. R., Mahowald, A. P.: Scanning electron microscopy of Drosophila embryos. Dev. Biol. **50**, 95—108 (1976)

Ursprung, H.: Fragmentierungs- und Bestrahlungsversuche zur Bestimmung von Determinationszustand und Anlageplan der Genitalscheiben von *Drosophila melanogaster*. Wilh. Roux' Arch. **151**, 504—558 (1959)

Wieschaus, E.: Clonal analysis of early development in *Drosophila melanogaster*. Thesis, Yale Univ. (1974)

Wieschaus, E., Gehring, W.: Clonal analysis of primordial disc cells in the early embryo of *Drosophila melanogaster*. Dev. Biol. **50**, 249—263 (1976 a)

Wieschaus, E., Gehring, W. J.: Gynandromorph analysis of the thoracic disc primordia in *Drosophila melanogaster*. Wilhelm Roux's Archives **180**, 31—46 (1976 b)

Wieschaus, E., Szabad, J.: The Development and Function of the Female Germline in *Drosophila melanogaster*: A Cell Lineage Study. Dev. Biol. (in press)

Wildermuth, H.: Differenzierungsleistungen, Mustergliederung und Transdeterminationsmechanismen in hetero- und homoplastischen Transplantaten der Rüsselprimordien von *Drosophila*. Wilh. Roux' Arch. **160**, 41—75 (1968)

Würgler, F. E., Ulrich, H.: Radiosensitivity of embryonic stages. In: Ashburner, M., Novitski, E. (Eds.): The Genetics and Biology of Drosophila, Vol. 1 C. New York: Academic Press 1976

Zalokar, M.: L'ablation des disques imaginaux chez la larve de *Drosophile*. Rev. Suisse Zool. **50**, 232—237 (1943)

Zalokar, M.: Anatomie du thorax de *Drosophila melanogaster*. Rev. Suisse Zool. **54**, 17—53 (1947)

Zalokar, M., Erk, I.: Division and migration of nuclei during early embryogenesis of *Drosophila melanogaster*. J. Microsc. Biol. Cell. **25**, 97—106 (1976)

Cell Lineage and Differentiation in Drosophila

Antonio Garcia-Bellido and Pedro Ripoll

*Centro de Biologia Molecular, C.S.I.C., Facultad de Ciencias, U.A.M.
Madrid, Spain*

I. Introduction

Embryonic differentiation is usually monitored by the appearance of cell, or tissue-specific enzymes or their products or structural proteins. These phenotypic traits correspond to the activities of specific genes, and therefore, in the long run, differentiation has to be explained in terms of selective gene activities. The mechanisms by which particular genes are expressed in certain cell types and not in others remain one of most appealing problems of modern biology.

We will analyze here two of the manifold properties associated with the phenomenon of cellular differentiation. The first property is that differentiated cells appear to differ by several gene products from other differentiated cell types. It is debatable whether this pleiotropy of expression results from the cascade activation of different genes or whether batteries of genes are simultaneously activated by one or a few gene signals. Stated as a question: Is differentiation a continuous or a discontinuous process? The second property is that states of differentiation are considered to be irreversible, i.e., the activation of certain genes in a given cell type appears to be maintained by cell heredity. If this is so, it is debatable whether it is due to functional loops of gene activity or to alterations at the chromosomal level. Stated as a question: Is differentiation in embryonic systems an open system like that of metabolic genes in bacteria, or a closed one?

In this paper we will analyze these problems with reference to the development of *Drosophila* and in particular to that of its imaginal discs. Paradoxically, most of our knowledge of the processes involved in the establishment of cell lineages and in cell differentiation is based on observations of cuticular structures, despite the fact that these structures result from the secretion of chitin, which is the last event in the life of epidermal cells. The existence of different types of single cell structures and of their arrangements into different spatial patterns and of mutants that modify them, enables us to ask questions about events occurring before that final event. Transplantation experiments, clonal analysis, and genetic analysis have uncovered the existence of steps of genetic decisions that take place during the development of these cells. Our aim is to discuss available evidence to see how the genetic information present in the individual cells is singled out, processed, and implemented in supracellular terms during development.

II. Cell Lineages

A. Embryonic Segmentation and Germ Layer Segregation

There is an increasing body of evidence that suggests that the first differential genetic decisions, possibly defining embryonic segments, take place in the blastoderm cells. This evidence derives from both transplantation and defect experiments and from clonal analysis.

Experiments of ligation of preblastoderm embryos have shown that whereas ligation in early cleavage stages disturbs the pattern of segmentation, at later cleavages it has lesser effects, and if performed at the time of blastoderm formation it has no effect on segmentation (Schubiger, 1976). This finding is of special relevance, for it indicates that segmental differentiation does not result from a clonal segregation of cleavage nuclei and that it does not reflect a fixed organization laid down by the mother on the unfertilized egg. It strongly suggests that segmental organization results from an interaction of more general physical properties of the egg that are expressed on the egg surface and that determine the fate of the incoming nuclei. We do not know the nature of these cortical determinants, nor the kinds of interactions occurring between them and the genes of the migrating nuclei. Only in the case of the germ cells has it been possible to relate the determinants with morphologically and functionally distinguishable organelles in the pole plasm (Illmensee et al., 1976). These organelles have been traced back to the oocytes and found to be already functional at stage 13. However, in view of the results of the ligation experiments mentioned above, it would be misleading to postulate a similar particulate distribution of cortical determinants for different segments or different germ layers.

Transplantation of cleavage nuclei into unfertilized eggs (Illmensee, 1972) or into eggs in early cleavage stages (Zalokar, 1971) has demonstrated that these nuclei are totipotent. However, the heuristic value of these results, with respect to the problem of genetic determination, is poor, because transplantation of nuclei from cells of blastoderm embryos or of gastrula embryos (Illmensee, 1973) and possibly from differentiated tissues (see Illmensee, this volume) to the same cleavage embryos, reveals a similar unrestricted capacity to support normal development. An operational difference between nuclei and cells becomes obvious in cell transplantation experiments. Illmensee (1976, and this volume) has shown that ectopic transplantation of blastoderm cells leads to clones that differentiate according to their origin. The data available so far suggest that the specificity of determination of these cells may correspond to segments or tissues, and for imaginal cells possibly even to particular disc qualities.

This result is in line with others obtained with dissociated cells of early embryos. In these experiments, genetically marked cells from anterior and posterior halves of early embryos were mixed either inter se (Schubiger et al., 1969) or with cells of whole embryos (Chan and Gehring, 1971). They were then cultured in adult flies and tested for their presumptive adult cuticular structures by transplantation into metamorphosing hosts. Both experiments showed that anterior halves give rise to structures from anterior imaginal discs and posterior halves, to structures of posterior discs. Moreover, mosaic structures corresponding to anlagen located in the middle of the embryo were also found. The first authors used slightly dissociated 10-h donor embryos (beginning of organogenesis). The latter authors mixed highly dissociated 3-h old embryos (cellular blastoderm). In both cases it could be argued that only intact blastoderm regions are capable of self-determination later. In fact, experiments in which single, or clumps of a few, dissociated cells from 3-, 5-, and 7-h-old embryos were mixed and grown in a feeding layer of irradiated cells (Garcia-Bellido and Nöthiger, 1976) showed that only those derived from the oldest embryos (advanced gastrulation) were capable of differentiating recognizable adult cuticular structures. The results of Illmensee (l.c.), seem to settle this question: They suggest that single blastoderm cells can express specific qualities of determination.

Of less heuristic value to the question of cell determination are those experiments based on the disturbance of the normal development by UV irradiation (Geigy, 1931; Nöthiger and Strub, 1972) or by pricking or cauterization (Howland and Child, 1935; Howland and Sonnenblick, 1936; Bownes and Sang, 1974). This is so because the lack of an effect can be interpreted as being due to either the lack of any determination, or to the occurrence of regulatory events that compensate for the damage, and also because positive effects could be due to disturbances of an unspecific nature that would affect development at later stages.

The results in the preceding discussion suggest that whereas cleavage nuclei are totipotent, when they become incorporated into cells at the time of blastoderm formation, the cells express very specific properties of different developmental pathways.

Clonal analyses support and extend previous conclusions. Data from gynandromorphs indicate that in any region of the adult cuticle, two landmarks can be of different sex and thus can be derived from different clones initiated at the first zygotic divisions (Sturtevant, 1929; García-Bellido and Merriam, 1969) or later, but before blastoderm formation (Gelbart, 1973). Moreover, the probability of any two landmarks of a given disc being differently marked can be higher than that of two landmarks from different discs. These data suggest that there is no clonal segregation prior to the formation of the imaginal discs.

When the cleavage nuclei reach the egg cortex they probably respond to spatial coordinates. Formally we may distinguish two types of coordinates: One would determine segmental position (anterior–posterior axis), the other (dorsoventral axis) the germ layer (Fig. 1). We do not know whether these responses occur simultaneously and independently throughout the egg surface or whether they appear in a sequential fashion. Clonal analysis and mutant analysis will probably soon help us to uncover the logic of the process. This is now becoming possible with the use of enzymatic cell markers (Janning, 1974). So far, however, most of our knowledge derives from the analysis of adult cuticular cell markers. In the case of imaginal disc derivatives it seems that discs corresponding to the same segment become segregated from others of neighboring segments very early in development (Wieschaus and Gehring, 1976b; Steiner, 1976; Lawrence and Morata, 1976b). In gynandromorphs, by coupling chromosome loss with the creation of a *Minute* mosaicism (see Sec. B), it has been shown that there is not much growth in the blastoderm cells prior to the segregation of segments, as defined by their imaginal discs (Ferrús and García-Bellido, 1977).

We do not know when imaginal disc cells become segregated from other anlagen or cell types of the different germ layers. Previous analyses pertain only to that sample of the blastoderm cells that will become incorporated into the imaginal discs and can thus be unequivocally recognized after metamorphosis as adult cuticular structures. Thus, the fact that dissociated cells, or probes of single blastoderm cells, can develop outside their normal environment as imaginal disc cells, suggests that the segregation for such developmental pathways has already taken place. It is possible, however, that such segregation occurs later in development from cells of a cell lineage, say epidermis, not yet specified for imaginal disc qualities (Fig. 1).

Clones originating from mitotic recombination, initiated at known developmental stages, enable us to ascertain the time interval in which clonal segregations take place. Moreover such an analysis allows the detection of possible patterns of sequential segregations. Clones initiated after blastoderm formation are restricted to derivatives of particular imaginal discs (Bryant and Schneiderman, 1969; Bryant, 1970; García-Bellido et al., 1973, 1976) or to the nests of dorsal histoblasts of the abdominal segments (García-Bellido and Merriam, 1971b; Guerra et al., 1973). This observation was taken as an indication that imaginal discs correspond to independent cell lineages from very early in development. The timing of the segregation of the cell lineages has been recently worked out in more detail. In the thoracic segments, whereas clones initiated in the blastoderm stage (3 h) may embrace cuticular elements of discs of the same segment (dorsal and ventral discs), they do not include imaginal discs of different segments (Wieschaus and Gehring, 1976b). Clonal analysis has shown that, at that time, this segregation also applies to anterior versus posterior compartments within the same segment (Steiner, 1976; Lawrence and Morata, 1976b). The segregation between dorsal and ventral discs, in both anterior and posterior compartments, occurs later on in development: before 7 h (advanced gastrulation) in the thoracic discs (Wieschaus and Gehring, 1976a).

A similar blastodermal clonal segregation separating the analia from the genitalia, in the genital disc, has recently been found (Dübendorfer, 1977). Data from comparative anatomical studies, however, indicate that although in *Drosophila* both anlagen develop within the same imaginal disc they actually derive from two different embryonic segments. Following this rationale, clonal restrictions separating the different cephalic segments were expected (Baker, 1978). The adult head of insects is considered to be built up out of five to seven embryonic segments, each, except for the acron, giving rise to a different set of ventral appendages (Snodgrass, 1935; Baker, 1978). In *Drosophila*, the adult head derives from the fusion of three pairs of imaginal discs: the clypeolabrum disc (Gehring and Seip-

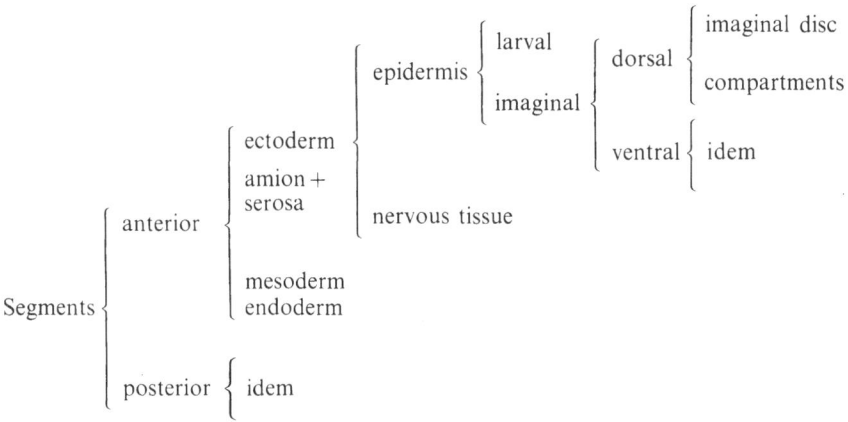

Fig. 1. Hypothetical segregation of cell lineage during embryogenesis

pel, 1967), the eye-antennal disc that contains the eye, the antenna, and the maxilar palp, and the labial (proboscis) disc. We do not know where the anlage of the clypeolabrum imaginal disc is situated. The anlagen of the two latter discs are widely separated in the blastoderm (Hotta and Benzer, 1972) and consequently no clones embracing both are expected. However, clones including antenna and maxilar palp, possibly belonging to different segments, have been found. Clones also may cross from the antenna to the head and eye until at least the first instar (Wieschaus and Gehring, 1976b; Baker, 1978). Thus, except possibly for the head capsule, segments and ventral and dorsal anlagen within a segment become separated by clonal restrictions early in embryonic development.

Clonal analysis in the hemipteran *Oncopeltus* (Lawrence, 1973a) shows that the abdominal segmentation, at least for the external cuticle, becomes defined very early in embryogenesis, around blastoderm.

Fate maps obtained from gynandromorph data do not provide any clue as to when the determination of the different anlagen takes place in development. However, consideration of these fate maps shows that the locations of the presumptive cells for the different embryonic anlagen (imaginal discs, larval epidermis, central nervous system, mesoderm, and endoderm) have the same topologic relationships (García-Bellido and Merriam, 1969; Hotta and Benzer, 1972; Janning, 1974; Kankel and Hall, 1975, see Janning, this volume) as those deduced for the blastoderm stage from embryologic studies (Poulson, 1950). An estimation of the number of primitive cells of the different anlagen, based on observed frequencies of mosaicism in gynanders, is difficult for most of the afore-mentioned anlagen. However, as for as the imaginal discs of the thorax are concerned, the data suggest that they contain numbers of cells that correspond well to those deduced from the fraction of the total blastoderm occupied by the anlagen in the gynandromorph fate map (Ripoll, 1972; García-Bellido and Ferrús, 1975; Wieschaus and Gehring, 1976a). Moreover, these data suggest that the anlagen of all the different imaginal discs of the thorax lie side by side, in a continuous cell layer. A generalization of these considerations to other anlagen would imply that determination for the different cell lineages may occur by sampling or splitting of the

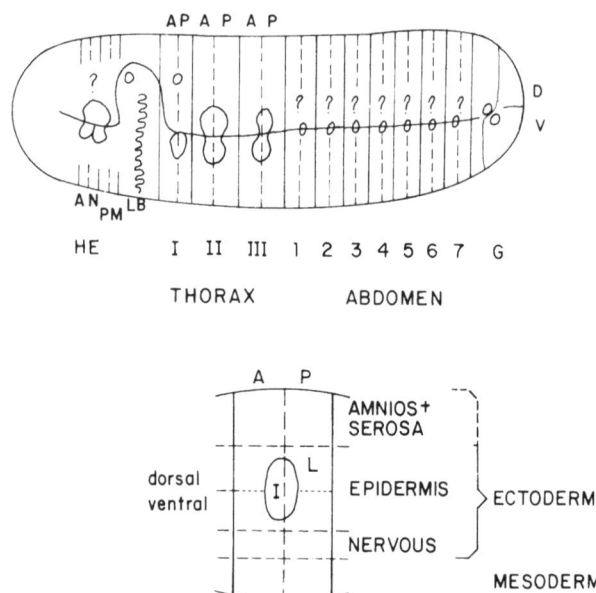

Fig. 2. Schematic representation of the organization of the embryo at around blastoderm time and hypothetical position of the anlagen of the different segments, germ layers, and cell lineages. *HE*, head; *AN*, antenna; *PM*, maxilar palpus; *LB*, proboscis or labium; *I–III*, thoracic segments; *1–7*, abdominal segments; *G*, genitalia and analia; *A*, anterior; *P*, posterior; *D*, dorsal; *V*, ventral; *I*, imaginal; *L*, larval

homogeneous population of blastoderm cells into subpopulations: i.e., into compartments (see below). We do not know how much, if any, cell division takes place between blastoderm and the segregation of the different cell lineages, nor if this segregation occurs simultaneously for all the different cell lineages or in a dichotomous and sequential fashion. Data from descriptive embryology already support the existence of pathways of segregation of the different cell types. Thus, it is possible that larval and adult (imaginal) ectoderms derive from a common cell pool, and that neuroblasts result from differential mitosis of ectodermal cells. In fact both analgen are adjacent in the blastoderm fate map (Hotta and Benzer, 1972; Figs. 1 and 2).

B. Imaginal Disc Compartimentalization

The existence of independent cell lineages during the development of the imaginal discs was first inferred from observations of clonal restrictions on the adult cuticular landscape. Becker (1957) noticed that clones initiated after the first instar would respect a line separating the dorsal from the ventral half of the eye. A similar line separating the dorsal from the ventral surface of the wing along the margin was described by García-Bellido (1968), Bryant (1970), and García-Bellido and Merriam (1971a). Another line of restriction separating thorax from wing was also described by Bryant (1970). Lines of restriction, however, could result from a real segregation of cell lineages or non-specific morphogenetic processes, such as cell migration, anlage fusions, intercalary cell death, etc. Making use of the property of non-*Minute (M⁺/M⁺)* clones of cells, resulting from mitotic recombination in heterozygous *Minute (M/M⁺)* anlage, to overgrow their neighboring cells (Morata and Ripoll, 1975), these uncertainties could be overcome and the demonstration of the existence of segregations of cell lineages was made possible.

The *Minute* technique thus uncovered the existence of lines of clonal restrictions delimiting discrete populations of cells called compartments (García-Bellido et al., 1973, 1976). The first restriction in the dorsal mesothoracic (wing) disc subdivides it into two compartments, an anterior and a posterior one. M^+ clones may be composed of several thousands of cells and almost fill either compartment but will respect a fixed anterior–posterior demarcation line. Data from Steiner (1976) and Lawrence and Morata (1976b) have shown that this segregation already exists at the blastoderm stage. The presumptive cells of both compartments lie side by side in the gynandromorph fate map (Ripoll, 1972; García-Bellido and Ferrús, 1975). Male–female boundaries in gynanders can separate any two landmarks within either compartment, indicating that their cells have no clonal relationship. Moreover, the distances in the fate map between landmarks within a compartment can be larger than those separating neighboring landmarks at both sides of the demarcation line. Both findings suggest that the primitive population of blastoderm cells, which will later become wing imaginal disc cells, are topologically split into two adjacent subpopulations called "polyclones" by Crick and Lawrence (1975), corresponding to an anterior and a posterior compartment. These cell populations will remain separate but side by side during subsequent development, and each will differentiate a fraction, either anterior or posterior, of the otherwise continuous cuticular pattern of the adult wing.

During subsequent development, and once the presumptive wing disc cells become incorporated in the imaginal disc, new compartment alizations arise. They subdivide gradually the proliferating cells of the disc into new compartments (Fig. 3) and separate a proximal (notum) from a distal (wing) region and a dorsal (notum and wing) from a ventral (pleura and wing) region during the first or early second larval instar. At the beginning of the third instar new subdivisions affect the notum and pleura and separate a proximal from a distal region in the wing blade (García-Bellido et al., 1973, 1976). Similar analyses on other imaginal discs have uncovered the occurrence of compartmentalization in the dorsal metathoracic disc (García-Bellido, 1975a; Morata and García-Bellido, 1976), the thoracic legs (Steiner, 1976), the head (Baker, 1978) and the genital disc (Dübendorfer, 1977). It thus appears to be a general property, at least of imaginal disks, that as proliferation proceeds different subpopulations appear, corresponding to different cell lineages, which define characteristic developmental pathways.

We will now discuss some general properties of compartments, which may be relevant to an understanding of their role in development.

1. Compartments are supracellular units of development. Although they arise in groups of cells having different clonal origins they confer upon them a common quality. This quality is of an abstract nature: anterior, dorsal, etc., shared by all the cells of the compartment in an identical way. Any cell within a compartment

can differentiate any structures belonging to it. Thus $Minute^+$ clones can almost fill a compartment and can certainly extend over any region of it. This regulative property of compartments is operationally equivalent to that found in the morphogenetic fields of classical embryology (García-Bellido, 1975a, p.180).

2. Compartmentalization occurs by a binary process. In each compartmentalization step the previously homogeneous cell population becomes subdivided into two different subpopulations. The two newly originated cell lineages are alternatives: anterior vs. posterior, dorsal vs. ventral, etc. These partitions are binary, possibly because each developmental alternative is associated with a genetic decision; i.e., a gene ("selector") becomes activated in one cell lineage but remains inactive in the alternative one (see below). There is one apparent exception to this rule: If segmentation is operationally identical to compartmentalization, the simultaneous appearance of segmental characteristics in the blastoderm cannot occur by binary partitions. However, we will see that genetic considerations show that the different segmental pathways may in fact represent different alternatives to a single pathway (see Sec. II.C). Compartment boundaries, between segments and between compartments, may represent the limits of the field of action of selector genes.

3. Homologous compartmentalizations, both anatomic and genetic, take place in different segments, in different imaginal discs, and indifferent compartments within a disc (Fig. 4). Thus an early anterior–posterior compartmentalization is found on all of the dorsal and ventral thoracic discs. It is interesting to recall here the exception of the cephalic (eye-yntenna) disc, which apparently has only anterior characteristics (Baker, 1978). The segregation of dorsal and ventral discs, from a common anlage, seems to take place simultaneously in all three thoracic segments. In the cephalic disc the possibly homologous compartmentalization separating antenna from head proper, occurs later (Baker, 1978). A dorsoventral compartmentalization is initiated in all the thoracic discs and in the eye part of the cephalic disc at the same time.

4. There seems to be a limit to the number and extent of compartments within a disc. Clonal restrictions cannot readily be defined when there are few cell divisions left before metamorphosis. The smallest compartments found in the wing embrace more than a thousand cells. They include cells that may differentiate into different kinds of cuticular structures arranged in spatial patterns. These patterns probably arise by a mechanism other than compartmentalization (see Sec. III.A). Crick and Lawrence (1975) suggested that compartment boundaries play a role in defining limits to gradients of positional information.

5. Compartments in imaginal discs are associated with certain types of cell behavior that may be relevant to an understanding of the process as a whole. The suggestion that cells of neighboring compartments are maintained apart by cell recognition characteristics (García-Bellido and Merriam, 1971a; García-Bellido, 1975a) has been recently supported experimentally (Morata and Lawrence, 1975; Lawrence and Morata, 1976a). In the mutant *engrailed (en)*, which transforms

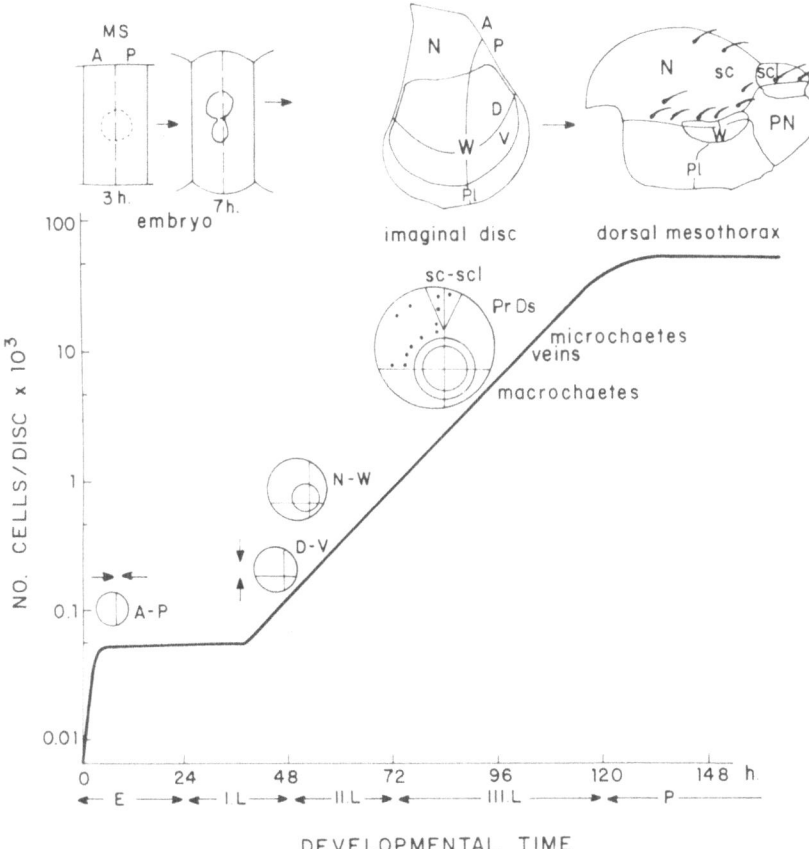

Fig. 3. Segregation of cell lineages in the development of the dorsal mesothoracic disc. *Above:* Organization of the anlage in embryos, in the imaginal disc, and in the adult fly (lateral view). *Below:* A schematic representation of the appearance of compartments and of the cell lineages for chaetes and veins during the proliferation phase of the disc. The size of the different compartments is proportional to the number of initial cells. *Pointing arrows* indicate opposite polarity of twin compartments. *A,* anterior; *P,* posterior; *D,* dorsal; *V,* ventral; *Pr,* proximal; *Ds,* distal; *MS,* mesothorax; *N,* notum; *W,* wing; *PL,* pleura; *SC,* scutum; *Scl,* scutellum; *PN,* postnotum

the posterior compartment of the wing into the anterior one, the anterior–posterior demarcation line becomes diffuse and disappears. Moreover, posterior *en/en* clones in phenotypically wild-type *(en/+)* anlagen will cross the anterior–posterior demarcation line. This finding suggests that the *engrailed* wild-type gene labels the cells of the posterior compartment so that they are different from those of the anterior one in order to prevent the two from mixing along the anterior–posterior border line. Experiments with other systems and other mutants (see García-Bellido and Lewis, 1976; Morata and García-Bellido, 1976) seem to allow the generalization that differences in cell recognition may be associated with the appearance of compartments and with their maintenance as separate entities.

6. The splitting of a homogeneous population into two is associated, in some cases, with the creation of opposite polarities. The pattern distribution of the cuticular elements in the anterior compartment of the wing disc is the mirror image of that of the posterior. Similar considerations apply to the relation between the dorsal and the ventral compartments in the same disc (García-Bellido et al., 1976). There are, however, no indications of such a change in polarity between the proximal and the distal compartments, perhaps because with the two previous ones the possible spatial symmetries are saturated. We do not know if this phenomenon is general and extends to other discs.

Since cells of the same compartment can mix with each other, organs that have identical counterparts on either side of the embryo and that maintain contact along the midline throughout development, should merge because they have the same recognition characteristic. The fact that they do not merge suggests that they are prevented from doing so by polarity. The imaginal discs of the first legs, whose anlagen are on opposite sides of the blastoderm, fuse early in development and remain fused until metamorphosis. Although clones may cross between them in early stages, they do not cross after the second larval instar (Steiner, 1976). Similarly, clones may cross the midline of the genitalia early in development but not later, although the lateral halves remain fused throughout development (Nöthiger and Dübendorfer, personal communication). The same probably applies to single bilateral organs in internal tissues, because clear left–right mosaic borders are found in gynandromorphs (Janning, 1974). These considerations suggest that polarities are instrumental in maintaining the cells of bilaterally symmetric compartments apart. Probably the same mechanism is used to prevent cells from different segments from mixing. Perhaps anterior–posterior alternating polarities are important for the segregation of segments, preventing them from merging before specific determination occurs, and even afterward.

7. The splitting of the primitive cell population into two subpopulations follows certain numerical rules. The two polyclones are, at the moment of their initiation, of different sizes. Thus in the wing disc the number of cells incorporated into the anterior compartment is about twice that incorporated into the posterior one. In a similar way the two homologous dorsal polyclones contain about twice the number of the two ventral ones and similarly 2:1 is the ratio between the cells of the wing and the notum compartments (García-Bellido et al., 1976). Similar conclusions can be drawn from the data of the leg discs (Steiner, 1976). We do not know whether this ratio is general, but it seem to hold that compartments are of different sizes at the moment of their initiation. However, there is no fixed number of cells characteristic of a given compartmentalization step. Thus the number of cells in the anterior dorsal mesothoracic anlage is larger than that in the anterior dorsal metathoracic one (Morata and García-Bellido, 1976), and the number of dorsal cells in the anterior compartments is different from the number of dorsal cells in the posterior one (García-Bellido et al., 1976). These differences suggest that the number of cells in a compartment at the moment of its initiation is under some specific control related to the particular developmental pathway in which it occurs.

To summarize: Compartmentalization appears to be a mechanism by which different cell lineages are segregated from previously homogeneous cell populations. It occurs in a sequential and progressive fashion and delimits, probably by means of specific genes, the characteristic steps of developmental pathways. It is possible that such developmental properties as are found in imaginal discs could be more general. Compartmentalization could be instrumental in the segregation of segments, germ layers, and imaginal disc anlagen.

The actual mechanism of compartmentalization is associated with cell properties such as polarity, cell recognition, segregation of a finite number of cells into polyclones, etc, which still are poorly understood.

C. Genetic Control

The existence of mutations that affect morphogenesis affirms our assumption of a genetic control of development. Yet the mechanisms by which genes exert this control has so far escaped our understanding. The discovery by Stern that most mutations express themselves autonomously in cells has provided a new strategy of studying gene control of development (see Stern, 1954). When genes express themselves in individual cells it is possible to see how the behavior of the cells is altered by the mutations. From these alterations we can deduce which cellular parameters are affected by which mutations and consequently define at least operationally their wild-type functions. Many recent studies in *Drosophila* have been made using this approach. Clonal analysis of the imaginal discs has provided us with a description of the cellular parameters of their normal development. The observed clonal restrictions, their topologic organization into compartments, and the logic of their sequential appearance, represent a developmental framework within which genetic questions can be asked. Mutant analysis can now be coupled with clonal analysis. Thus, gene organization and gene interaction, basic to our understanding of control mechanisms, are open to our analysis.

As we will see later, morphologic phenotypes seem to result from the combined effect of several gene functions. Therefore it is difficult (1) to deduce from the mutant phenotype the altered wild-type functions without knowing the combinatorial code and (2) to know the effect of the total lack of gene function without a genetic analysis of the mutant locus. Most of the known mutations seem to represent disruptive alterations of development, i.e., they cause effects not related to their primary wild-type gene function. Fortunately the homeotic mutants represent an exception. They appear to substitute one developmental pathway with another, and thus may correspond to control genes. Homeotic mutations have been the subject of recent reviews (see Postlethwait and Schneiderman, 1974; Ouweneel, 1976; García-Bellido, 1975a, b, 1977a), and so we will only summarize here their most relevant properties.

Some homeotic mutants specifically affect individual segments or compartments. Individually they lead to the transformation of a certain organ into another specific one (Fig. 4). Remarkably most of the segmental transformations are toward mesothorax. For the ventral appendages, *Antennapedia (Antp)* and *arista-*

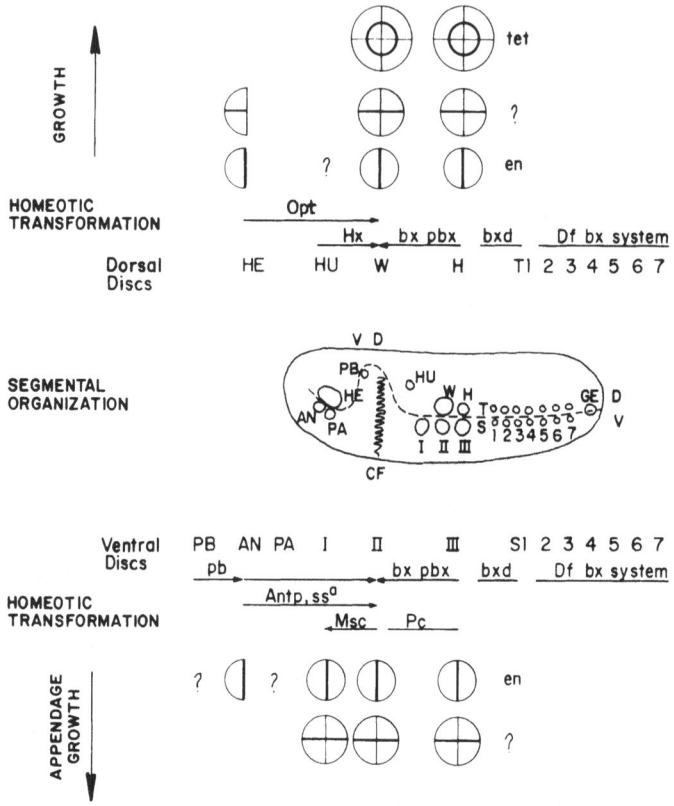

Fig. 4. Some homeotic mutants of *Drosophila* and segments and compartments they affect (description of the mutants in text). *Arrows* indicate the direction of transformation. The anlagen in the fate map of the blastoderm are: *HE*, head; *HU*, humerus; *W*, wing; *H*, haltere; *PB*, proboscis or labium; *PA*, palpus (maxilar); *I, II, III*, thoracic legs; *1–7*, abdominal segments; (*T*, tergites; *S*, sternites) *G*, genitalia; *CF*, cephalic furrow. Longitudinal line separates: *D*, dorsal from *V*, ventral anlagen

pedia (ss^a) transform the antenna, and *proboscipedia (pb)* the proboscis, into a second thoracic leg (see Lindsley and Grell, 1968 for further description of the mentioned mutants). Other homeotic mutants, *Multiple sex comb (Msc)*, *Polycomb (Pc)*, etc, however, lead to the transformation of the second and third thoracic legs into those typical of the first segment. For the dorsal appendages, *ophthalmoptera (oph)* changes the head-eye capsule, and *Hexaptera (Hx)* the humerus or dorsal prothorax, into dorsal mesothorax. Mutants in the *bithorax (bx and pbx)* locus change both dorsal and ventral appendages of the metathorax and abdominal segments back into the corresponding mesothoracic organs. In all of these changes, anatomic coordinates other than the affected one remain invariant. Thus, these mutants are specific for ventral or dorsal organs, and thus retain unchanged anterior and posterior or proximal and distal coordinates, i.e., *bithorax (bx)* affects the anterior compartments and *postbithorax (pbx)* the posterior compartments. Other homeotic mutants also change individual compartments:

Thus *engrailed (en)* changes posterior into anterior wherever these compartments exist: in both dorsal and ventral thoracic appendages. Double mutants cause additive transformations: as for example in the combinations *Antp Pc, bx pbx, bx en, pbx en,* etc.

The interpretation given by E. B. Lewis (1963, 1964) for the transformations caused by the *bithorax* mutants can be summarized as follows: The mutant phenotype results from the lack of function of a gene product necessary to prevent that transformation. Thus the function of the wild-type gene of *bithorax* is required to prevent the anterior metathorax from being converted into the anterior mesothorax and that of the gene ss^a to prevent the antenna anlage from shifting to a ventral mesothoracic pathway, etc. The logical consequence of this interpretation is that the wild-type genes of these mutants act by opening up a new pathway as an alternative to the other, the "archetypal" one, to which development will shift in the mutant condition (García-Bellido, 1975a). The archetype for the segments seems to be the mesothorax, that for compartments, the anterior, dorsal, proximal one (notum). Since these mutants affect only one of the anatomic coordinates and are additive in their effects, we postulated that the different pathways are defined by a combination of signals each under the control of a gene that we have called "selector gene" (García-Bellido, 1975a).

In this sense mutant selector genes are specific for segments or compartments: these are the realms of action of their wild-type alleles. Under this criterium, not all the homeotic mutations correspond to selector genes. Transformations between regions of the same appendage, otherwise considered as duplications, or even between appendages, can result from nonspecific alterations of the normal developmental system (see Sec. III.C). It would not be surprising that mutants causing insufficiency of gene product or an excess of an unwanted one would lead to homeotic transformations. In these cases the line of symmetry of the duplication or the extent of the transformation may not correspond to known segmental or compartmental boundaries. It is therefore, important to analyze homeotic mutants both genetically and developmentally in order to know whether they are specific and restricted to compartments and so may be considered as alleles of genes controlling the corresponding cell lineages.

A functional correspondence between homoeotic genes and segments or compartments is clear in some cases. The *bithorax* system is the best analyzed genetically (Lewis, 1963, 1964). Among the point mutations of the *bithorax (bx)* locus the expressivity of the transformation varies. In the most extreme allele, bx^3, almost equivalent to the deficiency for the locus (amorphic condition), the transformation is restricted to the anterior metathoracic compartment but affects all of its derivatives. In $bx^3/Df(bx)$ (Df: deficiency for the locus) flies, the demarcation line separating wing and haltere tissue in the metathorax coincides with the anterior–posterior demarcation line. In the *postbithorax* locus the allele *pbx* corresponds to an almost amorphic condition, and the transformation in homozygous flies is again restricted to the posterior compartment. The double mutant $bx^3 pbx/Df$ has a perfectly complete second mesothoracic segment in place of the metathoracic one. A similar phenotype appears in *Ultra-bithorax (Ubx)* mutants, whether point mutations or chromosome rearrangements, and in *Df (bx, pbx)* flies. Although all are homozygous lethal, their transformation can be studied in larvae and in adult structures in mitotic recombination clones (Morata and García-Bellido, 1976). It is interesting to note that the larvae display, in addition to the described transformation of the imaginal disc, segmental transformation in the larval cuticular pattern (spiracles, cuticular hooks), in the mesoderm derivatives and probably also in the central nervous system (Lewis, personal communication). The latter observation indicates that the *bithorax* genes are required in all the germ-layer derivatives of the metathoracic segment.

The behavior of other homoeotic mutants seems to fit a similar scheme to that found in *bithorax*. They may appear as alleles that correspond to hypomorphic or amorphic mutations, which cause incomplete transformations. However these transformations can increase in expressivity under experimental conditions. The expression of the transformation is as a rule stronger in flies heterozygous for the mutant and its deficiency than in homozygous condition [as ss^a in ss^a/Df(ss)]; or in homozygous condition but in the presence of genetic modifiers (as *en* in combination with *Minutes*, Lawrence and Morata, 1976a) or raised under higher or lower temperatures, as *oph* (Postlethwait, 1974) ss^a (Vogt, 1964b; Grilliatti and Suzuki, 1971) and *en* (Morata and Lawrence, 1976a). In other cases new alleles in a locus may uncover a much wider spectrum of transformation for that locus than previously thought, such as *Nasobemia* compared to other *Antp* mutants in the *Antp* locus (Denell, 1973). The limit of these transformations, however, seems be restricted to either entire appendages or entire compartments, an indication of the limit of action of their wild-type alleles.

Analyses in gynandromorphs and in mitotic recombinant clones have shown that the transformations caused by the *bithorax* alleles are cell autonomous (Lewis, 1963; Morata and García-Bellido, 1976). Production of mutant clones results in a morphogenetic mosaic of structures of both segments, in which the topologic characteristics of the transformed clone correspond to the location at which it appears. These topologic homologies probably reflect the fact that both meso and metathoracic dorsal discs share the same compartmentalizations. In fact, marked clones show the same clonal restrictions in the normal metathoracic and normal mesothoracic discs. Mitotic rate and mitotic orientation of the proliferating cells are also the same in both wild-type imaginal discs. However, cell affinities, which are different between wild-type mesothoracic and metathoracic discs (García-Bellido and Lewis, 1976), are also different in the mutant *(bx³)* recombinant cells (Morata and García-Bellido, 1976). This is an indication that during development the proliferating cells express differential cell behavior properties corresponding to their genetic constitution. These differences must appear early in development because the mutant constitution of the cells expresses itself early in the organization of the primitive metathoracic anlage. The number of cells early in this anlage is different in the wild-type metathoracic and mesothoracic segments but is found to be identical in both anlagen in *bx³ pbx/Ubx* flies (Morata and García-Bellido, 1976). Thus, the function of the *bithorax* genes seems to be required from very early stages of development and throughout subsequent development to maintain the metathoracic pathway in the proliferating cells.

Clonal analyses performed with other homoeotic mutants yield similar results. The genetic change after mitotic recombination is immediately followed by homoeotic changes in the clones as seen during the proliferation period in mitotic orientation (*en*, García-Bellido and Santamaría, 1972), in mitotic rates (*ss*^{a}, Roberts, 1964; Postlethwait and Girton, 1974) and in cell affinities (*en*, Morata and Lawrence, 1975) and also in the corresponding homoeotic cuticular pattern seen after metamorphosis. Again these studies have shown that the products of the wild-type alleles are required in successive cell generations during development, to retain the normal developmental pathway.

Genetic analysis of the homoeotic mutants uncovers several types of genetic behavior. This analysis is most important for an understanding of the mode of action of the corresponding wild-type genes. The phenotype of some mutations in a homozygous condition is similar to that of the deficiency of the locus, as seen in mitotic recombinant clones, when the deficiency is zygotic lethal. These correspond to hypomorphic or amorphic mutations (e.g., *bx*, *bx*^{3}, *pbx*) and are as a rule recessive. Amorphic point mutations or deficiencies are occasionally dominant (*Ubx*, *Df(bx)*), but haplo-insufficient loci are rare in *Drosophila* (Lindsley et al., 1972). Dominant mutations, not corresponding to deficiencies, are, however, relatively common (*Antp*, *Msc*, etc.), and an interpretation of the function of their wild-type alleles in mostly lacking. In the *bithorax* system one such mutation is *Contrabithorax* (*Cbx*). It causes the transformation of mesothorax to metathorax. The genetic behavior of *Cbx* suggests that it corresponds to an overproducer mutation, i.e., to the derepressed condition of the *bithorax* genes: *Cbx* is dominant over several doses of wild-type *bithorax* loci. Its phenotype can be explained as being due to the presence of *bithorax*$^+$ products in the mesothorax, which will lead to the repression of the mesothoracic pathway in the same way as it does in the wild-type metathoracic segment (Lewis, 1964; Morata, 1975). Thus, *Cbx* can be considered operationally similar to an 0 mutation (Lewis, 1964). *Rg-pbx* is one of several mutants that map far outside of the *bithorax* locus, and have a phenotype similar to that caused by mutations in this locus. It causes the appearance of spots of mesothoracic tissue in the posterior metathorax. This mutant is also dominant over several wild-type doses of its locus but decreases its expressivity with increasing doses of wild-type *bithorax* genes (Lewis, 1968); it behaves as a neomorphic mutation and may correspond to a mutation in a gene with regulatory effects upon *bithorax*, operationally similar to a superrepressor mutation in bacteria. However, it is conceivable that the phenotype results from a nonspecific interference with the expression of the *bithorax* genes. If this is the case, its wild-type function could be unrelated to the control of the metathoracic pathway. The existence of mutations in different loci that show the same phenotype suggests that they may affect the same developmental pathway, and opens the question of the functional interaction between their wild-type alleles. Thus we cannot safely interpret the mode of action of the wild-type genes of *ss*^{a} and *Antp*, of *Msc* and *Pc*, etc, because we do not know to which genetic system they belong.

The anatomic specificity of homoeotic mutations raises the question of how homoeotic genes are expressed in certain developmental pathways but not in others. Again the *bithorax* system is a model system where, to a certain degree, these questions are being analyzed. Phenocopies of *bithorax* alleles can be induced by ether treatment of wild-type blastoderm embryos (Gloor, 1947). The transformation seen in these phenocopies is patchy, and clonal analysis has shown that the phenocopy effect is maintained clonally in the progeny of individual cells

(Capdevila and García-Bellido, 1974). Genetic analysis has further shown that the effect is not on the products of the *bithorax* structural genes, nor upon cytoplasmic constituents of the affected cells, but rather upon the mechanism of activation of the *bithorax* locus. Several considerations suggest that ether acts by disturbing the positional signals for segmentation, leading to mosaics in which some cells have the *bithorax* system repressed in what is otherwise the metathoracic anlage (*bithorax* phenocopies) or to others in which the *bithorax* system is derepressed in a mesothoracic anlage (*Contrabithorax* phenocopies) (Capdevila and García-Bellido, 1974, 1977). Apparently, position is registered in a cell-autonomous way and leads to the initiation and establishment of a mesothoracic or metathoracic pathway both in normal flies and in those bearing phenocopies. It is therefore interesting to know if cell sensitivity to phenocopies is affected in mutations that map outside the *bithorax* locus. It has been found that the presence of *Rg-pbx* in embryonic cells increases the frequency of phenocopies regardless of whether it comes from the maternal or paternal gamete (Capdevila and García-Bellido, 1978). This suggests that *Rg-pbx* interferes with the mechanism of activation of the *bithorax* system and not with the expression of its products. In fact in *Rg-pbx* flies, the patchy *postbithorax* phenotype has a clonal origin. Moreover the removal of the mutant allele during the proliferation period by mitotic recombination does not change the mutant phenotype of the resulting clones [see García-Bellido, (1977a); García-Bellido and Capdevila (1978) for further discussion].

These findings, if they represent a general rule, suggest that the initiation of a developmental pathway and its maintenance are two different genetic operations. In the former, position is translated into genetic terms by the interaction of the elements, called "inductor," "activator" gene, and „selector" gene (García-Bellido, 1975). In the metathoracic pathway these would correspond to positional information in the egg cortex, the wild-type alleles of *Rg-pbx* and the *bithorax* system, respectively. The maintenance of this pathway is probably only under the control of the *bithorax* genes, for we have seen that the removal of *Rg-pbx* during the proliferation phase of the anlage has no effect on the phenotype. Obviously mutations in the genes coding for inductors or activators or selectors will have homoeotic phenotypes. The phenotypes will vary depending on the nature of the mutation and on the type of control (positive or negative) exerted upon the selector genes.

Selector genes are probably required for the registering of position and the initiation and maintenance of specific developmental pathways in the zygote or embryo. This is strongly suggested by the spatial restrictions of the mutant phenotype, and by the temporal coincidence of the first effects of the mutations and of the appearance of the corresponding compartments. The logic behind the combinatorial sequence of compartments and the additive effects of selector mutations also supports this idea. This interpretation, which is derived from the analysis of imaginal disc development, may have a general validity for other cell lineages. The segregation of segments, of blastoderm cells into germ layers, of germ layers into different tissues, and of imaginal discs all conform to the same picture. They seem to occur in groups of cells and may possibly occur by binary partitions under the effect of selector genes. We have seen that gene activation occurs within definite spatial coordinates. We do not know the nature or the distribution of the postulated inductor molecules. The general problem of positional information has so far escaped genetic analysis. This is mainly because most mutations studied to date are more easily explained as failures in the interpretation of positional information rather than in the actual mechanism of specification of positional information. For one thing, they only express their phenotypes in the embryo that carries them. Perhaps from the study of mutants with maternal effects more information about gene function defining positional coordinates in the egg will be obtained. The mutant *bicaudal* is interesting in this context. Homozygous females produce eggs that, independently of the zygotic genetic constitution, show a transformation of anterior segments of the embryo into posterior ones (Bull, 1966). This transformation would suggest the existence of a primitive anterior versus posterior genetic decision under the control of its wild-type allele. However, this conclusion contradicts the assumption that the first genetic decisions already distinguish particular segments (see above). It is possible that this mutation nonspecifically disrupts the cytoplasmic organization of the egg in a way similar to that of UV irradiation (Kalthoff, 1971) or centrifugation (Schubiger, personal communication). Genetic and developmental analyses of early embryogenesis may reveal the existence of sequential genetic steps. If this is the case, the pattern complexity of segments and germ layers may result from the genetic programming of cell lineage segregations rather than from the existence of different positional cues for the qualities of individual cells (Figs. 1 and 2).

III. Morphogenesis and Pattern Formation

A. Size and Shape

From the establishment of the imaginal disc anlage to the beginning of metamorphosis, cell proliferation is associated with morphogenesis. The different imaginal discs can be distinguished, prior to metamorphosis, not only by their anatomic position but also by their specific size and shape. Since individual imaginal disc cells are of the same size and histologic appearance, the different sizes of the discs correspond to a difference in cell numbers. The final adult shape of the cuticular derivatives of the imaginal discs depends in part on differences in cell density, as for example between the notum and the wing blade (García-Bellido and Merriam, 1971a). It is debatable whether the final shape also results from morphogenetic movements taking place after mitoses have ceased or whether it is a direct reflection of the position of the cells in the premetamorphic disc (Fristrom and Fristrom, 1975).

A comparison of the relative position of the presumtive cells in the embryonic anlage of imaginal discs and the actual position of these structures in the adult, precludes the existence of major morphogenetic movements (García-Bellido and Merriam, 1969; Ripoll, 1972). Clonal analysis has shown that all the cuticular elements of a clone remain side by side during proliferation. Thus minor movements of cells relative to one another, if they exist, are of little relevance in overall morphogenesis. Therefore the shape and size of imaginal discs reflect the behavior of the proliferating cells during previous development.

Clonal analysis suggests that there is a more or less constant relationship between the numbers of cells in the anlage and those of the final organ. As a rule the frequency of mosaicism in gynandromorphs is correlated with the size of the adult appendage (see García-Bellido and Merriam, 1969; Nöthiger, 1972). This suggests that the average number of divisions is similar in all discs. In the wing disc there seem to be 10 divisions (García-Bellido and Merriam, 1971a), and a similar number was estimated for the haltere disc (Morata and García-Bellido, 1976). In all discs studied, cell proliferation is exponential and intercalary and affects all the cells of the anlage. In the wing the clone size dispersion is no larger than three divisions (García-Bellido and Merriam, 1971a), although for the scorable structures of the eye facets this dispersion is larger (Becker, 1957). Differences in clone size are found when comparing different regions within the same disc (Becker, 1957; Bryant, 1970; García-Bellido and Merriam, 1971a). Temporal variations in cell sensitivity to the induction of mitotic recombination between regions exist, which suggests the existence of a variable number of target cells and/or that of regional differences in mitotic activity.

A comparison of the gynandromorph map of presumptive structures with that of the actual distances, in terms of cells, in the adult suggests the occurrence of regional differences in the preferential orientation of growth (Ripoll, 1972). The direct observation of clones in these regions confirms this assumption. Moreover, the existence of indented clone borders suggests that clone shape is primarily determined by the orientation of mitoses rather than by mechanical stretching during evagination (García-Bellido and Merriam, 1971a).

Cell dissociation and mixing experiments have shown that the imaginal disc cells possess differential recognition properties characteristic not only of the different imaginal discs but also of their different regions (Nöthiger, 1964; García-Bellido, 1966a, b; Tobler, 1966; García-Bellido, 1972).

In summary, the imaginal disc cells show during the proliferation phase, behavioral properties that are characteristic of the different developmental pathways. These properties cannot be a direct result of the gene products of selector genes. It has been proposed that these behavioral properties of differentiation reflect the activity of other types of genes ("realisator genes"), which are under the control of selector genes (García-Bellido, 1975). These behavioral properties probably result from the interaction of several distinct and identifiable cell functions, each amenable to detection by mutational analysis. Mutations affecting these functions do not necessarily correspond to the genes controlling them. However, it is hoped that when these functions are identified, we will be able to study their interactions and eventually understand the logic of their genetic control.

Morphogenetic mutants are logical candidates for the study of cell behavior lesions. It is expected that alterations in cell behavior properties will be reflected at the clonal level. However, very few viable morphogenetic mutants, other than homoeotics, have been studied at that level. Most of the morphogenetic mutants analyzed are zygotic lethals. Their morphogenetic mutant character has been detected upon dissection of the larva and subsequent characterization of the transplants (Stewart et al., 1972; Shearn et al., 1971) or directly by the changes observed in clonal parameters in homozygous spots resulting from mitotic recombinations (Ripoll and García-Bellido, 1973; Ripoll, 1977). More recently, morphogenetic alterations have been directly detected in *Minute* clones, which is a way to isolate morphogenetic mutants, irrespective of their lethal or viable character in zygotes (Ferrús and García-Bellido, 1976). Not surprisingly, these mutants are all zygotic lethals.

Since the final goal of this analysis is to ascertain the nature of the function of the wild-type genes, we have to know the genetic nature of the mutant allele. If cell behavior results from the effect of the interaction of several genes, individual wild-type functions can only be ascertained when we know the effect of the total absence of these functions. An incomplete absence of wild-type function, as a result of a hypomorphic allele, for example, might hinder the identification of the normal function. The analysis of phenotypes becomes even more complex when the existence of pleiotropic effects is taken into account. Since pleiotropy cannot be disregarded even at the cellular level, it is not surprising that most mutations appear at a first glance as disruptive mutations and that they give us no clear hints as to the nature of the affected wild-type gene function.

These shortcomings of the analysis of mutations are relevant when trying to describe tissue and stage specificity of a given mutation. Thus, mutants that appear to affect imaginal discs or even some sets of them may only reflect the existence of different thresholds of requirements (Shearn et al., 1971). Therefore any conclusion about tissue or disc specificity should be preceded by a comparative analysis of alleles closer to the amorphic condition, and a study at the cellular level in somatic spots. Even at this level the developmental properties of the system under consideration are again relevant. Thus, autonomous cell lethals (Ripoll and García-Bellido, 1973) in the wing imaginal disc that survive in the tergites may simply reflect the existence of cell behavior peculiarities in one system but not in the other (Morata and Ripoll, 1975; Ripoll, 1977). Thus, conclusions, derived from the occurrence of lethals specific for different stages (embryogenesis, larval, and pupal development) or different tissues (imaginal discs, etc.), about the existence of different sets of genes at work at these system, should be reconsidered.

The fraction of lethals that alter the normal clonal parameters in mitotic recombination spots is small (Ripoll, 1977). Of 86 lethals of the first chromosome, a large fraction (37) lead to normal gynandromorphs and 6 of them only produce erratic and slight malformations in the hemizygous male tissue. Among 43 lethals, nonviable in male tissue in gynandromorphs, studied in epidermal cells, 18 had normal clonal parameters, 19 had slower cell division rates in the wing, but only 7 of them were inviable in both wing and tergites. From the remaining six mutants, two showed abnormal cuticular differentiation and four abnormal morphogenesis in clones (see below). From this small sample it derives that (1) about 10% of the genome, at least, is necessary for the viability of epidermal cells (similar estimates were obtained for temperature-sensitive cell lethals: Arking, 1975), and (2) only 4/86 (5%) correspond to genes required for normal cell behavior in functions necessary for disc morphogenesis.

Two main supracellular properties are associated with morphogenesis: size and shape. As seen above the size of the imaginal discs is a direct consequence of the number of cells. Mutations, malnutrition, or premature metamorphosis lead to smaller individuals with smaller cuticular organs, and to patterns with a reduced number of elements. However, these variations, being of a disruptive nature, are difficult to interpret and possibly have little bearing on the problem of control of size. Implants of first instar imaginal discs grown in the adult reach a maximal size, which corresponds to the maximal one in the larva in situ (García-Bellido, 1965). In gynandromorphs consisting of *Minute* and non-*Minute* tissue, the size of the *Minute* territory is normal if flies pupate with a delay like that which occurs in *Minute* flies, but smaller when pupation occurs at normal non-*Minute* time. Moreover, the M^+ territory of these mosaics is wild-type in size, even when pupation occurs with the *Minute* delay (Ferrús and García-Bellido, 1977).

This fixed size does not depend upon a fixed number of cell divisions per clone. In fact M^+/M^+ cells growing in a M/M^+ anlage may go through many more rounds of cell division and yet attain a normal final size (Morata and Ripoll, 1975). The difference in size between males and females is reflected autonomously in gynandromorphs: In mosaic wings the number of triple row elements depends largely on the sex of the cells (García-Bellido, unpublished). A mutation [1(1)ts 1126] that decreases the rate of cell division and viability under restrictive temperature conditions, may also affect the final size of mosaic organs (Simpson, 1976). In these cases the size disturbance must not necessarily correspond to compartment boundaries. Mutations that change cell size may also alter the final adult size. Two lethal alleles *1(3)Me 25* and *1(3)Me 109* (locus: 3.36) that cause an increase in cell size (Fig. 5a) show in mosaics an autonomous increase in the size of the mutant region (see Fig. 2 in Ferrús and García-Bellido, 1975). On the other hand, the *1(3)Me 10* (3-43) that cause the autonomous differentiation of cells smaller than normal, compensates by extra cell divisions in mosaics to attain a final normal size, and an endoduplicated pattern of chaetes (Fig. 6c). Thus size may be controlled by mechanisms based on counting the number of cells, rather than by physical distances.

One such mechanism for counting the number of cells could be mitotic waves. Schatz (1951) and Stumpf (1956) reported the existence of two mitotic waves in the prepupal wing. In the first one, mitoses are oriented in the longitudinal, in the second, in the transverse axis of the wing. Analyses of mutations, and of their heat-induced phenocopies, which change the size and shape of the wing into either a "broad" or a "lanceolate" shape, showed a correlation between these phenotypes and the suppression of one of these waves (Schatz, 1951). The importance of compartment borders as references for size and shape has been emphasized by Crick and Lawrence (1975). One observation is that in mosaics of some mutations mutant territories that touch the wing margin may lead to extra growth and shape deformations, whereas they have no effect when occupying only internal territories. This applies to mutants such as *engrailed* (Lawrence and Morata, 1976a), *Costal* (Whittle, 1976), and *Lyra* (Santamaría, 1973).

As mentioned above, clonal analysis of normal development has indicated that shape may be the consequence of the existence of preferential mitotic orienta-

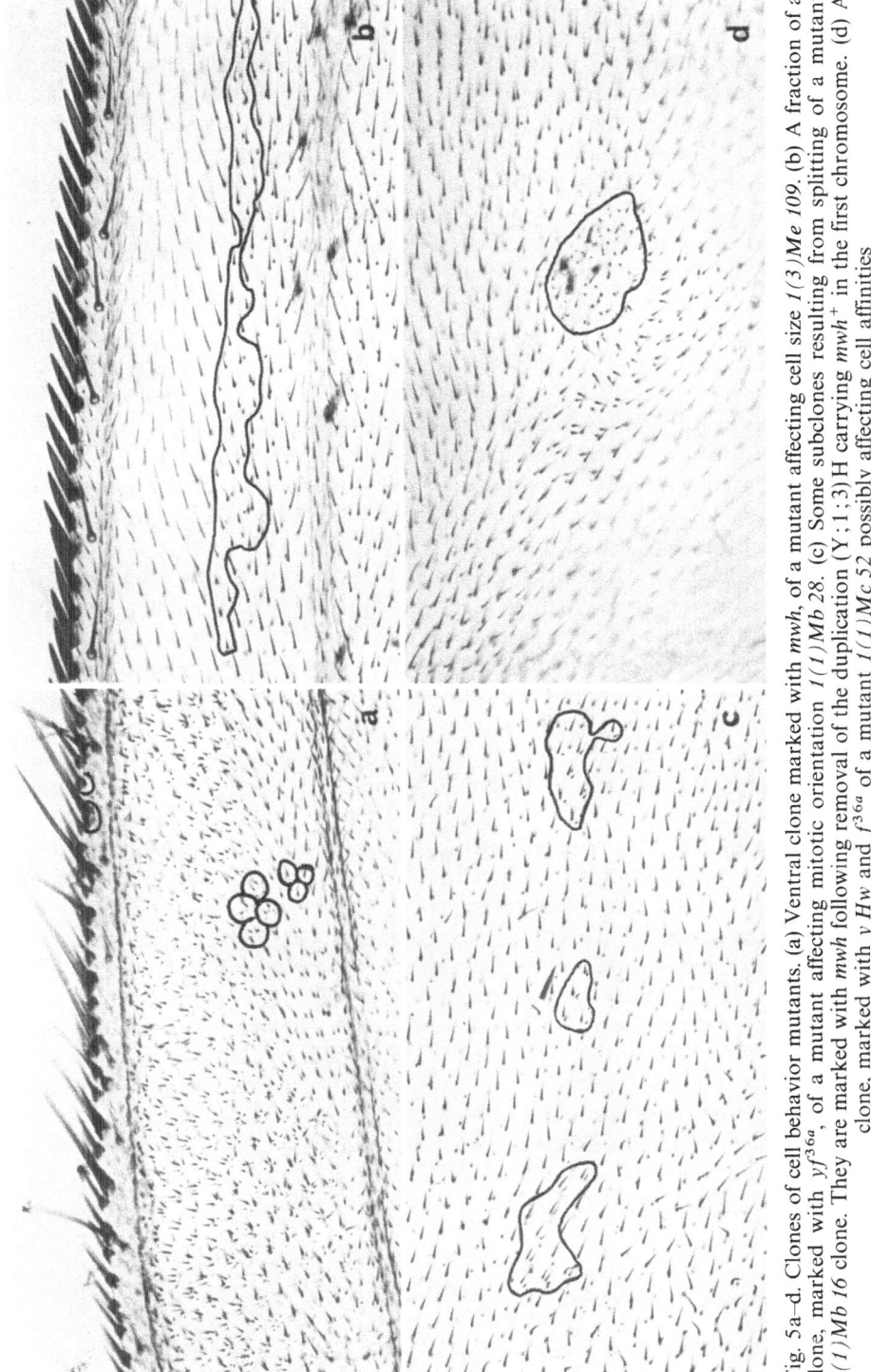

Fig. 5a–d. Clones of cell behavior mutants. (a) Ventral clone marked with *mwh*, of a mutant affecting cell size *l (3) Me 109*. (b) A fraction of a clone, marked with *yf³⁶ᵃ*, of a mutant affecting mitotic orientation *l (1) Mb 28*. (c) Some subclones resulting from splitting of a mutant *l (1) Mb 16* clone. They are marked with *mwh* following removal of the duplication (Y; 1; 3) H carrying *mwh⁺* in the first chromosome. (d) A clone, marked with *y Hw* and *f³⁶ᵃ* of a mutant *l (1) Mc 52* possibly affecting cell affinities

tions (see Postlethwait and Schneiderman, 1971). Three different mutants *1 (1) Mb 28* (1-15), *1 (3) Me 51* (3-11), and *1 (3) Me 55* (3-15) lead to the appearance of elongated clones, thirty to fifty times longer than wide (Fig. 5 b). Also wings that are mosaic for these mutants are slightly deformed by the presence of large clones: This is clearly visible when the clones embrace the wing margin. This finding suggests that clone-shape abnormalities in internal clones are compensated for by the surrounding cells. Yet, in entire wings failures in the orientation of the last few cell divisions may have consequences on the final shape (Schatz, 1951).

Another cell property expected to be relevant during morphogenesis is cell adhesion. The clonal behavior of some scalloping mutants such as *cut, (ct⁶)*, *Beadex (Bx^J)*, *vestigial (vg)*, and *Lyra (Ly)* and of their phenocopies suggests that the lack of tissue in the borders of the wing and in other compartment borders may result from general failures in cell adhesion (Santamaría, 1973; Santamaría and García-Bellido, 1975). The frequent occurrence of split clones in spots, resulting from mitotic recombination, in a lethal *(1(1)Mb 16, 1.-0.5)*, probably reflects others kinds of failures in cell adhesion (Fig. 5 c; Ripoll, 1977).

Another lethal *(1(1)Mc 52; prox. to forked)* is more likely to affect cell affinities, because clones of homozygous cells show a tendency to segregate from the surrounding wild-type cells (Ripoll, 1977; Fig. 5 d).

The morphogenetic mutants discussed previously seem to affect cell properties whose existence was previously inferred from the analysis of normal development. Other morphogenetic mutants produce abnormalities that so far escape our understanding in terms of individual cell behavior properties. Most of the lethals shown to lead to abnormal morphogenesis in clones are known to die in the homozygous condition in embryos or early larvae. Possibly the corresponding wild-type genes are required on all of the developmental systems and in imaginal disc development from very early stages. It is hoped that a better understanding of the pleiotropic effects of these mutations, and of the causal relationship of the processes they affect, will emerge from the analysis of more such mutations and of their interactions in double mutant combinations. Moreover, this analysis may uncover new cell functions relevant to morphogenesis. Whether these functions are under the control of specific genes or whether they represent physiologic crises differentially affected by disruptive mutations is not known. If they are genetically controlled, what characterizes different developmental pathways is probably the modulation of these functions. Thus pattern specificity may result from the combinatorial effect of "realisators" under the control of a combination of "selector" signals. The properties exhibited by the proliferating cells relative to the final morphogenesis cannot yet be explained in terms of an inventory of gene products; however, their combined effects define a "certain differentiation." We will call it "proliferative differentiation" in contrast to that taking place at the end of the proliferation phase, i.e., "terminal differentiation."

B. Terminal Differentiation

We will consider here possible mechanisms responsible for the spatial distribution of cells into morphologic patterns. The problem of pattern formation has been reviewed by several authors in recent years with different approaches (see

Tokunaga, this volume, for further references). It is obvious from such reviews that we still lack a comprehensive view of the causal mechanisms involved in the generation of patterns. In the present discussion we will concentrate on the problem of the developmental origin and genetic control of differentiation of the cuticular elements and see how far it can account for their spatial positioning. Cuticular patterns can be seen as a characteristic distribution of cuticular elements and are therefore visible only after terminal differentiation at metamorphosis. The elements of those patterns are single-cell derivatives of the epidermal cell layer, identifiable as chaetes, sensillae, and trichomes of different sizes and shapes.

Formally speaking the entire cuticle of the adult constitutes a single pattern. However, early experimental analysis established that the different imaginal discs contribute to this continuous pattern by making fixed and characteristic parts of it (Zalokar, 1943). Fragmentation experiments allowed the construction of fate maps of the presumptive elements in the premetamorphic imaginal discs (Vogt, 1946a; Hadorn, 1953; Schubiger, 1968; etc.). Cell dissociation experiments of the discs revealed that identifiable elements could be found isolated from the normal context in the differentiated reaggregates (Hadorn et al., 1959; Nöthiger, 1964; see García-Bellido, 1972). Clonal analysis has shown that there are only a few, residual divisions, between the time of explantation of the imaginal disc or its fragment, and the onset of the process of cuticle formation in situ (García-Bellido and Merriam, 1971a). The same analysis has shown that the mother cells for chaetes are genetically committed before pupariation (García-Bellido and Merrian, 1971c; see below). Phenocopy experiments and mutant analysis (see below) finally confirm that statement. Thus, metamorphosis seems to lead merely to the cuticular expression of a pattern of cell differentiation already defined in the imaginal discs during the cell proliferation phase.

Cuticular patterns are not only specific to different imaginal discs but also to different compartments. Patterns do not seem to be independent of cell lineages, in the sense that different cell lineages contribute with a fixed inventory of elements to the whole organ pattern. Thus, the dorsoventral demarcation line of the wing always runs between the medial and ventral rows of the triple row of chaetes, the anterior-posterior demarcation line always runs on the anterior side of the fourth vein of the wing, etc. Moreover in mutants that distort the pattern the particular modifications also are characteristic for the different compartments. As we have seen, this is true even in small patches of homeotically transformed tissue resulting from mitotic recombination. In such morphogenetic mosaics the pattern of the homoeotic tissue corresponds to the inventory of the combination of selector genes at work in the compartment in question; however, the spatial distribution of elements also corresponds to the *position* of the cells within that compartment.

As we have seen from the problem of shape and size (Sec. III.A), the formation of patterns is probably also a question of cell interaction. To understand this idea, we need a deeper knowledge of the different properties that the cells exhibit during the proliferation stages. At the present time we can only accumulate information about the causal mechanism of differentiation of the pattern elements and of their modification caused by mutations, with the hope that a clearer picture of the general rules of pattern formation will emerge.

One of the most conspicuous elements of the cuticular pattern is the chaetes. They probably arise in *Drosophila*, as in other insects, from two differential mitoses, the first one separating a nervous from an epidermal pathway, the second giving rise to a trichogen (shaft) and a tormogen (socket) in the epidermal pathway, and to a centripetal nerve, and possibly to a neurilemma or glia cell in the nervous pathway (Stern, 1938; see Lawrence, 1966). There are variations to this scheme, and so the chaetes of the medial and ventral rows, as well as those not bearing a socket at both sides of the dorsoventral demarcation line along the wing margin are not innervated (García-Bellido and Santamaria, 1978). Chaetes have different sizes, perhaps associated with different degrees of endopolyploidy as in other homologous organs in insects. In the notum there are two classes that are clearly distinguishable and that are called macrochaetes and microchaetes. Comparative chaetotaxy in different *Drosophila* species has revealed that the evolution of most macrochaetes is not correlated with that of the microchaetes (Sturtevant, 1970). Moreover, mutants such as *achaete (ac)* and *scute (sc)*, which probably belong to the same pseudoallelic system (García-Bellido, 1978; Fig. 6a, b), affect the pattern of chaetes: *achaete* removes mainly the microchaetes in the acrostical region, *scute* mainly the macrochaetes. Both alleles affect in a similar and specific way chaetes of head and legs. The genetic deficiency for the entire *achaete-scute* system removes all the innervated chaetes of the body, leaving only those that are not innervated. It is possible that these genes are required in the process of segregating the nervous from the epidermal pathways (García-Bellido and Santamaria, 1978). It is interesting to note that loci of the same genetic system code for functions that are expressed in different positions. We have seen a similar topographic specificity in the case of the pseudoalleles of the *bithorax* system.

There seem to be temporal differences in the segregation of cell lineages and in the differentiation of different chaetes. The existence of a perdurance period, within which the removal of the wild-type allele by mitotic recombination has no effect on the phenotype of the homozygous mutant cells, is an indication that the wild-type products of this gene have already been released (García-Bellido and Merriam, 1971c). Mitotic recombination with the *Df (ac, sc)* has shown that macrochaetes enter the perdurance period about 40 h before pupariation, whereas microchaetes do so at about 20 h and *hairy* extrachaetes 8 h before pupariation (García-Bellido and Merriam, 1971c; García-Bellido and Santamaria, 1978). This result indicates when different cell lineages of chaetes become segregated. The genetic commitment to the chaete pathway occurs several cell divisions before the remaining epidermal cells of the disc cease dividing. Clonal analysis has shown that for the microchaetes of the notum this difference is of at least two divisions. Possibly shortly afterward the differential mitoses take place (García-Bellido, 1971a). The period sensitive to phenocopies, which may interfere with the actual differentiation (García-Bellido and Merriam, 1971a; García-Bellido and Santamaria, 1978; Poodry, 1975) and to temperature shocks in the mutant *shibire* (Poodry et al., 1973) occur around pupariation. Trichogen and tormogen cells are first visible histologically 15 h after pupariation (Lees and Waddington, 1942).

Clonal analysis has shown that there are no clonal relationships between different macrochaetes (Sturtevant, 1929) and this possibly applies to the microchaetes as well. In the notum chaetes are determined in situ, according to the possition they occupy in relation to the cell population as a whole (Stern, 1954).

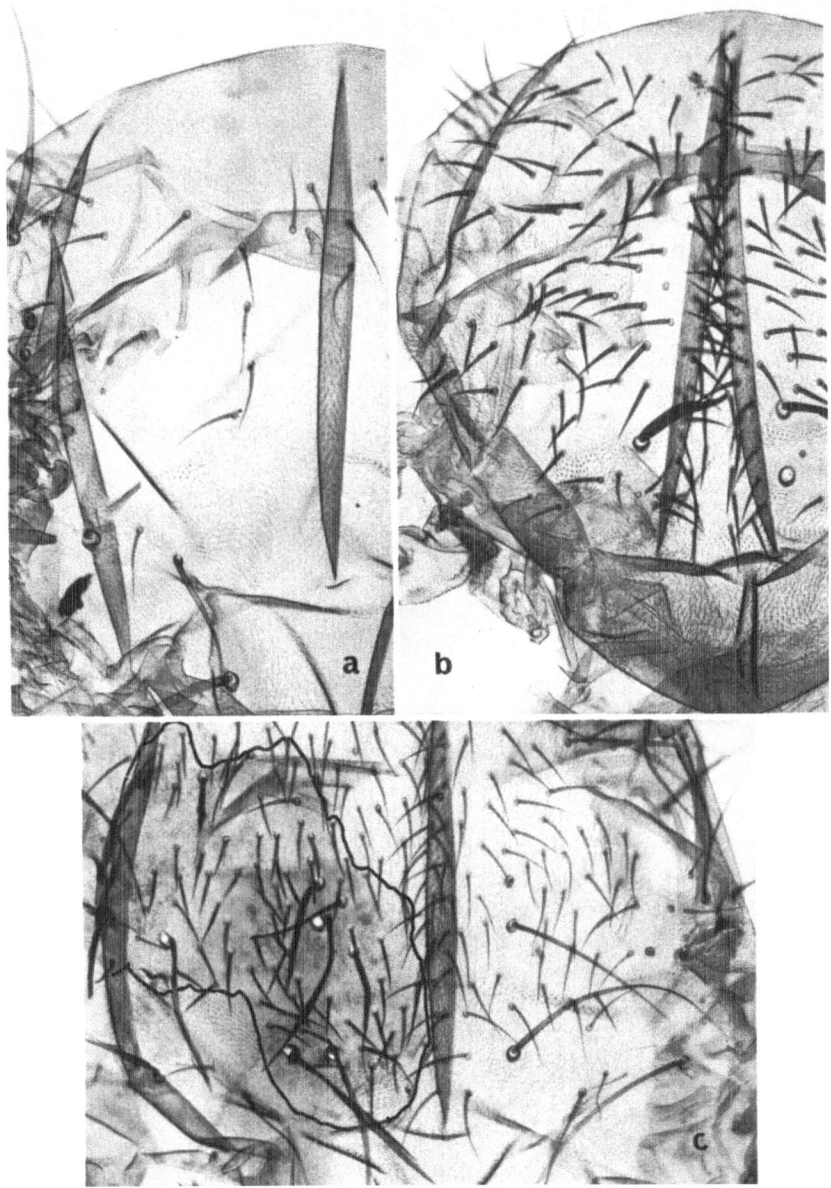

Fig. 6 a–c. Pattern mutants. (a) *achaete* (*In(1)y³ᴾᴸsc⁸ᴿ*) notum. (b) *scute* (*In(1)sc⁸ᴸsc⁴ᴿ*) notum. (c) A clone marked with *mwh* and *jv* of a mutant, *1(3)Me 10* causing endoduplicated pattern of chaetes. Wild-type pattern in the *right side* of the notum in (c)

We do not know the mechanisms of this effect of position, and several factors may intervene. The analysis of morphogenetic mutants in mosaics will hopefully throw some light on this problem. Stern (1954) already showed that, in *achaete* mosaics, the position of a chaete is defined by a maximal probability around a certain site. The mutant *1(3)Me 10* causes a diminution of cell size and higher cell density

(Ferrús and García-Bellido, 1976). In clones of genetically marked cells this mutant expresses autonomously, giving a higher density of chaetes but maintaining the same relative distances between them, as measured in numbers of cells (Fig. 6 c). On the other hand in mutants that cause the appearance of extrachaetes, these may appear distributed over regular distances (*Polichaetoyd, hairy*), but also concentrated in certain regions (*Groucho, shibire, Tuft*) with no clonal relationships between them. Explanations based on competition for morphogens that would lead to regular positioning (Wigglesworth, 1940; Lawrence, 1973 b) should take into account these variations.

Apparently the position of chaetes in the wing margin presents a simpler situation. The scalloping mutants *cut* (ct^6), *Beadex* (Bx^J), and others cause nicks in the wing margin. This possibly happens as a result of failures in the adhesive properties of the cells at both sides of the margin, and as a consequence, chaetes in this region fail to differentiate (Santamaria and García-Bellido, 1975). Thus, cell contact could be a prerequisite for cell differentiation. We stated previously that in morphogenetic mosaics of *bx* in the haltere, wing structures differentiate autonomously according to the position in which they appear in the haltere. When these clones touch the dorsoventral margin of the haltere, they differentiate wing marginal chaetes, dorsal and medial, or ventral, depending on the compartment in which they are found. Thus, possibly both wing and haltere share the same signals to elicit the response of the wing cells to differentiate into marginal chaetes (for more details see Morata and García-Bellido, 1976).

The latter hypothesis also seems to be appropiate for the mechanism of differentiation of chaetes on the veins. In *Drosophila*, the veins are void of chaetes, but the mutants *hairy (h)* and *Hairy wing (Hw)* lead to the appearance of extrachaetes along the veins, following a characteristic pattern. These chaetes are innervated (Stern, 1938). In double mutant combinations of *h* with mutants that suppress veins, *veinlet (ve)* and *radius incompletus (ri)*, or add extra veins, *plexus (px)*, the *h* chaetes follow the course of the mutant veins, i.e., they are absent in the areas affected by *ve ri* that are without veins, and appear in the extra veins of *px*. The same happens in clones of these mutants even when these affect only small stretches of veins (García-Bellido, 1977 b; Fig. 7 a, b). It seems then that the expression of *h* depends on the genetic constitution of the cells forming the vein.

As is true elsewhere, chaetes in the tergites have no common clonal origin but neither do they seem to differentiate in situ according to their position. When clones are marked with mutants that label trichomes as well as chaetes, the chaetes are frequently found singled out and outside the clone of marked trichomes. Reciprocally, nonlabeled chaetes frequently appear within the continuous patch of marked trichomes (García-Bellido and Merriam, 1971 b). Apparently cells determined to differentiate into chaetes can migrate away from their clones. Whether this occurs at random or because of specific cell properties is not known. Also the number of chaetes on the tergite can be altered by damaging the larval tergite over which the adult histoblast will spread. In this way adult tergites appear with no chaetes or with several times the normal number of chaetes distributed at regular distances (Santamaria and García-Bellido, 1972; see García-Bellido, 1972).

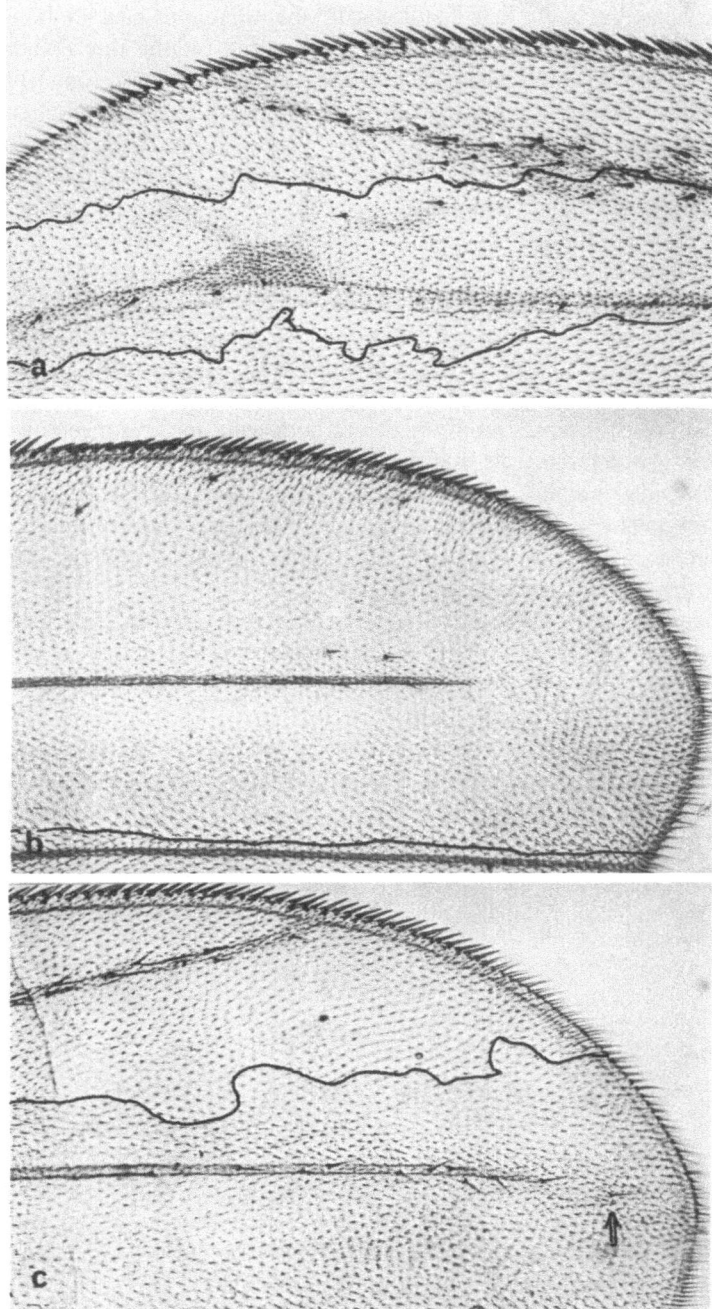

Fig. 7 a–c. Clones of vein mutants. (a) A dorsal clone of *pwn px^{72}h* cells causing extra venation in the ventral surface. (b) A dorsal clone of *mwh ve h ri* cells causing suppresion of veins *II* and *III* in both surfaces and concomitantly of the *h* extra chaetes in these veins. (c) A dorsal clone of *mwh ve* cells in a *h* fly, causing suppression of the tip of vein *III* but not of the *h* chaetes *(arrow)* of the corresponding ventral surface

The same applies to the mechanism involved in the differentiation and patterning of veins. Veins are first histologically identifiable as lacunae between the wing surfaces in the prepupal wing after evagination of the disc (Waddington, 1940). This is the first time when the two wing surfaces, which belong to different compartments, contact each other during development. The results of cauterization experiments in the prepupal wings lead to the suggestion (Braun, 1940; Lees, 1941) that the course of the veins is determined at that time and that the ventral surface is under the inductive influence of the dorsal one. Clonal analyses with the vein suppression mutants *ve* and *ri*, and the extra vein mutant *px* have shown that the mutant condition is expressed autonomously if the clones occupy the dorsal surface but nonautonomously if they are in the ventral one (Fig. 7a, b). However, in clones that were mutant for both *h* and one of the mutants affecting the veins, the expression of the *h* phenotype was suppressed in dorsal clones but not in the corresponding region of the ventral wing surface (Fig. 7c). Since *h* chaetes are determined prior to pupariation, i.e., before both wing surfaces meet, its presence in the ventral surface suggests that the course of the veins was laid down autonomously and independently in the cells of the ventral wing surface also. Analysis of vein formation shows that between the genetic determination to initiate a given differentiation pathway and its expresion, modifying mechanisms of cell interactions may intervene (García-Bellido, 1977 b).

A similar inductive mechanism has been shown to occur in the appearance of bracts, which are modifications of normal epidermal cells found at the base of certain chaetes and which develop under their influence (Tobler et al., 1973; see García-Bellido, 1972 for discussion).

There are mutants that seem to affect one process of differentiation or another (macrochaetes, microchaetes, veins) and in particular regions, but they leave the rest of the pattern elements unaffected. It appears as though not only the differentiation of the different elements of the pattern but also their distribution in different patterns, have different genetic controls. The overall pattern, as we see it, again seems to be the result of a combination of additive signals having independent genetic controls.

C. Regeneration

Experiments of fragmentation and transplantation of imaginal discs have led to the construction of maps of presumptive cuticular structures (Bodenstein, 1941; Vogt, 1946a; Hadorn et al., 1949). The imaginal discs of mature larvae behave as a mosaic of presumptive structures when directly transplanted into metamorphosing hosts. However, it was observed that the inventory and pattern organization of these fragments changed if they were cultured for some time in vivo, before transplantation to the metamorphosing host. Three different kinds of results can be distinguised operationally: either the fragments show duplicated structures (duplication) or else new structures appear, which do not correspond to the fate map of the original fragment but belong either to the inventory of the same disc (regeneration) or to that of a different imaginal disc (transdetermination, Hadorn, 1967). Clonal analysis (Gehring, 1967; Postlethwait et al., 1971;

Nöthiger, 1972) or thymidine labeling (Wildermuth, 1968) established that in all cases the new structures are derived from the cells of the original fragment, and more particularly from its edge (Kröger, 1958), which underwent extra proliferation.

Fragmentation and culture experiments have since been performed to study both the qualities transferred by cell heredity and the mechanism of pattern formation in regenerates. It has been shown that there is a topologic organization, typical for each imaginal disc, that determines whether the fragment will duplicate or regenerate (Schubiger, 1971; Bryant, 1971). In duplication, the level of the cut on the disc corresponds to the line of symmetry of the adult pattern. Most imaginal discs correspond to appendages. The distribution in these discs of potentialities to duplicate or to regenerate follows an angular gradient: Thus, some sector—fragments duplicate, whereas others regenerate. There is also a proximo-distal (radial) gradient, in which proximal fragments will regenerate, distal ones will duplicate (see Bryant, 1975; French et al., 1976). We do not know the cellular basis for these rules, but they indicate the existence of cell properties related to cell communication and cell polarity. These same properties are probably also at work during the cell proliferation phase of normal development.

Cuts of early larval discs performed in situ lead to the same results as those observed in explanted fragments (Bryant, 1971). Also following X-ray irradiation of early larvae (Postlethwait and Schneiderman, 1973) and cauterization of blasto-derm embryos (Bownes, 1975), flies appear with a given frequency of morphologic abnormalities that can be explained by the same rules. In these cases the most obvious effect is that of duplication. The line of symmetry of the duplication may run through virtually any line in the pattern. Since the number of cells in the disc at these stages is very low, it is obvious that the individual cells cannot be specified for the details of the final pattern. Thus these duplications probably do not result from the death of specifically determined cells but from a general disturbance caused by the X-rays. In fact X-rays are known to cause, even in low doses, alterations in the orientation of the mitotic spindle (Kröger, 1957) and in cell adhesion (Morata and García-Bellido, 1973). Similar disturbances in the pattern were found following treatment of isolated discs with sodium salts (Ha-dorn and Fritz, 1950). Moreover, mutants are known that cause duplications, probably as a consequence of generalized damage, and eventually cell death (Russell, 1974). It is therefore interesting to recall that compartmentalization is probably associated with the creation of opposite polarities in the two new cell populations, and that this opposite polarity is maintaned even between adjacent cell populations of the same cell affinities (p. 128). Thus, duplications may result from disturbances in the organization of the growing anlage with the creation of subpopulations of cells with opposite polarities. These duplications do not contain all the elements of the pattern typical of entire compartments. We do not know why there is no regulation toward whole compartments, nor how such incomplete patterns are stable.

Apparently, during the process of duplication, at least as it concerns the cells of the cut-edge fragments, cell proliferation is associated with a loss of genetic specification for terminal differentiation. Clonal analysis has confirmed that there is no clonal relationship between identical structures in the old and in the duplicated regions (Wildermuth, 1968; Nöthiger, 1972). We do not know whether all the cell lineages (e.g., chaetes) are capable of such dedifferentiation or whether extra cell proliferation occurs only in cells not terminally committed. The presumed combination of selector genes at work in both primitive and duplicated regions is the same. However, structures belonging to compartments not represented at the wound surface must have appeared with respecification of selectors. This is obviously the case in regeneration. Thus, apparently, cells undergoing extra proliferation may erase previous genetic decisions and acquire new ones. We do not know how this occurs. Hopefully clonal analysis, will show whether the new genetic decisions are taken in polyclones and are stepwise as in normal development. Otherwise, we will be faced with the paradox that the genetic rules at work during normal development do not apply in regeneration. If regeneration occurs by compartmentalization, new questions arise. What is the stimulus that elicits initiation of new developmental pathways? How do cells recognize the absence of cells of neighboring compartments? How do the progeny of these cells enter precisely into the developmental pathways of neighboring compartments? Are there preferential directions of regeneration, related to activation or repression of selector genes?

Cell interactions between the original cells of the fragment and those generated by regeneration or duplication somehow determine the qualities of the new cells. Experiments in which single cells or small groups of cells are grown on a feeding layer of cells of another histotype have shown that the rules of duplication and regeneration are no longer valid in these experiments. Wing cells retain wing quality even if the feeding layer is of leg cells. Progeny of cells of the anterior wing disk compartments retain this anterior determination as do cells from the dorso-proximal (notum) compartments, although the latter may give rise to cells of the dorsal and ventral wing compartments, a thing they never do in cultured fragments (García-Bellido and Nöthiger, 1976). Thus, proliferation of isolated cells provides a system for the study of the rules of cell heredity under experimental conditions, other than regeneration from disc fragments.

Proliferating fragments occasionally give rise to cell types corresponding to other imaginal discs (Hadorn, 1967). This phenomenon, called transdetermination, demonstrates that the imaginal disc cells can change not only the genetic specification of compartmental restrictions but also that of segmental ones. Clonal analysis (Gehring, 1967) has shown that transdetermined structures can derive from cells that also produce structures of the disc of origin. This suggests that transdetermination does not occur in uncommitted cells. The same analysis suggests that the transdetermination event may affect several cells simultaneously: i.e., that it occurs in a polyclonal way. Analysis of the sequences of transdetermination from fragments of different discs reveals certain rules of probability and preferential directions of transdetermination (Hadorn, 1967). The sequence of these changes fits well with the idea that the imaginal discs cells may be specified by different combinations of genes (Kauffman, 1973, 1975). Unfortunately in these experiments no distinction was then made between the different compartmental origins of the imaginal disc fragments. It will therefore be interesting to know whether compartment specifications are also invariant or are elements of these combinatorial codes.

Contrary to regeneration, transdetermination appears to be a shift in developmental pathways that has no positional relationship to the fragment of origin. However, if the basic change in these events is primarily genetic, then both fall in the same category of phenomena. This would show that genetic specification is reversible. If developmental pathways are genetically controlled by different combinations of genes, then changes in the individual signals will lead to changes in the resulting developmental pathways that will not necessarily bear any relationship to either the sequence of these signals in normal development or to anatomic position.

Such reversibility raises the question of how developmental pathways are genetically controlled and how the genes responsible are maintained in an active state. Analysis of phenocopies of *bithorax* has shown that the initiation and the maintenance of activity of this gene are two different genetic operations (p. 134). In the first, both the position of the cell and the activity of other genes seem to be involved. For the maintenance of activity only the *bithorax* gene itself is needed and is responsible for this particular pathway. We do not know if this continuous activity in the metathorax and its repression in the mesothorax is due to conservative alterations in the DNA or is maintained by feed-back loops. Several considerations seem to suggest the second alternative (Capdevila and García-Bellido, 1978; see García-Bellido, 1977a). If this is the case, the eventual repression of its activity will again be a heritable commitment. If this is the general underlaying mechanism of transdetermination, due to subsequent repressions, transdetermination will finally lead to dead-end developmental pathways. It is interesting to note that all the data available support this hypothesis: The mesothoracic, anterior, dorsal, proximal (notum) compartment is one such dead-end (Hadorn, 1967). This compartment also seems to be the archetype compartment in the developmental pathways of imaginal discs (García-Bellido, 1975a, 1977a).

This reversibility of the genetic commitments applies only to imaginal disc cells. We do not know whether transdetermination occurs between other cell types and imaginal disc cells or between other histotypes themselves. The full potentiality of differentiated plant cells (Steward, 1958; see Meins, 1975) and of nuclei derived from differentiated animal tissues to support normal development (Gurdon, 1974), indicates that transdetermination is a possibility in all the developmental pathways. Experimentally, development appears, in genetic terms, as an open system. We have seen above that in *Drosophila* the behavior of transplanted nuclei and of transplanted cells is different. Whereas nuclei from gastrula cells are totipotent (Illmensee, 1973), blastoderm cells are already committed to grow and to differentiate into specific structures (Illmensee, 1976; and this volume). It may be that stable determination in the classical sense is not associated with chromosomal alterations. Probably, specific gene activity is maintained in a conservative way by the interaction of gene products synthesized in the xytoplasm and acting back upon the genes themselves in some regulatory way. These interactions are possibly disrupted by nuclear transplantation or during regeneration and transdetermination.

The meaning of the word determination should now be revised. In the classical sense, determination denotes an irreversible event that fixes a cell in a pathway for a particular type of differentiation. Determination is considered to be separated in time and through cell divisions from the actual differentiation. When applied to imaginal disc development in *Drosophila*, this has led to the operational discrimination between the initiation of a developmental pathway and its manifestation in terms of cuticular differentiation. However, we have seen (Sec. III.A) that the activation of selector genes is followed immediately by the expression of these genes, and this continues throughout the proliferative phase of the disc, defining characteristic cell behavior properties. Thus, cuticular differentiation is but the final expression of a number of differentiation events expressed during development. In this sense the concept of state of determination loses its operational value, as does consequently, the idea that determination is, at the genetic level, different from actual differentiation.

IV. Concluding Remarks

Clonal analysis has shown that the development of the imaginal discs proceeds by the progressive segregation of new cell lineages. These segregations occur by binary steps associated with the initiation of activity of a selector gene in one cell lineage and the repression in its alternative. Genetic analysis indicates that at the initiation of the selector, the position of the cell (inductor) and the activity of a regulator gene (activator) may be involved. During development new cell lineages appear, defining clonal developmental pathways. Comparative analysis has shown that the same selector genes might work in different pathways. Thus, as a whole, developmental pathways are defined by selectors acting in different combinations so that a few selector genes can account for a variety of developmental pathways. Genetic considerations suggest that this process may account for the segregation of segments and of germ layers in early stages of development. This scheme, based on inferences from a few analyzed systems, requires further genetic and clonal analysis in order to ascertain whether or not it is a general one.

The few cases in which clonal analysis has been applied to the study of regeneration and transdetermination indicate that the genetic specification of developmental pathways is reversible. As we have seen above, the signals emanating from the selector genes are implemented into actual differentiation events throughout the cell proliferation phase of development. They are first expressed in cell behavior characteristics ("proliferative differentiation") and in cuticular differentiation ("terminal differentiation") later. Thus, throughout development it is not only the determination but the actual cell differentiation pattern that emerges. Several considerations, genetic analysis of phenocopies among others, suggest that the genetic basis of the maintenance of a specific pathway depends on regulatory gene interactions upon selector genes. Therefore these can be altered by nuclear isolation or extra cell proliferation. Analysis of these gene interactions becomes one of the most appealing endeavours of a genetic analysis of development.

The genetic decisions taking place during normal development and during extra cell proliferation occur in a supracellular and positional context. Thus, the initiation of cell lineages and of transdetermined territories may occur simultaneously in several cells. The genesis of shape and size reflects the existence of cell functions at work during the cell proliferation period that coordinate the supracellular organization. Cell lineage initiation, but also proliferative and terminal differentiation depend on positional cues. Cell properties such as polarity and cell recognition seem to play an important role in these positional cues. The analysis of certain mutants has shown that several cell functions with supracellular effects can be operationally isolated. Hopefully from the study of these and other new mutants we will understand how the signals of selector genes are expressed in coordinate supracellular communities. It is possible that genetic analysis will provide us with the experimental variation necessary to understand the parameters of these functions, and even to discover new ones. We hope that the strategy of studying development as a consequence of the action of genes working within cells will, at least, open new perspectives. It will undoubtedly raise new problems.

Acknowledgement. The unpublished work reported here was supported by grants of CADC and CAICT 1539.

References

Arking,R.: Temperature-sensitive cell lethal mutants of *Drosophila*: Isolation and characterization. Genetics **80**, 519—537 (1975)

Baker,W.K.: A clonal analysis reveals early developmental restrictions in the head of *Drosophila*. Dev. Biol. **62**, 447—463 (1978)

Becker,H.J.: Über Röntgenmosaikflecken und Defektmutationen am Auge von *Drosophila* und die Entwicklungsphysiologie des Auges. Z. Indukt. Abstamm. Vererbungsl. **88**, 333—373 (1957)

Bodenstein,D.: Investigation on the problem of metamorphosis. VIII. Studies on leg determination in insects. J. Exp. Zool. **87**, 31—53 (1941)

Bownes,M.: Adult deficiencies and duplications of head and thoracic structures resulting from microcautery of blastoderm stage *Drosophila* embryos. J. Embryol. Exp. Morph. **34**, 33—54 (1975)

Bownes,M., Sang,J.H.: Experimental manipulation of early *Drosophila* embryos. I. Adult and embryonic defects resulting from microcautery at nuclear multiplication and blastoderm stages. J. Embryol. Exp. Morph. **32**, 253—272 (1974)

Braun,W.: The effect of punctures in the developing wings of several mutants of *Drosophila melanogaster*. J. Exp. Zool. **84**, 325—349 (1940)

Bryant,P.J.: Cell lineage relationships in the imaginal wing disc of *Drosophila melanogaster*. Dev. Biol. **22**, 389—411 (1970)

Bryant,P.J.: Regeneration and duplication following operations in situ on the imaginal discs of *Drosophila melanogaster*. Dev. Biol. **26**, 637—651 (1971)

Bryant,P.J.: Regeneration and duplication in imaginal discs. In: Cell Patterning, pp. 71—93. Ciba Found. Symp. 29. Amsterdam: Elsevier 1975

Bryant,P.J., Schneiderman,H.A.: Cell lineage, growth and determination in the imaginal leg discs of *Drosophila melanogaster*. Dev. Biol. **20**, 263—290 (1969)

Bull,A.L.: *Bicaudal* a genetic factor which affects the polarity of the embryo in *Drosophila melanogaster*. J. Exp. Zool. **161**, 221—242 (1966)

Capdevila,M.P., Garcia-Bellido,A.: Development and genetic analysis of *bithorax* phenocopies in *Drosophila*. Nature (Lond.) **250**, 500—502 (1974)

Capdevila,M.P., Garcia-Bellido,A.: Phenocopies of *bithorax*: clonal and genetic analysis. Wilhelm Roux's Archives (in press, 1978)

Chan, L. N., Gehring, W.: Determination of blastoderm cells in *Drosophila melanogaster*. Proc. Natl. Acad. Sci. USA **68**, 2217—2221 (1971)

Crick, F. H. C., Lawrence, P. A.: Compartments and polyclones in insect development. Science **189**, 340—347 (1975)

Denell, R. E.: Homoeosis in *Drosophila*. I. Complementation studies with revertants of *Nasobemia*. Genetics **75**, 279—297 (1973)

Dübendorfer, K.: Die Entwicklung der männlichen und weiblichen Genital-Imaginalscheibe von *Drosophila melanogaster*. Eine klonale Analyse. Inaugural-Dissertation Univ. Zürich, 1977

Ferrus, A., Garcia-Bellido, A.: Morphogenetic mutants detected in mitotic recombination clones. Nature (Lond.) **260**, 425—426 (1976)

Ferrus, A., Garcia-Bellido, A.: Minute mosaics caused by early chromosome loss. Wilhelm Roux's Archives **183**, 337—344 (1977)

French, V., Bryant, P. J., Bryant, S. V.: Pattern regulation in epimorphic fields. Science **193**, 969—981 (1976)

Fristrom, D., Fristrom, J. W.: The mechanism of evagination of imaginal discs of *Drosophila melanogaster*. I. General considerations. Dev. Biol. **43**, 1—23 (1975)

Garcia-Bellido, A.: Larvalentwicklung transplantierter Organe von *Drosophila melanogaster* in Adultmilieu. J. Ins. Physiol. **11**, 1071—1078 (1965)

Garcia-Bellido, A.: Pattern reconstruction by dissociated imaginal disc cells of *Drosophila melanogaster*. Dev. Biol. **14**, 278—306 (1966a)

Garcia-Bellido, A.: Changes in selective affinities following transdetermination in imaginal discs of *Drosophila melanogaster*. Exp. Cell Res. **44**, 382—392 (1966b)

Garcia-Bellido, A.: Cell lineage in the wing disc of *Drosophila melanogaster*. Genetics **60**, 181 (1968) (abstract)

Garcia-Bellido, A.: Pattern formation in imaginal disks. In: Ursprung, H., Nöthiger, R. (eds.): Results and Problems in Cell Differentiation, Vol. V, pp. 59—91. Berlin-Heidelberg-New York: Springer 1972

Garcia-Bellido, A.: Genetic control of wing disc development in *Drosophila*. In: Cell Patterning. Ciba Found Symp. 29, pp. 161—182. Amsterdam: Elsevier 1975a

Garcia-Bellido, A.: Genetic control of imaginal disc morphogenesis in *Drosophila*. In: McMahon, D., Fox, C. F. (eds.): Developmental Biology ICN-UCLA Symposia. London: Benjamin 1975b

Garcia-Bellido, A.: Homoeotic and atavic mutations in insects. Am. Zoologist **17**, 613—629 (1977a)

Garcia-Bellido, A.: Inductive mechanisms in the process of wing vein formation in *Drosophila*. Wilhelm Roux's Archives **182**, 93—106 (1977b)

Garcia-Bellido, A.: Genetic analysis of the *achaete-scute* system of *Drosophila melanogaster*. (submitted) (1978)

Garcia-Bellido, A., Capdevila, M. P.: Initiation and maintenance of a developmental pathway. In: Symp. Amer. Soc. Develop. Biol. (in press, 1978)

Garcia-Bellido, A., Ferrus, A.: Gynandromorph fate map of the wing-disk compartments in *Drosophila melanogaster*. Wilhelm Roux's Archives **178**, 337—340 (1975)

Garcia-Bellido, A., Lewis, E. B.: Autonomous cellular differentiation of homoeotic bithorax mutants of *Drosophila melanogaster*. Dev. Biol. **48**, 400—410 (1976)

Garcia-Bellido, A., Merriam, J. R.: Cell lineage of the imaginal discs in *Drosophila* gynandromorphs. J. Exp. Zool. **170**, 61—76 (1969)

Garcia-Bellido, A., Merriam, J. R.: Parameters of the wing imaginal disc development of *Drosophila melanogaster*. Dev. Biol. **24**, 61—87 (1971a)

Garcia-Bellido, A., Merriam, J. R.: Clonal parameters of tergite development in *Drosophila*. Dev. Biol. **26**, 264—276 (1971b)

Garcia-Bellido, A., Merriam, J. R.: Genetic analysis of cell heredity in imaginal discs of *Drosophila melanogaster*. Proc. Natl. Acad. Sci. USA **68**, 2222—2226 (1971c)

Garcia-Bellido, A., Nöthiger, R.: Maintenance of determination by cells of imaginal discs of *Drosophila* after dissociation and culture in vivo. Wilhelm Roux's Archives **180**, 189—206 (1976)

Garcia-Bellido, A., Ripoll, P., Morata, G.: Developmental compartmentalization of the wing disk of *Drosophila*. Nature (Lond.) New Biol. **245**, 251—253 (1973)

Garcia-Bellido, A., Ripoll, P., Morata, G.: Developmental compartmentalization in the dorsal mesothoracic disc of *Drosophila*. Dev. Biol. **48**, 132—147 (1976)

Garcia-Bellido, A., Santamaria, P.: Developmental analysis of the wing disc in the mutant *engrailed* of *Drosophila melanogaster*. Genetics **72**, 87—104 (1972)

Garcia-Bellido, A., Santamaria, P.: Developmental analysis of the *achaete-scute* system of *Drosophila melanogaster*. Genetics **88**, 469—486 (1978)

Gehring, W.: Clonal analysis of determination dynamics in cultures of imaginal disks in *Drosophila melanogaster*. Dev. Biol. **16**, 438—456 (1967)

Gehring, W., Seippel, S.: Die Imaginalzellen des Clypeo-labrums und die Bildung des Rüssels von *Drosophila melanogaster*. Rev. Suisse Zool. **54**, 637—712 (1967)

Geigy, R.: Erzeugung rein imaginaler Defekte durch ultraviolette Eibestrahlung bei *Drosophila melanogaster*. Wilhelm Roux' Arch. **125**, 406—447 (1931)

Gelbart, W. M.: A new mutant controlling mitotic chromosome disjunction in *Drosophila melanogaster*. Genetics **76**, 51—63 (1973)

Gloor, H.: Phänokopie-Versuche mit Äther an *Drosophila*. Rev. Suisse Zool. **54**, 637—712 (1947)

Grigliatti, T., Suzuki, D. T.: Temperature-sensitive mutations in *Drosophila melanogaster*. VIII. The homoeotic mutant, ss^{a40a}. Proc. Natl. Acad. Sci. USA **68**, 1307—1311 (1971)

Guerra, M., Postlethwait, J. H., Schneiderman, H. A.: The development of the imaginal abdomen of *Drosophila melanogaster*. Dev. Biol. **32**, 361—372 (1973)

Gurdon, J. B.: The control of gene expression in animal development. Oxford: Clarendon Press 1974

Hadorn, E.: Regulation and differentiation within field-districts in imaginal discs of *Drosophila*. J. Embryol. Exp. Morph. **1**, 213—216 (1953)

Hadorn, E.: Dynamics of determination. Symp. Soc. Dev. Biol. **25**, 85—104 (1967)

Hadorn, E., Anders, G., Ursprung, H.: Kombinate aus teilweise dissoziierten Imaginalscheiben verschiedener Mutanten und Arten von *Drosophila*. J. Exp. Zool. **142**, 159—175 (1959)

Hadorn, E., Bertani, G., Gallera, J.: Regulationsfähigkeit und Feldorganisation der männlichen Genital-Imaginalscheibe von *Drosophila melanogaster*. Wilhelm Roux' Arch. **144**, 31—70 (1949)

Hadorn, E., Fritz, W.: Veränderungen am transplantierten weiblichen Geschlechtsapparat von *Drosophila melanogaster* nach Behandlung der Imaginalscheiben in Salzlösungen. Rev. Suisse Zool. **57**, 477—488 (1950)

Hotta, Y., Benzer, S.: Mapping of behaviour in *Drosophila* mosaics. Nature (Lond.) **240**, 527—535 (1972)

Howland, R. B., Child, G. P.: Experimental studies on development in *Drosophila melanogaster*. I. Removal of protoplasmic materials during late cleavage and early embryonic stages. J. Exp. Zool. **70**, 415—427 (1935)

Howland, R. B., Sonnenblick, B. P.: Experimental studies on development in *Drosophila melanogaster*. II. Regulation in the early egg. J. Exp. Zool. **73**, 109—125 (1936)

Illmensee, K.: Developmental potencies of nuclei from cleavage, preblastoderm, and syncytial blastoderm transplanted into unfertilazed eggs of *Drosophila melanogaster*. Wilhelm Roux' Arch. **170**, 267—298 (1972)

Illmensee, K.: The potentialities of early gastrula nuclei of *Drosophila melanogaster*. Production of their imago descendants by germ-line transplantation. Wilhelm Roux' Arch. **171**, 331—343 (1973)

Illmensee, K.: Nuclear and cytoplasmic transplantation in *Drosophila*. In: Lawrence, P. A. (Ed.): Insect Development. Symp. R. Soc. London **8** (1976)

Illmensee, K., Mahowald, A. P., Loomis, M. R.: The ontogeny of germ plasm during oogenesis in *Drosophila*. Dev. Biol. **49**, 40—65 (1976)

Janning, W.: Entwicklungsgenetische Untersuchungen an Gynandern von *Drosophila melanogaster*. I. Die inneren Organe der Imago. Wilhelm Roux' Archiv **174**, 313—332 (1974)

Kalthoff, K.: Position of targets and period of competence for UV-induction of the malformation "double abdomen" in the egg of *Smitia* spec. (Diptera, Chironomidae). Wilhelm Roux' Archiv **168**, 63—84 (1971)

Kankel, D. R., Hall, J. C.: Fate mapping of nervous system and other internal tissues in genetic mosaics of *Drosophila melanogaster*. Dev. Biol. **48**, 1—24 (1975)

Kauffman, S.: Control circuits for determination and transdetermination. Science **181**, 310—318 (1973)

Kauffman, S.: Control circuits for determination and transdetermination: interpreting positional information in a binary epigenetic code. In: Cell patterning. Ciba Foundation Symp. 29, pp. 201—214. Amsterdam: Elsevier 1975

Kroeger, H.: Eine Analyse röntgeninduzierter Modifikationen in Flügelgeäder der Mehlmotte *Ephestia Kühniella*. Wilhelm Roux' Archiv **150**, 77—104 (1957)

Kroeger, H.: Über Doppelbildungen in die Leibeshöhle verpflanzter Flügelimaginalscheiben von *Ephestia Kühniella*. Wilhelm Roux' Archiv **150**, 401—424 (1958)

Lawrence, P. A.: Development and determination of hairs and bristles in the milkweed bug, *Oncopeltus fasciatus* (Lygaeidae, Hemiptera). J. Cell. Sci. **1**, 475—498 (1966)

Lawrence, P. A.: A clonal analysis of segment development in *Oncopeltus* (Hemiptera). J. Embryol. Exp. Morph. **30**, 681—699 (1973 a)

Lawrence, P. A.: The development of spatial patterns in the integument of insects. In: Counce, S. J., Waddington, C. A. (eds.): Developmental Systems: Insects, Vol. II. New York: Academic Press 1973 b

Lawrence, P. A., Morata, G.: Compartments in the wing of *Drosophila*: A study of the engrailed gene. Dev. Biol. **50**, 321—337 (1976 a)

Lawrence, P. A., Morata, G.: The early development of mesothoracic compartments in *Drosophila*. An analysis of cell lineage, fate mapping, and an assessment of methods. Dev. Biol. **56**, 40—51 (1976 b)

Lees, A. D.: Operations in the pupal wings of *Drosophila melanogaster*. J. Genet. **42**, 115—142 (1941)

Lees, A., Waddington, C. H.: The development of the bristles in normal and some mutant types of *Drosophila melanogaster*. Proc. Roy. Soc. Ser. B. **131**, 87—110 (1942)

Lewis, E. B.: Genes and developmental pathways. Am. Zool. **3**, 33—56 (1963)

Lewis, E. B.: Genetic control and regulation of developmental pathways. Symp. Soc. Dev. Biol. **23**, 231—251 (1964)

Lewis, E. B.: Genetic control of developmental pathways. Proc. 12th. Int. Congr. Genetics **1**, 96—97 (1968)

Lindsley, D. L., Grell, E. H.: Genetic variations of *Drosophila melanogaster*. Carnegie Inst. Wash. Publ. N. 627 (1968)

Lindsley, D. L., Sandler, L., Baker, B. S., Carpenter, A. T. C., Denell, R. E., Hall, J. C., Jacobs, P. A., Miklos, G. L. G., Davis, B. K., Gethman, R. C., Hardy, R. W., Hessler, A., Miller, S. M., Nozawa, H., Parry, D. M., Gould-Somero, M.: Segmental aneuploidy and the genetic gross structure of the *Drosophila* genome. Genetics **71**, 157—184 (1972)

Meins, F. Jr.: Cell division and the determination phase of cytodifferentiation in plants. In: Reinert, J., Holtzer, H. (eds.): Results and Problems in Cell Differentiation, Vol. 7: Cell Cycle and Cell Differentiation, pp. 151—175. Berlin-Heidelberg-New York: Springer 1975

Morata, G.: Analysis of gene expression during development in the homoeotic mutant Contrabithorax of *Drosophila melanogaster*. J. Embryol. Exp. Morph. **34**, 19—31 (1975)

Morata, G., Garcia-Bellido, A.: Behaviour in aggregates of irradiated imaginal disk cells of *Drosophila*. Wilhelm Roux' Archiv **172**, 187—195 (1973)

Morata, G., Garcia-Bellido, A.: Developmental analysis of some mutants of the bithorax system of *Drosophila*. Wilhelm Roux's Archives **179**, 125—143 (1976)

Morata, G., Lawrence, P. A.: Control of compartment development by the *engrailed* gene of *Drosophila*. Nature (Lond.) **255**, 211—221 (1975)

Morata, G., Ripoll, P.: Minutes: Mutants of *Drosophila* autonomously affecting cell division rate. Dev. Biol. **42**, 211—221 (1975)

Nöthiger, R.: Differenzierungsleistungen in Kombinaten, hergestellt aus Imaginalscheiben verschiedener Arten, Geschlechter und Körpersegmente von *Drosophila*. Wilhelm Roux' Archiv **155**, 269—301 (1964)

Nöthiger, R.: The larval development of imaginal disks. In: Ursprung, H., Nöthiger, R. (eds.): Result and Problems in Cell Differentiation, Vol. V, pp. 1—34. Berlin-Heidelberg-New York: Springer 1972

Nöthiger, R., Strub, S.: Imaginal defects after UV-microbeam irradiation of early cleavage stages of *Drosophila melanogaster*. Rev. Suisse. Zool. **79**, 267—279 (1972)

Ouweneel, W. J.: Developmental genetics of homoeosis. Adv. Genet. **18**, 179—248 (1976)

Poodry, C. A.: A temporal pattern in the development of sensory bristles in *Drosophila*. Wilhelm Roux's Archives **178**, 203—213 (1975)

Poodry, C. A., Hall, L., Suzuki, D. T.: Developmental properties of *shibire^{ts1}*: A pleiotropic mutation affecting larval and adult locomotion and development. Dev. Biol. **32**, 373—386 (1973)

Postlethwait, J. H.: Development of the temperature sensitive homoeotic mutant *eyeless-opthalmoptera* of *Drosophila melanogaster*. Dev. Biol. **36**, 212—217 (1974)

Postlethwait, J. H., Girton, J.: Development of antennal-leg homoeotic mutants in *Drosophila melanogaster*. Genetics **76**, 767—774 (1974)

Postlethwait, J. H., Poodry, C. A., Schneiderman, H. A.: Cellular dynamics of pattern duplication in imaginal discs of *Drosophila melanogaster*. Dev. Biol. **26**, 125—132 (1971)

Postlethwait, J. H., Schneiderman, H. A.: A clonal analysis of development in *Drosophila melanogaster*: Morphogenesis, determination, and growth in the wildtype antenna. Dev. Biol. **24**, 477—519 (1971)

Postlethwait, J. H., Schneiderman, H. A.: Pattern formation in imaginal discs of *Drosophila melanogaster* after irradiation of embryos and young larvae. Dev. Biol. **32**, 345—360 (1973)

Postlethwait, J. H., Schneiderman, H. A.: Developmental genetics of *Drosophila* imaginal discs. Ann. Rev. Genet. **7**, 381—433 (1974)

Poulson, D. F.: Histogenesis, organogenesis, and differentiation in the embryo of *Drosophila melanogaster*. In: Demerec, M. (Ed.): Biology of Drosophila. New York: Wiley 1950

Ripoll, P.: The embryonic organization of the imaginal wing disc of *Drosophila melanogaster*. Wilhelm Roux' Archiv **169**, 200—215 (1972)

Ripoll, P.: Behaviour of somatic cells homozygous for zygotic lethals in *Drosophila melanogaster*. Genetics **86**, 357—376 (1977)

Ripoll, P., Garcia-Bellido, A.: Cell autonomous lethals in *Drosophila melanogaster*. Nature (New Biol.) **241**, 15—16 (1973)

Roberts, P.: Mosaics involving *aristapedia*, a homoeotic mutant of *Drosophila melanogaster*. Genetics **49**, 593—598 (1964)

Russel, M. A.: Pattern formation in the imaginal discs of a temperature sensitive cell lethal mutant of *Drosophila melanogaster*. Dev. Biol. **40**, 24—39 (1974)

Santamaria, P.: Control genetico de la morfogenesis del borde del ala de *Drosophila melanogaster*. PhD. Thesis. Univ. Madrid. (1973)

Santamaria, P., Garcia-Bellido, A.: Localization and growth of the tergite Anlage of *Drosophila*. J. Embryol. Exp. Morph. **28**, 397—417 (1972)

Santamaria, P., Garcia-Bellido, A.: Developmental analysis of two wing scalloping mutants, *ct^6* and *Bx^J*, of *Drosophila melanogaster*. Wilhelm Roux's Archives **178**, 233—245 (1975)

Schatz, E.: Über die Formbildung der Flügel bei Hitzemodifikationen und Mutationen von *Drosophila melanogaster*. Biol. Zbl. **70**, 305—353 (1951)

Schubiger, G.: Anlageplan, Determinationszustand und Transdeterminationsleistungen der männlichen Vorderbeinscheibe von *Drosophila melanogaster*. Wilhelm Roux' Archiv **160**, 9—40 (1968)

Schubiger, G.: Regeneration, duplication and transdetermination in fragments of the leg disc of *Drosophila melanogaster*. Dev. Biol. **26**, 277—295 (1971)

Schubiger, G.: Adult differentiation from partial *Drosophila* embryos after egg ligation during stages of nuclear multiplication and cellular blastoderm. Dev. Biol. **50**, 476—488 (1976)

Schubiger, G., Schubiger-Staub, H., Hadorn, E.: Mischungsversuche mit Keimteilen von *Drosophila melanogaster* zur Ermittlung des Determinationszustandes imaginaler Blastome im Embryo. Wilhelm Roux' Arch. **163**, 33—39 (1969)

Shearn, A., Rice, T., Garen, A., Gehring, W.: Imaginal disc abnormalities in lethal mutants of *Drosophila*. Proc. Natl. Acad. Sci. USA **68**, 2594—2598 (1971)

Simpson, P.: Analysis of the compartments of the wing of *Drosophila melanogaster* mosaic for a temperature-sensitive mutation that reduces mitotic rate. Dev. Biol. **54**, 100—115 (1976)

Snodgrass, R. E.: Principles of Insect Morphology. New York: McGraw Hill 1935

Steiner, E.: Establishment of compartments in developing leg imaginal discs of *Drosophila melanogaster*. Wilhelm Roux's Archives **180**, 9—30 (1976)

Stern, C.: The innervation of setae in *Drosophila*. Genetics **23**, 172—173 (1938)

Stern, C.: Two or three bristles. Am. Sci. **43**, 213—247 (1954)

Steward, F. C.: Growth and organized development of cultured cells. III. Interpretations of the growth from free cell to carrot plant. Am. J. Bot. **45**, 709—713 (1958)

Stewart, M., Murphy, C., Fristrom, J. W.: The recovery and preliminary characterization of X chromosome mutants affecting imaginal discs of *Drosophila melanogaster*. Dev. Biol. **27**, 71—83 (1972)

Stumpf, H.: Die Richtungen der Teilungsspindeln auf dem Puppenflügel von *Drosophila* im Verlaufe der Mitosenperiode. Biol. Zbl. **75**, 17—27 (1956)

Sturtevant, A. H.: The *claret* mutant type of *Drosophila simulans:* a study of chromosome elimination and of cell lineage. Z. Wiss. Zool. **135**, 323—356 (1929)

Sturtevant, A. H.: Studies on the bristle pattern of *Drosophila*. Dev. Biol. **21**, 48—61 (1970)

Tobler, H.: Zellspezifische Determination und Beziehung zwischen Proliferation und Trans-determination in Bein- und Flügelprimordien von *Drosophila melanogaster*. J. Embryol. Exp. Morph. **16**, 609—633 (1966)

Tobler, H., Rothenbühler, V., Nöthiger, R.: A study of the differentiation of bracts in *Drosophila melanogaster* using two mutations, H^2 and sv^{de}. Experientia **29**, 370—371 (1973)

Vogt, M.: Zur labilen Determination der Imaginalscheiben von *Drosophila*. I. Verhalten ver-schiedenaltriger Imaginalanlagen bei operativer Defektsetzung. Biol. Zbl. **65**, 223—238 (1946a)

Vogt, M.: Zur labilen Determination der Imaginalscheiben von *Drosophila*. II. Die Umwand-lung präsumptiven Fühlergewebes in Beingewebe. Biol. Zbl. **65**, 238—254 (1946b)

Waddington, C. H.: The genetic control of wing development in *Drosophila*. J. Genet. **41**, 75—139 (1940)

Weber, H.: Grundriß der Insektenkunde. Stuttgart: Fischer 1966

Whittle, J. R. S.: Clonal analysis of a genetically caused duplication of the anterior wing in *Drosophila melanogaster*. Dev. Biol. **51**, 257—268 (1976)

Wieschaus, E., Gehring, W.: Gynandromorph analysis of the thoracic disc primordia in *Drosophila melanogaster*. Wilhelm Roux's Archives **180**, 31—46 (1976a)

Wieschaus, E., Gehring, W.: Clonal analysis of primordial disc cells in the early embryo of *Drosophila melanogaster*. Dev. Biol. **50**, 249—263 (1976b)

Wigglesworth, V. B.: Local and general factor in the development of "pattern" in *rhodnius prolixus* (Hemiptera). J. Exp. Biol. **17**, 180—200 (1940)

Wildermuth, H.: Autoradiographische Untersuchungen zum Vermehrungsmuster der Zellen in proliferierenden Rüsselprimordien von *Drosophila melanogaster*. Dev. Biol. **18**, 1—13 (1968)

Zalokar, M.: L'ablation des disques imaginaux chez la larve de *Drosophila*. Rev. Suisse Zool. **50**, 232—237 (1943)

Zalokar, M.: Transplantation of nuclei in *Drosophila melanogaster*. Proc. Natl. Acad. Sci. USA **68**, 1539—1541 (1971)

Genetic Mosaic Studies of Pattern Formation in Drosophila melanogaster, with Special Reference to the Prepattern Hypothesis

CHIYOKO TOKUNAGA

Department of Molecular Biology, University of California, Berkeley, CA, USA

I. Introduction

During the early period of *Drosophila* genetics, it was Sturtevant who recognized the importance of genetic mosaics for the field of developmental biology. His early studies with genetic mosaics provided later researchers with phenomena, ideas, and methods, such as: the analysis of cell lineage relations between organs or specific sites of a gynander or a genetic mosaic by statistically assessing the frequency of genetically marked organs or sites (1929). He also recognized two kinds of interactions (autonomy and nonautonomy) between the different genotypes in a genetic mosaic, as an important clue for studies of gene action (1920, 1927, 1932). Thus, he contributed much to the progress not only of developmental biology, but also of biochemical genetics and behavioral genetics.

Stern's discovery of the somatic crossing-over phenomenon (mitotic recombination) in *Drosophila* (1936) also contributed to the progress of developmental biology by furnishing a powerful method to provide genetic mosaics with marker genes on any chromosome. Finally, Becker's finding that mitotic recombination can be induced at desired times during development by X-ray (1956, 1957) allowed all developmental stages to be studied. Thus, the methodology of mosaic studies was brought to its present state.

However, Stern's main involvement in this field was in problems of pattern formation, as analyzed through the use of genetic mosaics in situ and genetic methods. He proposed a unifying view, the "prepattern" hypothesis, after his studies of bristle pattern on the mesothorax by means of gynandric mosaics of *achaete*[1] (1954). Since then much work on genetic mosaics has been done by his associates and others to find the hypothetical "prepattern mutant," mostly for bristle patterns. The rarity of such mutants, as well as the fact that even hom-

---

[1] For a complete description of mutants. see Lindsley and Grell (1968).

oeotic mutants generally behave autonomously in mosaics, suggested that the prepattern was invariant for equivalent regions of mutants of widely different bristle patterns and for different regions of a single individual when the mosaic cell was induced.

Later these findings, as well as those from other areas of developmental biology, were incorporated by Wolpert (1969) in his general proposal for the establishment and interpretation of "positional information" within fields of cells.

The accumulation of new knowledge frequently leads one to abandon or modify earlier working hypotheses, and the latter has been the case for the relation between the prepattern and the positional information hypotheses.

This short article concerns the classic attempts to find the role of genes in pattern formation in *Drosophila melanogaster* via studies of genetic mosaics and an evaluation of the prepattern hypothesis in the light of recent developments in this field.

II. The Prepattern Hypothesis
vis-a-vis the Positional Information Hypothesis

Stern was concerned with the problem of how a pattern gene acts to specify a spatial pattern of cellular differentiation. He chose to approach the problem by genetic mosaic studies, and the results led him to the prepattern hypothesis. It was elegantly explained by Stern (1968); thus, readers are strongly encouraged to read his account. The following quotes are excerpts from Stern's own descriptions of the prepattern concept (1954, 1968), with *italics* for emphasis.

"A prepattern is a descriptive term for any kind of spatial differentness in development. It may not have any morphogenetic consequences at all or it may even form the basis of a field or mosaic of fields. Thus, the meaning of the term prepattern is not the same as that of 'morphogenetic field.' *Development may be regarded as a sequence of prepatterns*, each one being a realized pattern as compared to its predecessor, and a new prepattern as the basis for its successor. At any stage in the sequence, *differential genic responses to a prepattern* will create a differential prepattern for the next stage. In the interrelationship of developmental processes, *prepattern does not solely depend on genic activity*. Factors of the *internal or external environment* amplify existent differentness or even create it. In the determination of pattern, the role of genes has a two-fold picture. In some cases, *different genotypes respond differently to the invariant prepattern*, in others *the genotypes create different prepatterns*. There is no basic difference between these two methods" (1954).

Stern further clarified his prepattern concept by introducing the term *singularity*. *Drosophila melanogaster* has 11 pairs of macrochaetae patterned on the mesothorax (see Fig. 10a). Among them the dorsocentral bristles are affected by a sex-linked gene *achaete (ac)*, which changes the length, and position of the bristles, and suppresses the differentiation mainly of the posterior dorsocentral bristles (pdc). The variations of these gene effects depend on the external and internal environments. The autonomous gene action found in *ac* gynandric mosaic spots covering the area of dorsocentral bristles, led Stern (1968) to the concept that "the difference in visible bristle pattern is the outcome of a differential response to the same prepattern by cells with different genotypes. A certain bristle is present when the cells differentiating it are *genetically competent* to respond to the *singularity of the prepattern* at the specific location. The bristle is absent when the cell for genetic reason lacks this competence. Preceding the determination of 11 pairs of cells, a mesothoracic disc possesses 11 pairs of locations, each of which, independently of others, gives a qualitatively unique signal for bristle production. These different qualities of the 11 different sites will be referred to as *singularities*. Different alleles endow a cell with specific properties which may be tuned to all these signals as in wild type, or to one or another assortment of them as in the mutant." Here, "the term *competence* refers to these genetically controlled capacities, that exist independently of epigenetically acquired ones." In other words, the cell's competence must be distinguished from its developmentally acquired specified state—which derives from the preceding prepattern, and also contributes to the subsequent prepattern as an internal environment, as well as influencing the cell's competence.

"If the different alleles cause different prepatterns of the discs, one would expect new types of such prepatterns in mosaic discs, some of whose parts contain one kind of prepattern-forming allele and whose other parts contain an allele responsible for a different prepattern. The new types of overall organization should result in new bristle patterns, and one could hope that the analysis of such new patterns would lead to some understanding of the control of prepattern in genetically homogeneous individuals."

This expectation of finding prepattern mutants was not fulfilled by the genetic mosaic experiments of the following decade, except in a few cases. Most of the pattern mutants tested, including homoeotic mutants such as *spineless-aristapedia* (*ss^a*, Roberts, 1964) or *bithorax* pseudoallelic mutants (Lewis, 1963, 1964), revealed that the mosaic disc does not make a completely new type of overall organization. Instead, each genotype tissue develops autonomously, independent from the other, as if responding to an invariant prepattern singularity according to the competence of the cell genotype, after the time of mosaic cell initiation.

Stern (1968) summarized these general results and evaluated the mosaic studies of pattern differentiation as follows: "The autonomy of bristle differentiation in mosaics for *ac* and *scute (sc)* led to the concept of prepattern and genetic control of competence. The autonomy of secondary sex comb formation in *engrailed (en)* revealed an unknown singularity in the forelegs of Drosophila. The autonomy of differentiation of homoeotic mutants (such as *ss^a*) showed that the determination of the features characteristic for each antenna versus leg occurs separately for each of these properties. The decision on what is to be realized out of a variety of alternatives is made *at each specific region:* there is no overall organization of an imaginal disc that would tend to decide on the development of the type of appendage as a whole." *The general results seemed to indicate that a prepattern change might be lethal or at least compensated for, as Stern mentioned.* The search for causal genetic factors in the initiation of the terminal bristles and sex comb patterns as well as other patterns in *Drosophila* had revealed only

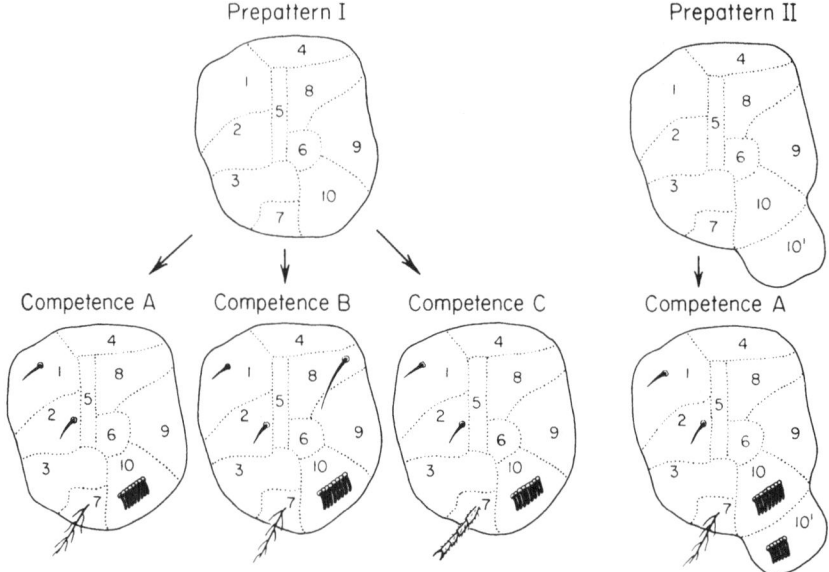

Fig. 1. General diagram of prepattern and competence interactions. Prepattern *I* consists of 10 different singularities. Three genotypes *A*, *B*, and *C* respond to prepattern *I*, each one being competent for specific differentiation at the various singularities, thus generating different final patterns. None of the three responded to the singularities 3, 4, 5, 6, and 9. A mutant prepattern *II* differs from *I* by having an enlargement (*10′*) of the area of singularity 10. Genotype *A* reacts to prepattern *II* by differentiation of a duplicate structure in area 10–10′. (After Stern, 1968)

differences in the competence of cells but not differences in their prepattern singularity. Also the fact that the few exceptional "prepattern" genes, exemplified by *eyeless-Dominant* (*ey^D*) changed the prepattern singularity *only quantitatively*, seemed to require from researchers a new approach to solving pattern formation problems in *Drosophila*. Figure 1 illustrates a summary of the relations between the prepattern and competence of cell genotype, based on the experimental data in mosaic studies (Stern, 1968).

Wolpert (1969) then proposed the "positional information hypothesis" as a possible universal mechanism by which the translation of genetic information into spatial patterns of differentiation could be achieved. According to him, the main proposals are the following:

(1) The cells in a developing system may have their position specified with respect to one or more points in the system. This specification of position is *positional information*. Cells which have their positional information specified with respect to the same set of points constitute a *field*. Thus, the identification of such reference points is of great importance. (2) Positional information largely determines, with respect to the cell's genome and developmental history, the nature of its molecular differentiation. The general process whereby positional information leads to a particular cellular activity or molecular differentiation will be termed the *interpretation of the positional information*. The specification of positional information in general precedes and is independent of molecular differ-

entiation. (3) *Polarity* may be defined in relation to the points with respect to which a cell's position is being specified; it is the direction in which positional information is specified or measured. (4) Positional information may be universal. (The same mechanisms that specify positional information may be operative in different fields within the same organism as well as in quite different organisms from different genera or even phyla) (5). Classical cases of pattern regulations are largely dependent on the ability of cells to change their positional information in an appropriate manner and to interpret the change.

According to this concept, the singularity of a prepattern is somewhat similar to the positional information specified at a particular location. The competence of a cell is expressed by the interpretation of the positional information. The concept of singularity or positional information is still highly speculative in the absence of information about its biochemical and physiologic basis. A difference of emphasis between the two theories lies in the definition of the area in which a specific pattern is formed. The positional information hypothesis emphasizes a field as a group of cells with a common reference point and the same options for interpretation, thus implying the mechanism by which the field was generated. In contrast, the prepattern hypothesis simply refers to a prepattern as any chosen ensemble of singularities, without reference to the mechanism by which singularities were positioned and without implying boundaries to the chosen area. Still another substantial difference exists between the concept of a prepattern singularity and a specified positional value, as can be further illustrated: The former suggests that factors other than position may be involved. For instance, some substrates which lead to the final bristle formation (e.g., macrochaeta or sex comb tooth) might be included in some singularity, whereas singularities for other structures might include quantitatively or qualitatively different substrates (Stern, 1956b). While the prepattern of all three pairs of legs had the same sex comb singularity, it was considered that either the intensity of this singularity differs from one leg to the next in a graded way, or the competence differs in a graded way (Fig. 2; Tokunaga and Stern, 1965). It should be pointed out here that, in these alternative explanations, the concepts of epigenetically varying prepattern and epigenetically varying competence were introduced. Both "may be regarded as subordinate to those of prepattern and competence since they serve only to *specify* variable aspects of the two main features" (Stern, 1968). In contrast, the positional information concept puts the weight more on the cell's genotype to interpret with precision a rather uniform positional information mechanism to achieve the variety of pattern differentiation in different regions. This modification from the prepattern concept is seemingly reasonable in view of the general findings of the autonomous gene action and invariant prepattern in the mosaic studies. However, this extremely universal working hypothesis still awaits further tests of its validity.

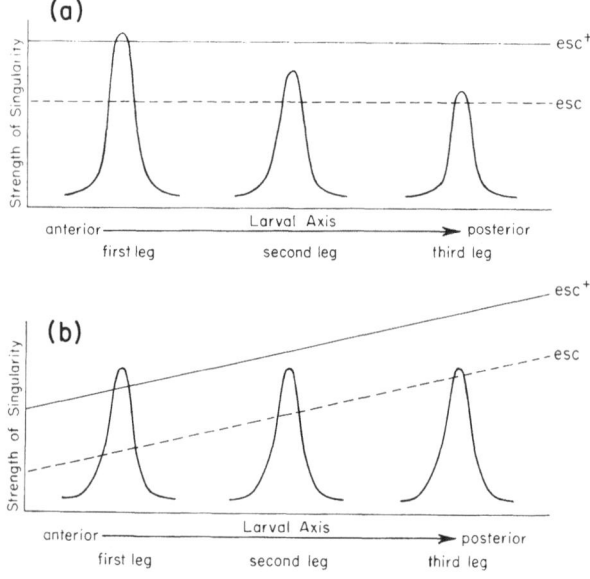

Fig. 2a and b. Two diagrams illustrating alternative modes of the action of genetic and epigenetic factors in the development of homoeotic structures (sex combs) on 2nd and 3rd legs. (a) Epigenetically variable strengths of prepatterns in the three legs and constant thresholds for esc^+ and esc action along the body axis for differentiation of 1st-leg structure. (b) Invariant strengths of prepatterns and epigenetically increasing thresholds for differentiation. (After Tokunaga and Stern, 1965)

III. Identification of Pattern Mutant Genotypes in Genetic Mosaics

In the study of pattern formation by genetic mosaics, one has to provide appropriate marker genes to identify the genotype of the pattern gene of the mosaic cells. The marker gene must not be a pattern-determining gene but a gene which exerts its effect whenever the differentiation of the cell permits its expression autonomously. For example, *yellow (y)*, *singed (sn)*, or *multiple wing hair (mwh)* sometimes in combination with *javelin (jv)* have been some of the most frequently used marker genes in the past, and for internal organs, the aldehyde oxidase mutant is one of the currently studied marker genes (Janning, 1972, 1975).

For the interpretation of the mosaic data, it is vitally important to know the advantages and disadvantages present in the method used to produce genetic mosaics. Therefore, I would like to refer briefly to those points which are important in the two most popularly used methods for the production of genetic mosaics in situ in pattern formation studies.

The basic principles of these two methods are chromosome elimination and mitotic recombination; both have been known for more than four decades.

Zygote

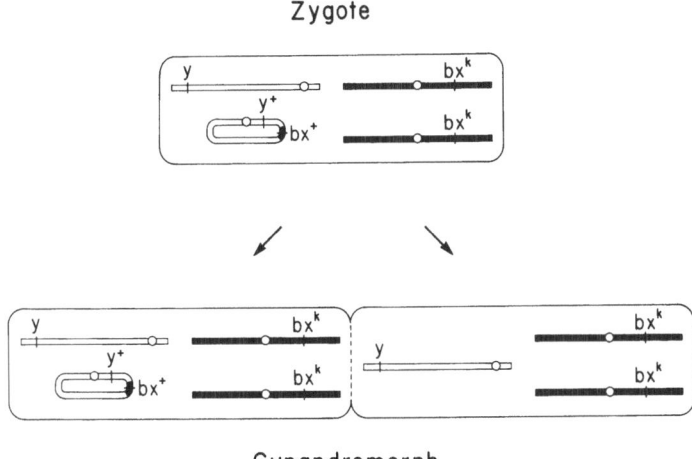

Gynandromorph

Fig. 3. Diagrams illustrating a genetic method to obtain gynandric mosaics for the y, bx mutants through the loss of a special ring-X chromosome with y^+ and bx^+ alleles. *White*, X chromosome; *black*, 3rd chromosome; *empty circle*, kinetochore; bx^k, represents bx pseudoalleleic recessive mutants. (Modified from Lewis, 1963)

A. Chromosome Elimination

Classical prototype studies on cell lineage by Sturtevant (1929) were done with gynanders induced by the *claret* mutant in *D. simulans*, which eliminates maternal chromosomes in the egg, thus producing XO, XX gynanders and diplo, haplo-IV mosaics, among other chromosomally abnormal flies. Similarly, a meiotic mutant, *claret nondisjunctional*, and other mutants in *D. melanogaster* were also used for gynandric mosaics. The more popular method has been the utilization of the ring-X chromosome of *D. melanogaster*. It was originally discovered and studied by L. V. Morgan (1926, 1933) and was found to be useful for the production of gynanders due to the spontaneous elimination of the ring-X chromosome from the female chromosome set of the zygote, constructed to have one normal and one ring-X chromosome during the early zygotic mitosis, or cleavage stage. A high frequency of gynanders can be obtained by introducing a genetic change or by environmental manipulations, such as an additional inversion or aging the female (see Hannah, 1955; Hinton, 1955). These methods have in common, therefore, the facts that genetic mosaicism is initiated spontaneously at the very early zygotic division of the nucleus, and that the marker gene must be on either the sex chromosome (gynander) or on the 4th chromosome (haplo-IV), although the pattern gene need not originally be on one of these chromosomes. One of the early studies of this kind illustrated in Figure 3, by which method a *bithorax (bx)* pseudoallelic mutant's spot of XO cells marked with y gene was successfully produced among the non-*bx* XX female cells, which are non-*y* (Lewis, 1963).

B. Mitotic Recombination (Somatic Crossing-Over)

This phenomenon allows us to utilize marker genes not only on the X but also on all chromosome arms, depending only on the availability of suitable marker genes for the specific pattern gene under study. X-ray-induced mitotic recombination is the most frequently used method since it not only increases the frequency of mitotic recombination, but also allows control of the induced clone size and approximate time of recombination by X-irradiating during various stages of development. However, as Becker (1956) demonstrated, there can be a time lag of the actual mitotic recombination events between two homologous chromosome pairs even in one cell of the eye disc. The time lag of one cell division will induce a spot one-half the size as the result, even though the time of irradiation was the same. Also precautions should be taken in the use of this method, as identification of the genotype by the marker gene phenotype may be difficult for several reasons. For instance, it was found in the early days, mainly from data concerning the X chromosome, that mitotic recombination occurs more frequently in the proximal heterochromatic region of the chromosome arm than in the euchromatic chromosome region (see Becker, 1969). When the marker gene is placed distal to the pattern gene on the same chromosome arm, the mitotic recombination between the kinetochore and the pattern gene will give rise to a cell homozygous for the pattern gene and for the marker gene. When mitotic recombination occurs between the pattern gene and the marker gene, the resulting cell with the homozygous marker gene will be heterozygous for the pattern gene. Thus, the phenotype of the marker gene spot may not indicate the genotype of the pattern gene.

It was found later that the right arm of the 3rd chromosome (3R) does not follow the general rule of high mitotic recombination frequency between the kinetochore and marker gene relative to the frequency between the marker gene and more distal area. Proximal mitotic recombination, when marked with *Kinked* (*Ki*, 3–47.6) or *Multiple sex comb* (*Msc*, 3–48.0), is much less frequent than that of the distal mitotic recombination frequency (Garcia-Bellido, 1972; Tokunaga, 1972). Furthermore, it was found that the proximal and distal mitotic recombination ratio is changed depending on the larval age at the time of irradiation for the induction of mitotic recombination on the 2L chromosome marked with *Tufted* (*Tft*, 2–53.2) or *Bristle* (*Bl*, 2–54.8) and identified in mosaic spots of mesothorax (Tokunaga and Arnheim, 1966). A similar phenomenon was not found in abdominal tergite spots (Garcia-Bellido and Merriam, 1971 c), and the reason may lie in the fact that the histoblasts, which later develop into a tergite, arrest their mitotic activity during the larval X-irradiation period, and do not divide until pupation (see also Guerra et al., 1973). For more details concerning mitotic recombination, see the review by Becker (1976, and this volume).

Thus, precautions should be taken to eliminate uncertainties in the identification of genotypes and also in the interpretation of the mosaic data. The direct effect of X-irradiation on morphogenesis also cannot be neglected (Waddington, 1942; Schweizer, 1972; Postlethwait and Schneiderman, 1973).

Nevertheless, the availability of genetic manipulations for the production of genetic mosaics far outweighs these problems and is one of the advantages in the study of *Drosophila*. Sample schedules for such manipulations are shown in Figures 3 and 4.

IV. Genetic Mosaic Studies to Discover Prepattern Mutants

A. Autonomous Gene Action in Mosaics

The first experiment with mitotic recombination-induced mosaics in the search for a prepattern mutant focused on the *engrailed* (*en*, 2–62.0) mutant, which was originally recognized by its engrailed (scalloped) scutellum. The pleiotropic effects of *en* affect the wing and also the male first leg, introducing a secondary sex comb at the mirror image position of the normal (primary) sex comb on the first tarsal segment; that is, the distal area between the longitudinal bristle rows 1 to 3 (see Figs. 6 and 15 b). The question asked was whether the secondary sex comb can differentiate within a small *en* spot on the otherwise non-*en* mosaic male tarsus. A preliminary experiment with the bristle color marker gene *straw*^{3} (*stw*^{3}, 2–55.1) on the same chromosome arm with *en* locus was not as satisfactory as with the *y* marker gene (1–0.0), mainly because the yellowish color of the normal transverse row bristles made it difficult to distinguish the bristle color of *stw*^{3} from the non-*stw*^{3} transverse row bristles. Thus, two experimental schedules were used to provide the *y* phenotype for the induced *en* clone, as shown in Figure 4 a, b. Both experiments produced desirable mosaics, the latter (Fig. 4 b) yielding more mosaics than the former (Fig. 4 a) experiment, simply because of the viability difference based on the genetic constitution (Tokunaga, 1961).

The results were interpreted as autonomous *en* gene action for the differentiation of secondary sex comb teeth, since *even a small en spot on the appropriate site differentiated a tooth or teeth.* Thus, the interpretation of the data was that the prepattern for the secondary sex comb singularity was present in the posterior half of a non-*en* leg disc when the *en* cell was induced by mitotic recombination, but the non-*en* cells were not competent to form the secondary sex comb, whereas the *en* cells were. Thus, the anterior and posterior halves of the leg disc appeared to have identical prepatterns, in mirror image arrangement.

In Tables 1 and 2, the early genetic mosaic studies of pattern mutants on the thorax, wing, head, antenna, and legs in *D. melanogaster* are summarized (Table 2 includes also the recent studies). Most of them indicated the same conclusion: the presence of an invariant prepattern singularity (i.e., genetically unaffected), but a difference in competence of the mutant cell, as indicated experimentally by the autonomous gene action of the mosaic cells.

In these mosaic studies, the general findings of autonomous gene response to an invariant prepattern do not necessarily mean that all the successive prepatterns during the subsequent development of the mosaic cell after its initiation are also invariant between the two different genotype tissues. Possibly the trouble in finding mutants affected in a prepattern singularity through the use of mosaic studies comes from the basically regional and autonomous nature of development itself in one disc. Thus, if a prepattern mutant were to cause the change of a singularity at a certain developmental stage after the mutant cell was founded in a nonmutant disc, the singularity change would be restricted to the nearby region, most likely occupied by cells of the same genotype. Thus, it would appear that the gene acts regionally in the surviving fly, not changing the whole disc organization as described earlier. If this assumption is correct, then it should be possible to find

Table 1. Early genetic mosaic studies of pattern genes on thorax, wing and head structures

Mutant (locus)	Genotype of pattern gene mosaic		Mosaic origin [X-ray doses]	Pattern criteria	Pattern gene action	Author
	spot	background				
achaete (1–0.0)	*ac*	*ac/+*	Gynander	Missing dorsocentral bristles	Autonomy and conditional nonautonomy in bristle position	Stern (1954) Roberts (1961)
			Induced m.r., 48–72 h age [1600–1800r]	Displacement of anterior dorsocentral bristle	As above	Claxton (1969)
aristaless (2–0.01)	*al/al*	*al/+*	Induced m.r., 24–72 h age [1300r]	Direction of posterior scutellar bristle	Autonomy or non-autonomy depends on the growth pattern change on scutellum by *al* gene autonomously	Tokunaga and Stern (1969)
dumpy (2–13.0)	*dp/dp*	*dp/+*	Induced m.r., 24–96 h age [2000r]	Vortex formation, abnormal bristle arrangement	Autonomy and conditional non-autonomy in bristle direction and position	King (1964)
hairy (3–26.5)	*h/h*	*h/+*	Spontaneous m.r.	Extra hair on scutellum	Autonomy	Stern (1954)
Hairy-wing 49c (1–0.0)	$+/+$ Hw^{49c}/Hw^{49c}	$Hw^{49c}/+$	Induced m.r., 37–96 h age [1200–2400r]	Extra bristles on thorax	Nonautonomy of +, semi-autonomy of Hw/Hw on central area	Gottlieb (1964)
$l(1)N^B$ (1–3.0)	fa^{no}/fa^{no}	$l(1)N^B\ ffa^{no}$	Induced m.r., 46–70 h age [1500r]	Reduced bristle number on thorax	Autonomy	Arnheim (1967)
scute (1–0.0)	*sc*	*sc/+*	Gynander	Missing scutellar bristles	Autonomy of *sc*	Young and Lewontin (1966)
scute alleles	*sc*	*sc/+*	Gynander	Missing bristles on head, abdomen, scutellum (scutellum)	Autonomy of sc^{D2}, sc^5, sc^8	Merriam (1969)

Theta (1-duplication)	$M^+ \cdot \theta/M^+$	$M \cdot \theta/M^+$	Spontaneous m.r.	Differentiation and position of interalar bristle	Autonomy and conditional nonautonomy	Stern (1956a)
Tufted (2–53.2)	+/+	Tft/+	Induced m.r., 46–70 h age	Extra bristles on thorax	Autonomy	Arnheim (1967)
bithorax alleles (3–58.8)	bx/bx	bx^+; bx/bx	Gynander	Changes of metathorax structures to mesothorax	Autonomy (of $a = bx^3$, $e = pbx$, $C = Ubx$)[a]	Lewis (1963)
	bx/bx	bx^+; bx/bx	Induced m.r. late 3rd instar [900–1200r]	Changes of metathorax structures to mesothorax	Autonomy in a and e mutants	Lewis (1964)
cut⁶ (1–20.0)	ct^6	ct^6/+	Gynander	Scalloping of wing margin and abnormal marginal bristles	Autonomy in general and nonautonomy when dorsal and ventral genotype is different at the wing margin	Williams (1968)
scalloped (1–51.5)	sd	sd/+	Gynander	As above	As above	Williams (1968)
scute¹ (1–0.0)	sc^1	sc^1/+	Gynander	Missing ocellar bristle	Autonomy	Stern and Swanson (1957) Roberts (1964)
spineless-aristapedia (3–58.5)	ss^a/ss^a	ss^+; ss^a/ss^a	Induced m.r. 1st instar [1200r]	Changes of arista to tarsal structures	Autonomy and conditional nonautonomy	Roberts (1964)

m.r. = mitotic recombination; [a] = means represented by.

(a)

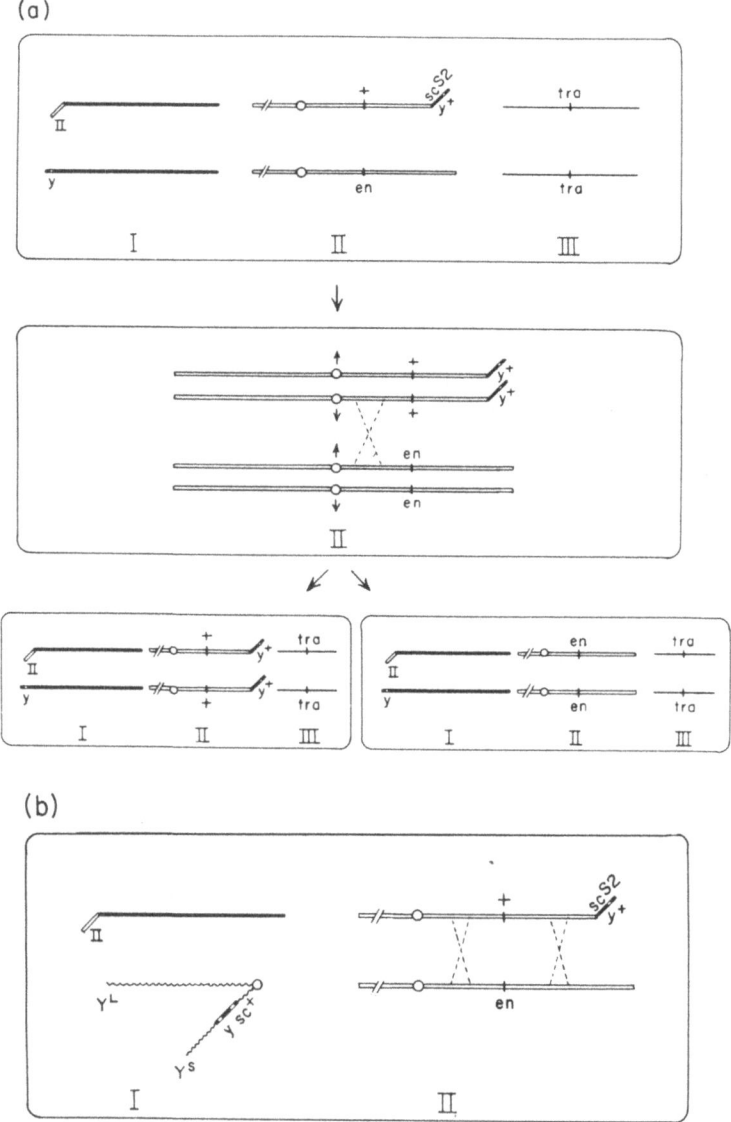

Fig. 4a and b. Diagrams illustrating the two experimental methods to obtain genetic mosaics for the *y; en* mutants through the use of induced mitotic recombination by X-ray. (a) Experiment *I. Top:* A cell with genetic constitution of first (*I*), second (*II*), and third (*III*) chromosome in heterozygous *en* larva before X-ray. *Center:* The position of mitotic recombination between the kinetochore and *en* locus is indicated by *dotted cross lines. Bottom:* Resulting two daughter cells after the mitotic recombination, left cell is non-*y*; non-*en*, right cell is *y*; *en*. The presence of *tra/tra* genotype changes the female phenotype to that of the male, thus the sex comb can differentiate. (b) Experiment *II*. Genetic constitution of the first and second chromosomes. *Black bar*, X chromosome; *wavy line*, Y chromosome; *Y^L*, long arm of the Y chromosome; *Y^S*, short arm of the Y chromosome; *white bar*, second chromosome; *black thin line*, third chromosome; *dotted cross lines*, possible positions of mitotic recombination on 2R (the *left* position is more frequent than the *right*); *empty circle*, kinetochore; *sc^{S2}*, scute bristle effect; *tra, transformer*. (After Tokunaga, 1961)

Table 2. Genetic mosaic studies of pattern genes on the leg

Mutant (locus)	Genotype of pattern gene mosaic		Mosaic origin [X-ray dose]	Pattern criteria	Pattern gene action	Reference
	spot	background				
XO genotype	XO	XX	Gynander	Sex comb teeth on first leg	Autonomy, conditional nonautonomy in tooth direction	Stern and Hannah, 1950; Tokunaga, 1962
sexcombless (1-rearrangement)	sx	sx/+	Gynander	Reduced sex comb on 1st leg	Autonomy	Mukherjee and Stern, 1965
	sx/sx	sx/+	induced m.r. 22–150 h age [1000–1800r]	As above	Autonomy	As above
engrailed (2–62.0)	en/en	en/+	Induced m.r. 34 94 h age [1500–1800r]	Secondary sex comb teeth on 1st leg	Autonomy	Tokunaga, 1961
extra sex combs (2L)	esc/esc	esc/+	Induced m.r. 27–130 h age [700–2400r]	Sex comb on 2nd and 3rd legs	Autonomy	Tokunaga and Stern, 1965
Multiple sex comb (3–48.0)	+/+	Msc/+	Induced m.r. 74–72 h age [1500r]	Sex comb on 2nd and 3rd legs, reduced sex comb size on 1st leg.	Autonomy	Tokunaga, 1972
eyeless-Dominant (4–2.0)	+/+	ey^D; +/+	Induced m.r., 0–72 h age [200–1500r]	Multiple sex comb	Nonautonomy	Stern and Tokunaga, 1967
four jointed (2–81) fj/fj	fj/fj	fj/+	Induced m.r., 24–72 h age [1200–1500]	Four jointed tarsi and their bristle pattern	Autonomy and conditional non-autonomy in tarsal segment formation and bristle pattern	Tokunaga and Gerhart, 1976
dachs (2–31.0)	d/d	d/+	As above	As above	As above	As above

examples of nonautonomy only at the border area where genotypes meet, at which location a mutant genotype, which changes the singularity during development, might influence the morphogenesis of the neighboring different genotype tissue. As will be shown later, this assumption seems to be correct in some cases of exceptional nonautonomy found in the mosaic studies.

Thus, it may be concluded that each genetic mosaic part develops according to its own prepattern successions, with the cell competence responding to specific singularities at each step, resulting in ever more diverse and refined differences in the resulting pattern. In other words, the succession of prepatterns involving the cell's competence in each step of mosaic disc development is supposed to be mosaic in nature, as far as any singularity change at a certain stage by a mutant is involved.

In addition to the main factor mentioned above, another factor which precluded the finding of prepattern mutants and led to the observation of the invariant prepattern singularity in most of the genetic mosaic studies, may be the time of the mosaic spot initiation relative to the time of pattern gene action for the specific pattern criteria studied during development. After finding the autonomous and region-specific nature of gene action for pattern genes, one realizes that the selection of bristle formation as a sole criterion for the search of prepattern mutants was unfortunate because of the fact that the final bristle determination occurs rather late in disc development as described in the following cases.

In the early studies with genetic mosaics induced by mitotic recombination at certain developmental stages, the pattern mutants studied were mostly concerned with bristle patterns on the thorax and on the leg, the latter often concerning sex comb teeth differentiation in a male. A bristle organ derives from four cells, after two cell divisions of a bristle mother cell. Lee and Waddington (1942) could first distinguish two cells (trichogen and tormogen) out of the four unit cells at 15 h after pupation by histologic studies. According to recent ultrastructural observations of the leg disc, the four-cell unit of a bristle organ is recognizable at 24 h after pupation, but not at the time of earlier observation at 5 h after pupation (Reed et al., 1975). Thus, bristle differentiation is a very late developmental event.

As shown in Tables 1 and 2, all the mosaic spots in the listed experiments were induced at least before 96-h larva, except in the study of *sexcombless (sx)* mosaics induced by X-ray. In this experiment, X-rays were applied between 22 h and 150 h after oviposition, the last being equivalent to the normal 128 h at $25°$ C, since the development of this specific genotype fly is slowed down. No mosaics were found among the specimens irradiated before 41 h of age, but it is not clearly stated whether the mosaic spot that was induced during the very late 3rd instar larval period was included among the mosaics analyzed in this report. In the study of the homoeotic mutant, *extra sex combs (esc)*, which differentiates the primary sex comb on all first tarsal segments of the three pairs of male legs (at the distal area between the longitudinal bristle rows 6 to 8, see Fig. 5 b–d), the X-irradiation of larvae older than 85 h did not produce genetically marked mosaic spots, although the irradiation was applied until the time of pupation (Table 2, 4th mutant). Instead, only the differentiation of teeth without the marker gene y was observed. Since similar y^+ teeth also differentiated in the earlier irradiated specimens, we assumed that this type of tooth was caused by some phenomenon other than mitotic recombination and, thus, excluded these cases from the genetic mosaic data (Tokunaga and Stern, 1965).

The autonomy of bristle pattern mutants may indicate that at least until 96 h the prepattern singularity for the future sex comb, or for the specific bristle differentiation, is still invariant. Furthermore, the autonomous differentiation of teeth by an *esc* mosaic spot does not necessarily mean that the *esc* gene did not change the later prepattern singularity for teeth differentiation after the *esc* mosaic cell was introduced. In gene combination studies with another extra sex comb gene, *esc^D*, we suggested that the *esc* may act as a late prepattern mutant. We based the suggestion on the assumption that *esc* provides the singularity for the secondary sex comb teeth on the 2nd and 3rd legs when in combination with the *en* gene, since *en* itself cannot differentiate these structures on the 2nd and 3rd legs (Tokunaga and Stern, 1965). A similar relation exists between *en* and *Msc* genes in differentiation of the secondary sex comb teeth on the 2nd and 3rd male legs (Tokunaga, unpublished), although each gene was found to act autonomously in the genetic mosaic studies for teeth differentiation (Table 2, 3rd and 5th mutants).

Garcia-Bellido and Merriam (1971a) reported that a mosaic spot induced earlier than 8 h before puparium formation (PF) behaves autonomously as far as the *Hairy-wing (Hw)*, *hairy (h)*, and *ac* gene actions on bristle formation in the wing disc. After this specific period of development, if the spot was initiated by X-ray-induced mitotic recombination, the differentiation of bristle was not changed despite the changed genotype of these mutant alleles, although marker genes used in these studies (*sta*, *sn^3*, *y* and *mwh*) manifested themselves after the change of bristle pattern genotypes. (They introduced the term "perdurance" to designate the persistence of a cellular developmental fate for several cell generations after the loss of the genetic basis for that cellular development.) Thus, *Hw*, *h*, or *ac* genes act before the critical period (8 h before PF) in the mesothoracic disc to assure the mutant phenotype, although the time of the determination of a bristle cell was suggested to occur 40 h before pupation. In contrast to the time of bristle determination in the wing disc, it occurs sometime during the pupal expantion for the growing imaginal cell population of the tergite (Garcia-Bellido and Merriam, 1971c).

The data for a similar time relation are not available for bristle or sex comb tooth differentiation on the leg, but one can surmise from the studies cited above that in tests of prepattern mutants, mosaic spots were initiated almost exclusively before the time of final irreversible bristle cell determination under normal conditions. The same is true for the gynandric mosaic studies.

The genetic mosaic studies with homoeotic *bx* mutants, induced by mitotic recombination during the embryonic and larval stages, also revealed autonomous mutant effects. The criterion for a small mosaic spot induced *near the end of the 3rd larval instar* was the differentiation of bristles marked with the *y* and *sn* genes. These experiments suggested that the wild-type alleles of at least the a gene (which changes anterior metathorax to anterior mesothorax, represented by *bx^3*) and the e gene (which changes posterior metathorax to posterior mesothorax, represented by *pbx*) continue to function late in larval development, preventing a mesothoracic-like transformation of even a few cells of the dorsal metathoracic imaginal disc

(Lewis, 1964). In other words, the *bx* gene acts autonomously, even in the late larval stages when its action is expressed in differentiating the characteristic bristle of mesothorax on metathorax.

Another homoeotic mutant *ss^a*, which transforms the arista into tarsal segments of the middle leg, showed autonomous gene action when the *ss^a* cells were induced during the first instar larva (Roberts, 1964). As will be discussed later, it was found that the final decision to differentiate into tarsal instead of antennal structures seems to be made between 72–96 h of age. Thus, for the reasons discussed above, it is not surprising to find generally autonomous gene action for *ss^a*.

B. Nonautonomous Gene Action in Mosaics

Because of the autonomous and region-specific nature of gene action generally found in genetic mosaic studies, it seemed hopeless to identify prepattern singularity changes by the use of genetic mosaics. However, thanks to several exceptional cases of nonautonomy found even in early studies of genetic mosaics, we were able to claim that the change of a prepattern singularity (or positional information specification) for pattern formation can occur. This section reviews this aspect of the genetic mosaic studies.

1. Bristle Direction and Location

Some cases of nonautonomy were expressed in mosaics under restricted conditions, and concerned the direction and the location of bristles. Although the majority of mosaic individuals showed autonomous direction or location in these cases, nonautonomy occurred when the tissue patch was specifically oriented with regard to the background tissue of the other genotype in the mosaic fly.

The first case was found in the abnormal direction of sex comb teeth, which autonomously differentiated in a narrow stretch of X0 male tissue extended to the sex comb site of otherwise XX female tissue (Stern and Hannah, 1950). This nonautonomy of teeth direction in a small X0 spot proved to be the result of differences of growth pattern between the X0 and XX tissue in a mosaic leg disc, as shown by the study of cell lineage relations in the bristle pattern of the first tarsal segment of the male first leg, by the use of genetic mosaics induced by mitotic recombination (Tokunaga, 1962). From this cell lineage study, it became clear that the sex comb in the male is a structure homologous with the distal transverse rows of the female first tarsus of the first leg. The main growth pattern difference in this segment between male and female is found in the distal transverse rows area (Figs. 5a, b and 6) where: (1) In the male, a sex comb structure differentiates instead of the last two distal transverse rows of bristles of the female. (2) The number of distal transverse row bristles or of sex comb teeth in the male is more than that of the corresponding transverse row bristles of the female. Thus, one can observe that, in the male, the number of distal transverse row bristles and of sex comb teeth has been increased. (3) *The orientation* of the sex comb suggests that the sex comb area of the male *shifts almost 90° C* from the transverse row

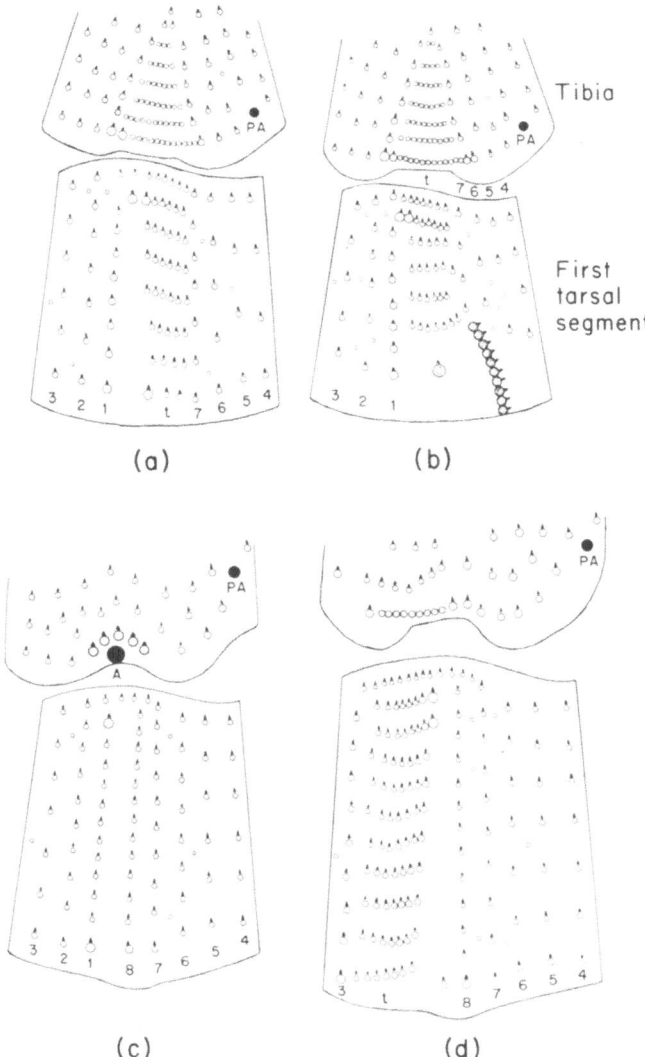

Fig. 5 a–d. Schematic bristle pattern of the first tarsal segment and distal part of tibia (left leg). (a) First leg, female; (b) first leg, male; (c) second leg, male; (d) third leg, male. *A*, apical bristle; *PA*, para-apical bristle; *C* central bristle; *cross-hatched*, sex comb teeth; *empty circle*, bractless bristle; *empty circle with triangle*, bracted bristle. *1–8*, longitudinal bristle rows; *t*, transverse row. [(a) After Stern, 1968; (b–d) after Tokunaga and Stern, 1965]

position. This shift is not likely to occur by the migration of each sex comb tooth cell, but rather is associated with the overall movement of this entire distal tissue area during morphogenesis of the tarsal segment.

    The growth pattern difference in this area between the male and female may be related to the *difference in tarsal length* which is observed in the wild-type fly. According to the data from the wild-type hybrid of Samarkand and Oregon

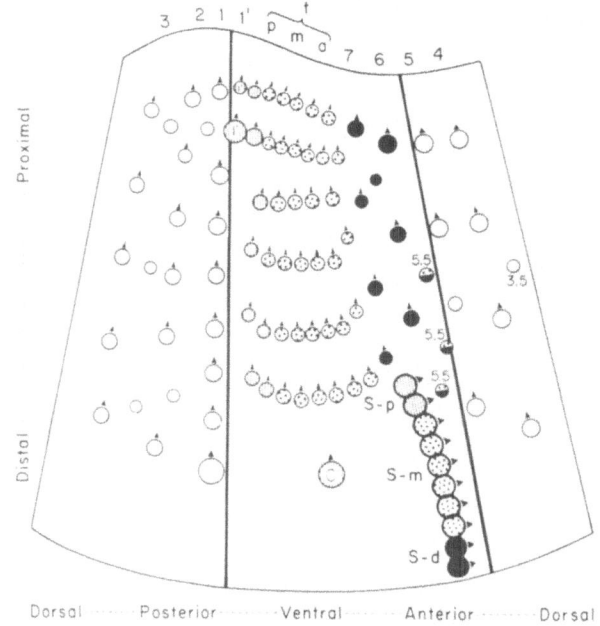

Fig. 6. Schematic representation of the cell lineage relationships on the first tarsus of the male first leg (*left*). The homology between the bristles of transverse rows (*t*) and sex comb (*S*) is shown by differently marked bristles in the area bordered by two proximodistal solid clonal lines between *1'* and *6*. *C*, central bristle; *S–p, S–m, S–d*, proximal, medial and distal sex comb teeth; *t(a, m, p)*, anterior, medial, posterior transverse row area; *1, 1', 2, 3, 4, 5, 6,* and *7*, longitudinal bristle rows; *3.5, 5.5,* and *circle*, bractless bristle; *circle with triangle*, bracted bristle. (Modified from Tokunaga, 1962)

stocks observed by Hannah-Alava (1958a), the female tarsal length is longer than that of the males in each of three pair of legs that were compared. When the male/female ratio of the length of the first tarsal segment is compared, the data are 0.89 for the 1st, 0.98 for the 2nd, and 0.97 for the 3rd leg, respectively, which indicates that the first tarsal segment with the sex comb is relatively short. The following examples, among others, also support this suggestion: (1) In *Msc* mutant male legs, the number of sex comb teeth of the first leg is distinctly smaller than that of control wild-type males (average 5.9 versus 10.75), and the length is longer than that of the wild type (average 0.19 mm versus 0.18 mm, and this difference is backed by comparisons of the ratio between the length of the first tarsus of the 1st and 2nd legs, and of the 1st and 3rd legs; see Tokunaga, 1966). (2) In combinations of the homoeotic mutant *esc, Polycomb (Pc)*, and *Extra sex comb (Scx)*—each differentiates extra sex comb teeth on the 2nd and 3rd legs in the corresponding position of the 1st leg sex comb—the expressivity in the differentiation of teeth and also transverse rows became high. The higher the synergistic effects of these gene combinations, the greater the transformation from the typical first tarsal segment of the 2nd or 3rd leg to that of the 1st leg, accompanied by shortening of the segment and increases of sex comb teeth and transverse rows as illustrated in Figure 8 (Hannah-Alava, 1958b).

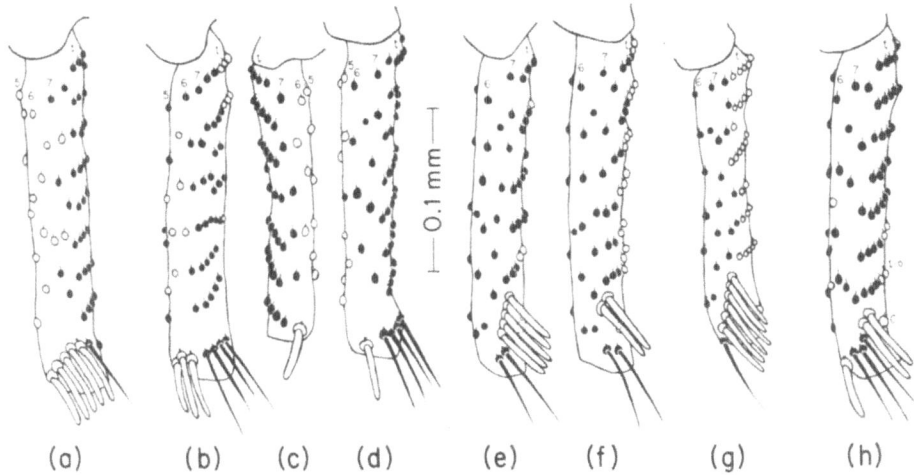

Fig. 7a–h. Gynandric first tarsal segment of first leg (anteroventral view). (a–d) Abnormal direction and position of sex comb teeth. Black bristles (XX) are present in or next to the sex comb area in the proximal or posteroventral position to the sex comb (X0). (e–g) Nearly normal direction and position of sex comb teeth. Black bristles are present in the sex comb area distal to the sex comb. (h) Abnormal direction of distal tooth and abnormal position of the last transverse row bristles of XX genotype. Black bristles are present between the proximal and distal teeth of the sex comb. *Black bristles*, non-*yellow*, XX; *white bristles*, *yellow*, X0; *black (or white) circle with or without bar*, position of the black-XX (or white-X0) bristle with or without bract. The bracts of sex comb teeth are indicated only when the phenotype of the bract differed from that of the tooth. (Modified from Tokunaga, 1962)

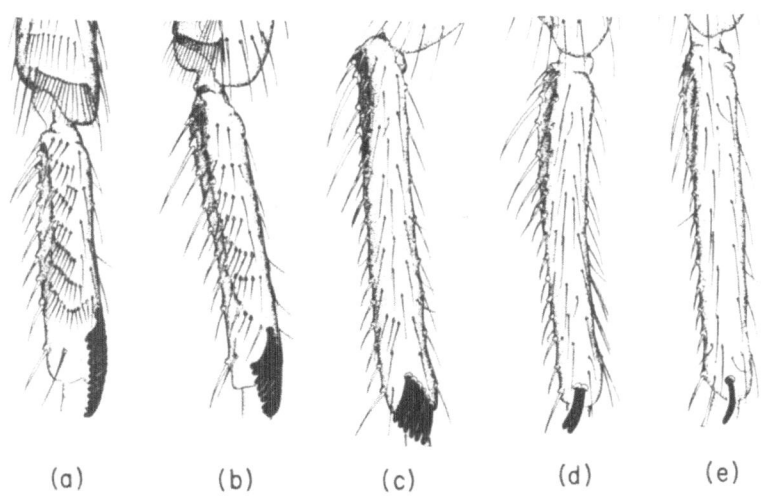

Fig. 8a–e. Illustration of stages in transformation of a male third leg into a first leg through changes in size and chaetotaxy of the tarsus (anteroventral view). (a and b) *esc; Scx;* (c) *Scx Pc;* (d and e) *Scx.* (Modified from Hannah-Alava, 1958b)

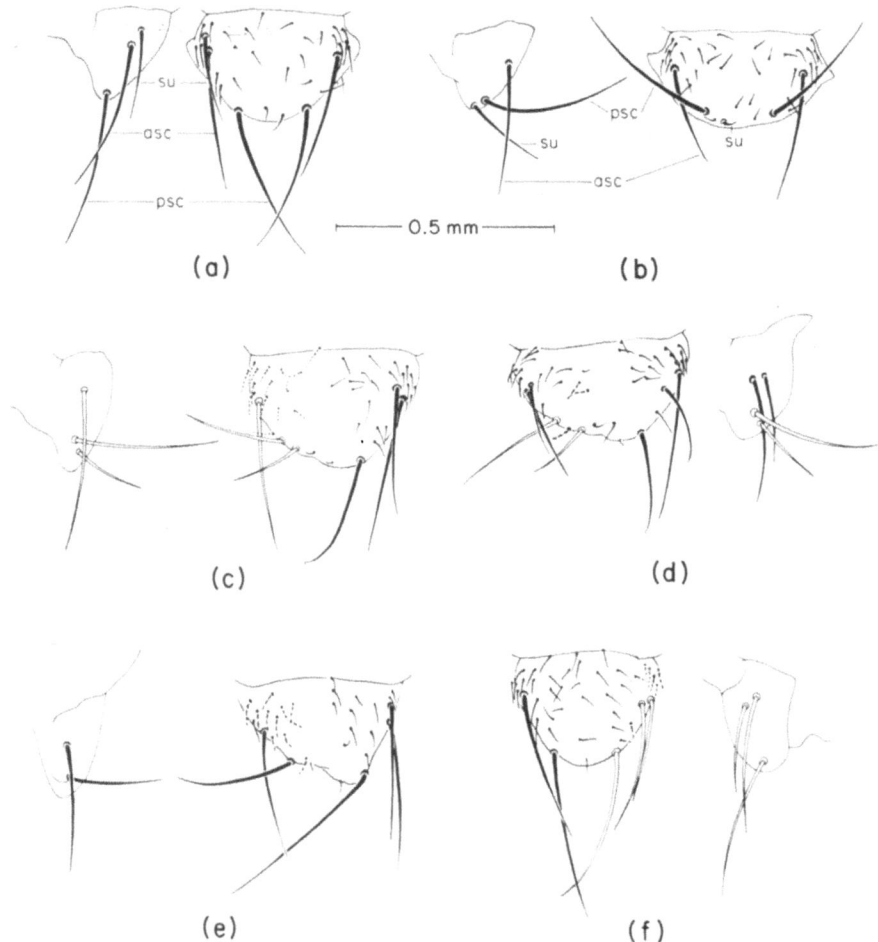

Fig. 9 a–f. Illustrations of the direction of the posterior scutellar bristle *(psc)* in *aristaless (al)* mosaics, depending on the growth pattern change due to the autonomous action of the *al* gene. (Lateral and dorsal view of scutellum.) (a) *y; h* male; (b) *y; al; h* male; (c–f) mosaic scutella with *y; al; h* spot. In mosaics, solid chaetae: non-*y*, non-*al*, *h*. Chaetae in *outline* or shown as *dotted lines: y; al; h*. Note the depression of the posterior edge of the left mosaic scutellar half in (c), (d), and (e) and the *al*-like direction of psc, as opposed to the lack of the depression in (f) with the normal direction of psc. (c and d) Autonomous expression of psc bristle (females). (e and f) Nonautonomous expression of psc bristle (males). *asc*, *psc*, anterior, posterior scutellar bristle; *su*, supernumerary bristle. (Modified from Tokunaga and Stern, 1969)

Thus, in gynandric mosaics the number of sex comb teeth depends on the size of the X0 genotype tissue located at the sex comb site; but the direction of teeth depends on the location of the small male spot in relation to the surrounding female tissue, as illustrated in Figure 7. Reciprocally, the location and direction of female transverse row bristles are also affected by their arrangement amidst male tissue as shown in Figure 7h. This is a clear example of a growth pattern difference based on tissue displacement, either by growth or movement.

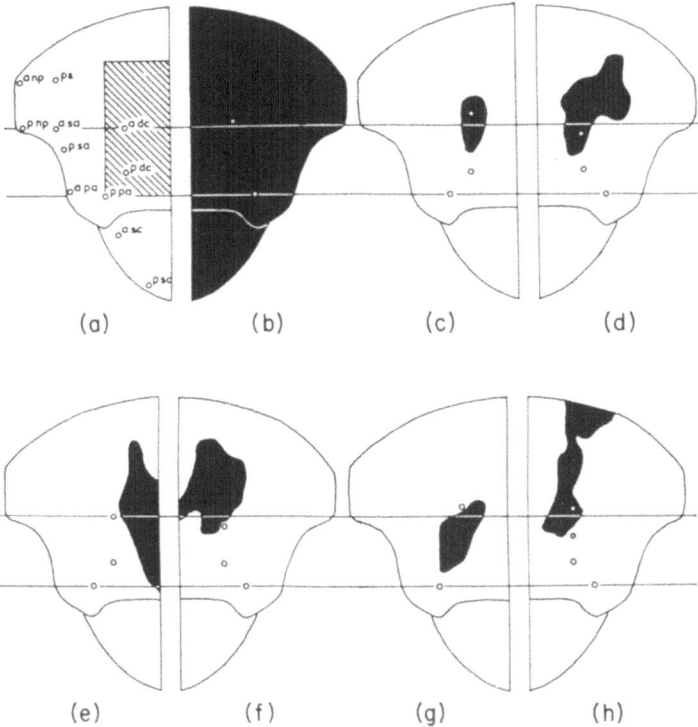

Fig. 10a–h. Half thoraces of non-*ac* *(white)* *ac* *(black)* and *ac* mosaic *(black* and *white)* flies illustrating the relative position of the dorsocentral and posterior postalar bristles. (a) The positions and abbreviated names of the mesonotal bristles. Only mosaics with *ac* tissue patches in or entering the hatched area were analyzed. (b) The average position of adc in *ac* flies. (c) A mosaic with *ac* adc anterior to the typical non-*ac* position. (d) A mosaic with *ac* adc posterior to the typical non-*ac* position. (e) A mosaic with non-*ac* adc. (f and g) Mosaics with *ac* tissue covering the typical non-*ac* adc site, and with displaced non-*ac* dorsocentral bristles. (h) Mosaic with supernumerary non-*ac* dorsocentral. *a-pdc*, anterior and posterior dorsocentrals; *a-psc*, anterior and posterior scutellars; *ps*, presutural; *a-pnp*, anterior and posterior notopleurals; *a-psa*, anterior and posterior supra-alars; *a-ppa*, anterior and posterior postalars. (After Claxton, Genetics **51**, 1969)

A similar relationship was found between bristle direction and growth pattern in the mosaic studies of the *aristaless* *(al)*; some of its pleiotropic effects include a shortened scutellum and changed direction of the posterior scuteller bristles (psc; Fig. 9b). A mosaic analysis has shown that the psc bristle direction is affected by a growth pattern change in the scutellum, not by the orientation of bristle cell, since, if a psc bristle with non-*al* genotype is surrounded by an *al* spot, its direction is changed to the *al* type (Fig. 9e). The opposite situation is also true in which the direction of an *al* bristle changes to normal in mosaics (Fig. 9f). That is, the direction of the psc bristle depends on the regional growth pattern of the surrounding posterior scutellum area (Tokunaga and Stern, 1969).

These studies indicated that certain genes could affect bristle pattern in an indirect way as a consequence of affecting the growth pattern of an area as their primary action. In mosaics, the interaction of areas differing in their growth pattern could thus lead to an abnormal direction or position of a bristle in tissue of either genotype, and thereby show nonautonomy just at the mosaic boundary. This effect is further illustrated by *dumpy (dp)* mosaics on the thorax. When the *dp* spot autonomously formed a vortex structure on the mesothorax, the adjacent non-*dp* tissue participated in vortex formation nonautonomously, and consequently, its bristle location and direction were affected (King, 1964).

Nonautonomy of bristle position was found earlier in some cases of *ac* mosaics and *Theta* mosaics in combination with *Minute*(1)*n* (Stern, 1954, 1956a). Claxton (1969) also studied *ac* mosaics, placing emphasis on the displacement of anterior dosocentral bristles (adc) by the *ac* gene (Fig. 10). Both of these authors and Roberts (1961) found a few cases of nonautonomy in bristle location of non-*ac* genotype near the borderline of the mosaic spot (see Fig. 10f, g). Stern proposed alternative explanations for this phenomenon. One is the spread of diffusible gene products from nearby tissue (1954), as in the cases of nonautonomy in facet color of *vermilion (v)* mosaic eyes (Sturtevent, 1920), and in *y* bristle color located next to the non-*y* spot on the gynandric abdominal tergite (Hannah, 1953). Another explanation proposed is that the production of a bristle at a specific location within the narrow dorsocentral line suppressed the production of a bristle close to it. When formation of a bristle at its normal position does not occur, then production of a bristle at a new location is possible (1968), a proposal equivalent to Wigglesworth's (1940, 1959) from studies on *Rhodnius*. As an additional possibility, the abnormal position of the bristle could be related to the differential growth pattern between the *ac* and non-*ac* tissue, as suggested by Claxton's (1969) data (e.g., cf. Fig. 10c–h). However, it may be difficult to prove such a relationship mainly because cell lineages on the mesothorax are variable from one sample to the next, especially in gynanders; so that the general relationships only become known by statistical treatment of many samples (Sturtevant, 1929; Noujdin, 1936; Murphy and Tokunaga, 1970; Garcia-Bellido et al., 1976). Moreover, recent mosaic studies with *ac* (Claxton et al., 1976) support the hypothesis of the diffusible gene products.

2. Quantitative Change of Bristle Pattern

The *Hairy-wing (Hw)* mutant differentiates a variable number of extra bristles (macro- and microchaetae) on the wing and thorax as well as on other parts of the fly. In genetic mosaic studies with the extreme allele Hw^{49c}, it was found that the Hw^{49c} gene changes the relative dimensions of the thorax on the central area studied, so that the differentiation of extra bristles occurred nonautonomously by non-*Hw* and semi-autonomously by Hw^{49c}/Hw^{49c} mosaic twin spots (Gottlieb, 1964). Thus, the effects of growth on bristle pattern formation were suggested by these nonautonomy phenomena. Later, the *Hw* gene action in mosaic studies on the scutellum was claimed to be autonomous (Garcia-Bellido and Merriam, 1971a). These results suggest regional differences in growth effect by *Hw* gene.

Another example of pronounced nonautonomy was found for sex comb differentiation in mosaics with wild-type spots on an *eyeless-Dominant (ey^D)* mutant

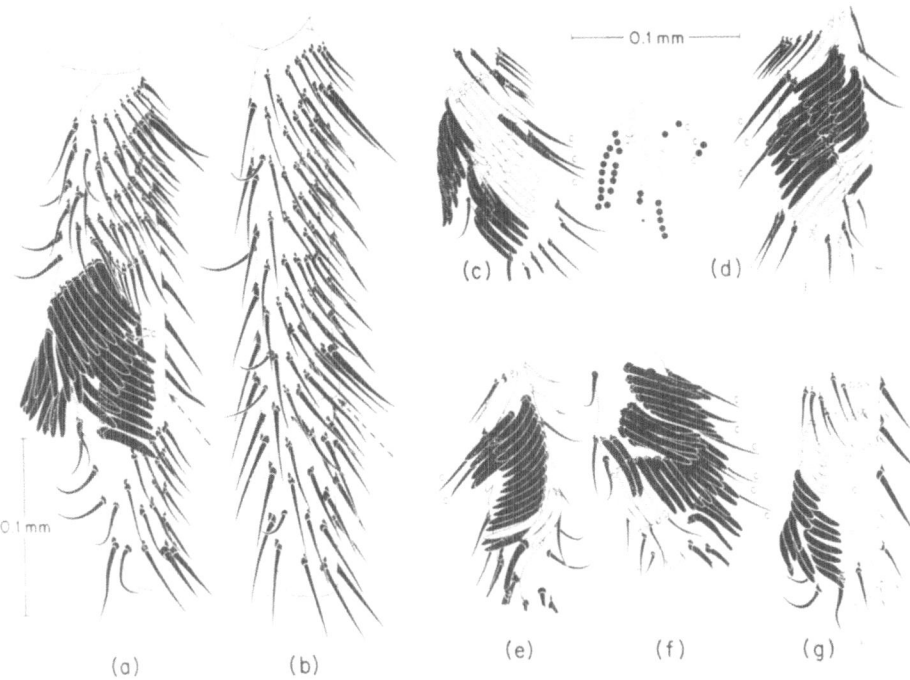

Fig. 11 a–g. Nonautonomy of + spot in ey^D mosaics differentiating ey^D type extra sex combs. First and second tarsal segments of first leg of ey^D male (a), and female (b), and sex comb site of five mosaics (c–g). In mosaics, solid chaetae or thick circle: non-y; ey^D. *Outlined chaetae or empty circle, y; non-ey^D; c*, central bristle; *arrow*, partially segmented site between the first and second tarsal segments. Mosaic (g) could be interpreted as autonomy of + or as low expressivity of ey^D gene since one occasionally encounters similar sex combs in ey^D males in which one row occupies the proximal but multiple rows the distal part of the sex comb.
[(b) After Stern, 1968; (a) and (c–g) after Stern and Tokunaga, 1967]

background. In such mosaics, the wild-type spot produced multiple rows of sex comb teeth, in accord with the surrounding ey^D phenotype of the male first tarsal segment (Fig. 11). This nonautonomy was interpreted as an enlargement of the tarsal region singular for sex comb formation, due to the action of the ey^D gene in the surrounding tissue, with the consequence that wild-type cells localized in the sex comb area encountered the enlarged singularity and responded by forming multiple rows of teeth rather than the normal single row (Stern and Tokunaga, 1967; Tokunaga, 1968, 1970). Thus, ey^D is regarded as a prepattern mutant, which changes the sex comb singularity quantitatively.

Since the ey^D mosaic case was exceptional in that autonomous gene action of the + allele was not proved, other explanations were presented by several authors. Wolpert (1971) suggested that positional information in the first tarsal segment is altered by an altered position of the field boundaries affected by the widening of the first tarsal segment. This, combined with our explanation (Stern and Tokunaga, 1967), suggests that the ey^D gene acts primarily to alter the dimensions of the tarsal region, and thereby secondarily affects sex comb specification.

Lawrence (1973a) suggested that the multiple rows of sex comb teeth in ey^D male could be explained by the speculation "that the key site of action of the mutant is the intersegmental membrane and from this stems the pattern change which results in the development of extra sex comb." This speculation was based on the idea developed by Piepho, Stumpf, Locke, and Lawrence in regard to the membrane separating abdominal segments in several insects other than *Drosophila*. These authors proposed that the boundaries specify the patterning of a segment by establishing and maintaining gradients used to specify position within the segment (see review by Lawrence, 1973a). Tarsal joint failure could also be a mechanism by which bristle pattern is altered in *Drosophila*, as will be shown later. In ey^D males the pronounced enlargement of the distal part of the first tarsal segment is accompanied by partial joint failure with the second segment, which in extreme cases is shortened and altered in its bristle pattern (Fig. 11a).

However, joint failure alone seems insufficient to cause multiple rows of sex comb teeth in ey^D. In the mutant *rotund (rn)*, for example, partial tarsal joints form and yet there are very few or no sex comb teeth (see Lindsley and Grell,

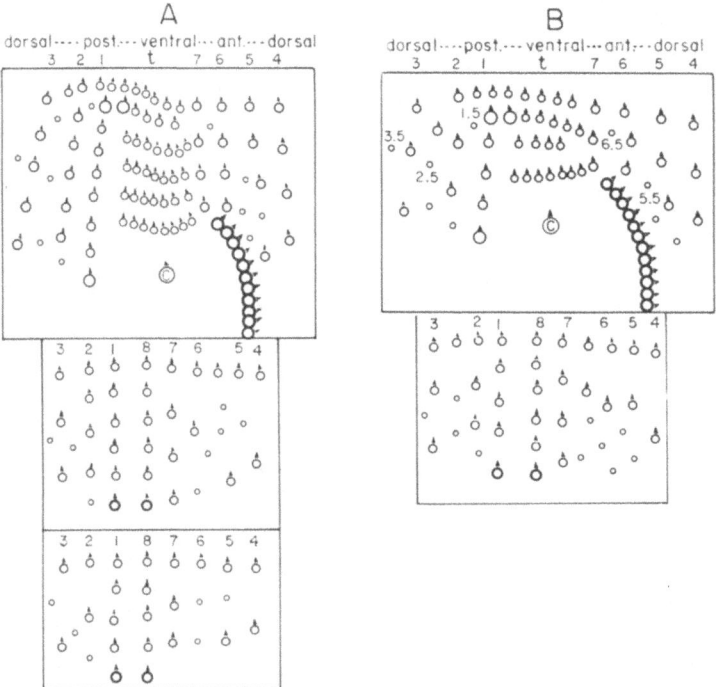

Fig. 12 A and B. Schematic bristle pattern of the proximal tarsal segments of $y; fj/sc^{S2}$ (A), and of $y; fj$ (B) male first legs. The number of bristles in each longitudinal row and the total number of transverse rows are averaged from 30 random samples. The $y; fj$ samples are those with complete third joint failure (between 2nd and 3rd tarsal segments). The bristle number of each t-row and the relative position of the bractless bristles are approximations, as are the length and width of the segments. *C*, central bristle; *t*, transverse rows; *1,2,3,– —,8*, longitudinal rows; *circle*, bractless bristle; *circle with triangle*, bracted bristle; *heavy circle*, sex comb tooth. (After Tokunaga and Gerhart, 1976)

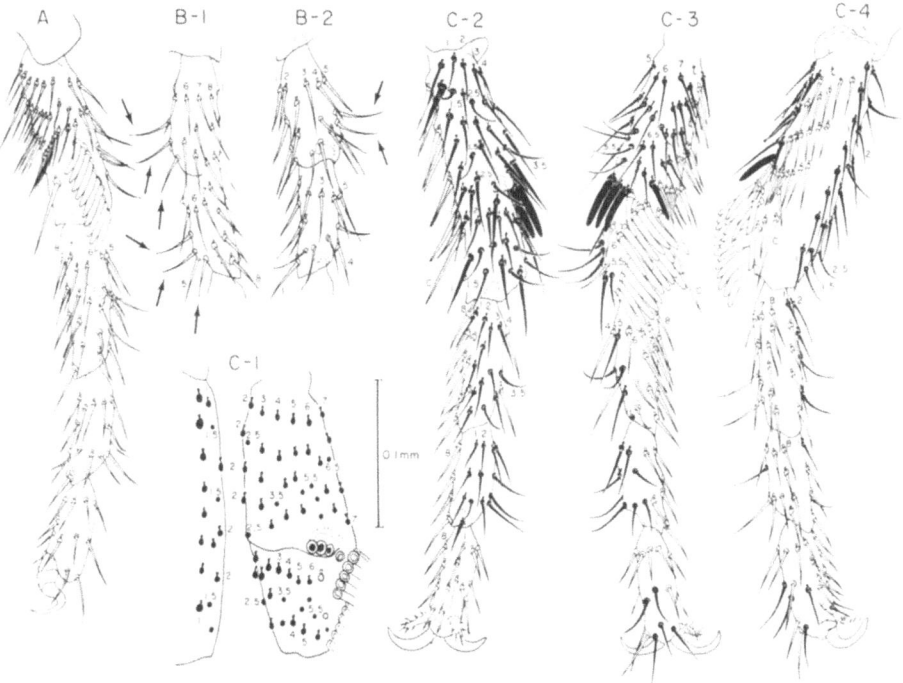

Fig. 13A–C. Partial joint failure and bristle pattern in the male first leg tarsi of *fj* and *d* homozygotes and of a *fj* mosaic. (A) Tarsus of *y; fj* with a small partial joint on the 2nd (2–3) segment. (B) A 2–3 segment of *y; d* with a well-developed partial joint. *Arrows* indicate increased bractless bristles. *B–1*, ventral view; *B–2*, dorsal view. (C) A rare type of *fj* mosaic in which two proximal non-*fj* segments correspond to one proximal *fj* segment. *C–1*, position of bristles on the dorsal *(right side)* and posterior *(left side)* area of the proximal tarsal segment. *C–2, C–3, C–4*, the same mosaic tarsi from different views. Note the nonautonomy of segment boundary formation by the non-*fj* region near the *fj* spot, and also the change of the bristle pattern. The similar nonautonomy is shown in Figure 14A. *Double circle*, *y; fj* tooth; *filled inner circle*, non-*y*, non-*fj* tooth; *open circle*, *y; fj* bristle; *filled circle*, non-*y*, non-*fj* bristle; *triangle*, bract; *C*, central bristle; *t*, transverse row; *1,2,3,——,8*, longitudinal row; *1.5, 2.5,—— —,6.5*, bractless bristle row; *white bristle*, *y; fj*; *black bristle*, non-*y*; non-*fj*. *Dotted area* indicates internal joint structure. These symbols also apply to Figure 14. (After Tokunaga and Gerhart, 1976)

1968; Tokunaga, unpublished). It is possible that excessive, localized growth induced by *ey^D* gene action on the first tarsal segment underlies joint failure. *Recently accumulated experimental data strongly suggest that cell death caused by ey^D gene at certain developmental stages is related to the phenomenon,* as J. Fristrom (1970) suggested earlier. This will be discussed in Section V.

3. Segmental Boundary Formation and Bristle Pattern

Recent studies (Tokunaga and Gerhart, 1976) on *four jointed* (*fj*, 2–81) and *dachs* (*d*, 2–31) mutants revealed that joint formation can be influenced by neighboring cells of different genotypes in mosaics. Here the mutants, both of which

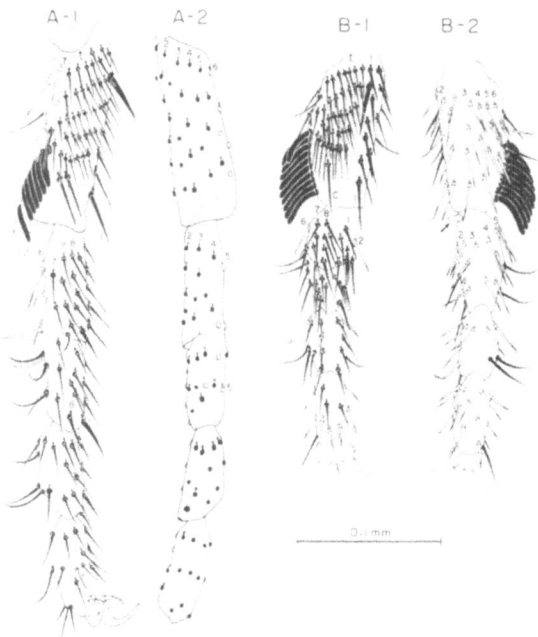

Fig. 14 A and B. Partial joint failure between the 2nd and 3rd tarsal segments and bristle pattern in the male first leg tarsi of *y; fj* mosaics. (A) A commonly observed type of mosaic leg. *A–1* ventral view; *A–2*, dorsal view (bristle position only). Note the bristle pattern change in relation to the nonautonomous failure of joint formation by non-*fj* as observed in Fig. 13 C. (B) Unusually short mosaic tarsi with a big *fj* mosaic spot. *B–1*, ventral view; *B–2*, dorsal view. Note the partial intersegmental boundary in *B 1*, which is accompanied by an increase in non-*fj* bristle numbers on both (proximal and distal) sides of the boundary. It should be noted that in mosaics illustrated in Figures 13 and 14, the tarsal shape was not bent, suggesting some compensating mechanism to keep the mosaic tarsi straight, despite differences in local growth. (After Tokunaga and Gerhart, 1976)

affect the tarsi, reduce them to four segments instead of the normal five. These experiments were designed to study the relationships between bristle patterns, segment formation, and growth patterns in mosaics, since earlier genetic mosaic studies had commonly shown growth pattern changes in examples of nonautonomy, as discussed above.

In contrast to the effect of ey^D, which increases the distal area of the first tarsal segment, the *fj* and *d* genes affect proximal tarsal segments by reducing their length. Thus, they differentiate a short first segment (reduced to ca. $^2/_3$ in length and in the total number of bristles in the proximodistal direction) and an adjacent single segment (2–3) instead of the 2nd and 3rd tarsal segments of the heterozygous fly (Fig. 12). The bristle pattern is changed in these two proximal segments, but the distal two tarsal segments are not affected in size and in bristle pattern as compared to the 4th and 5th tarsal segments in control heterozygous flies (Figs. 12 and 13 A). Among the mutant homozygotes, there are some that develop partial segmental boundaries in the middle of the 2nd (2–3) segment. In these cases, the bristle pattern is also changed on the areas between the proximal or

distal segmental borderlines and the partial segmental border. Homozygotes for
fj rarely differentiate this partial segment, but *d* does so with greater frequency
and with a greater degree of the partial segment formed (Fig. 13 B).

When genetic mosaics of *fj* or *d*, induced by mitotic recombination, were
studied, we found autonomy of segmental boundary failure in both mutant spots
between the 2nd and 3rd tarsal segment of nonmutant heterozygotes in the
majority of cases. Thus, the genes appear to act autonomously with respect to
failure of segmental boundary formation. Similarly, the nonmutant spot forms the
segmental boundary autonomously, but failed to form the segment at the area
located adjacent to the mutant spot, a case of *localized nonautonomy*. When this
happened, the nonmutant tissue differentiated the bristle pattern of its own genotype as if it belonged to a single segment, unlike the bristle patterns of the two
adjacent segments of the same genotype tissue where the partial segment was
formed (Figs. 13 C and 14 A).

This nonautonomy of wild-type tissue clearly suggests that: (1) The mutant
genes *fj* and *d* act in two ways in changing the bristle pattern. One is indirect
through the failure of joint formation; another is direct, resulting in a change of
the pattern in the longitudinal direction in specific tarsal segments, accompanying
the shortening of the tarsal length as mentioned earlier. The common characteristic of these gene effects is that of shortening the specific tarsal segment area, which
suggests the involvement of these genes with the growth pattern of the developing
leg disc. (2) The segment boundary is a key site to define the longitudinal bristle
pattern, as has been suggested for other insects (Lawrence, 1973 a).

Thus, according to the positional information hypothesis, the segmental
boundary could be regarded as a reference point of the proximodistal axis for
bristle pattern differentiation on the tarsal segment. If a tarsal segment represents
a field as Wolpert defines it, the field seems not to be fully regulative, since size
variance in bristle pattern of a special mosaic (Fig. 13 C, compare to the standard
bristle pattern in Fig. 12) seems to be apparent in an unusually long *fj* spot, which
occupies the ventral area of the proximal tarsal segment. Namely, the total number of transverse row units and sex comb teeth is increased more than the corresponding average numbers in *fj* homozygotes. If the transverse row area and sex
comb area are each considered as independent units of the bristle pattern under
discussion, then this bristle pattern change, in step with the change of length of the
tarsus, might be considered to match the French Flag Model of Wolpert (1969),
which is size invariant. However, the same consideration cannot be applied to the
ey[D] bristle pattern (Fig. 11 a), since the transverse rows in the male first leg were
not increased in step with the increase in the sex comb rows.

The discrepancy between the two cases in the compatibility with the positional information hypothesis may lie, among other possibilities, in the not yet
disclosed set or sets of points in the area studied. Earlier, the presence of a set of
hypothetical reference points in the leg was suggested by Hollingsworth (1964).
He considered the leg segments from femur to tarsi as bilaterally symmetric in
their basic bristle pattern and suggested the ventral midline (between bristle rows
1 and 8 in tarsus, see Fig. 5 c as an example) to be the source from which the
diffusion of one of the hypothetical morphogens occurs laterally. If this is the case,
genes affecting leg growth in the lateral direction may cause disturbances in terms

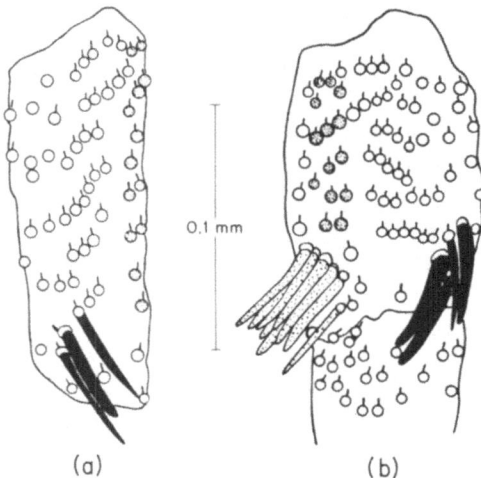

Fig. 15a and b. Male first tarsal segment differentiating secondary sex comb and secondary transverse rows by the effect of gene combinations of sx; en (a) and sx; en; ey^D (b). *Stippled circle with bar*, secondary transverse rows; *stippled bristles*, normal and intermediate teeth of secondary sex comb; *black bristles*, normal and intermediate teeth of primary sex comb. (After Mukherjee, 1965)

of the positional information specification in regard to this reference point. Possibly, ey^D might include such an effect since its additional growth proceeds in both longitudinal and lateral directions (in en; ey^D male legs, secondary sex comb teeth also increase in number, Tokunaga, unpublished). Likewise, the en phenotype, having a second and mirror image sex comb itself, may express itself in this category. The sx phenotype, having a reduced width of transverse rows and number of sex comb teeth, could be due to alterations with regard to this hypothetical lateral reference point and lateral axis, as viewed from the positional information theory. These are highly speculative considerations, but the synergistic effects found in the increase of the secondary sex comb teeth and in the appearance of secondary transverse rows in sx; en and sx; en; ey^D male tarsi (Fig. 15, Mukherjee, 1965) and the increase in secondary sex comb teeth in en; ey^D support these speculations.

Thus, at this point, one can only suggest that mutants such as ey^D, fj, d, or Hw may be associated with changes of regional growth pattern, changing the bristle pattern singularity or positional information during morphogenesis. In the next section, this suggestion will be considered further in relation to recent accumulated knowledge in related fields of study.

The basic consideration in pattern formation discussed so far has been concerned with epigenetic spacial differentness within a bounded multicellular area, represented by a hypothetical prepattern singularity or by specific positional information, and with competence as an expression of the individual genotype during development. The genetic mosaic studies aimed at finding mutations that change this spacial differentness lead to the following conclusions: (1) Regional growth patterns affected by mutants can alter the final bristle pattern of an area.

(2) Tarsal segment boundaries play an important role in bristle pattern formation.
(3) Tarsal segment boundary formation is affected by the growth pattern controlled by cell genotype.

V. Effects of Pattern Mutants on Growth Pattern and Boundary Formation

The following review in relation to the subjects discussed so far will center on mechanisms by which genes affect growth pattern and boundary formation in situ as revealed mainly by the use of genetic mosaics of pattern mutants.

A. Growth Pattern Control by Pattern Mutants

There are several factors known to be involved in the change of growth pattern in the imaginal discs of normal and mutant morphogenesis, such as movement of cells or tissues; mitotic activities, including the orientation and rate of cell division; cell death; size and shape of the cells; cell affinities; and hormonal controls. Genetic mosaic studies contributed to some of these findings, especially in the case of cell movement, such as illustrated earlier for sex comb differentiation, and in the following two categories, including the aspects of cell size and the shape of wing.

1. Cell Death

Mutants are known that affect the growth of specific organs and lead to structural deficiencies, e.g., *vestigial (vg)*, *scalloped (sd)*, and *Beaded (Bd)* mutants that cause reduced and incomplete wings. Two controversial explanations have been given for these defective growth patterns. Goldschmidt (1935) proposed the degeneration of a particular region, giving rise to the *vg* wing; Waddington (1940), on the other hand, suggested that the missing regions of *vg* wing simply failed to develop. After histologic studies with light and electron microscopes, Fristrom (1968, 1969) was able to show that Goldschmidt's hypothesis was correct in the wing mutants, *vg*, *Bd*, *cut (ct)*, and *apterous-Xasta (ap^{Xa})*, and in the eye mutants, *Bar (B)* and *eyeless (ey)*. In these mutant discs, she observed cell death occurring in specific areas at specific stages, and suggested that, in most cases, the degenerate areas of the disc could be correlated with the missing parts of the adult wing or eye. Postlethwait and Schneiderman (1973) observed malformations, including partially duplicated or deficient antennae, ocelli, or leg structures, following X-irradiation during late embryonic and early larval stages. Cell death by the direct effect of X-rays has been suggested as the cause of the abnormal development, and they interpreted these abnormal morphogeneses according to Bryant's (1971) proposed gradient model for duplication and regeneration of *Drosophila* leg imaginal discs after surgery followed by in situ growth (Fig. 16).

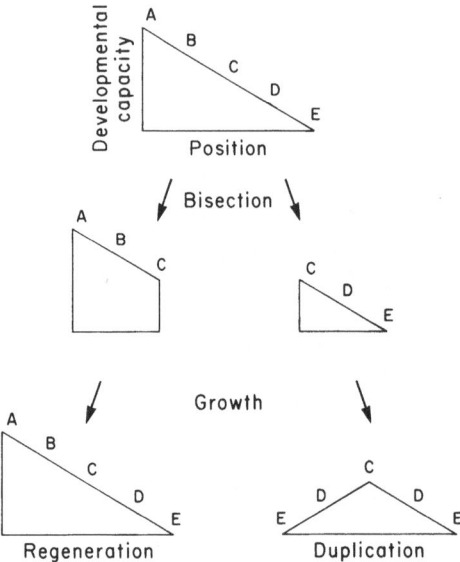

Fig. 16. The postulated gradient of developmental capacity and its response to bisection. (Modified from Bryant, 1971)

The findings of temperature-sensitive mutants by Suzuki and his associates (Suzuki, 1970, Suzuki et al., 1976, for review) and the use of the powerful mutagen ethyl methane sulfonate (EMS) (Lewis and Bacher, 1968) finally made possible the following genetic mosaic experiments, which clearly indicate morphogenetic and pattern changes resulting in situ from cell death during disc development.

Russell (1974) observed a high frequency of structural deficiencies and dupli-cations when the larvae of a cell-autonomous temperature-sensitive sex-linked mutant [*l*(1)*ts* 726], were subjected to *sublethal* heat treatments at the late 2nd and 3rd instar. By the use of gynandric mosaics induced by a meiotic mutant, *pal* (*paternal loss*, 2–35.7: see Baker and Hall, 1976), and temperature treatment, he observed abnormalities in the head (eye deficiencies only) and in the leg, with duplications in the *ts* mutant spot (X0) in several cases. More importantly, in two cases, he found duplicated elements in the nonmutant (XX) tissue. These are clear cases of nonautonomous effects of mutant tissue on the growth pattern of normal tissue. Therefore, it was proposed that the lethality of the mutant cells was the cause of the phenotypic malformations observed. Arking (1975) also obtained basically similar results by studying induced cell-lethal *ts* mutants of the X chro-mosome, including gynandric mosaic studies with *lethal* heat treatment. A similar conclusion was also derived from the studies concentrated on eight *ts* lethal mutants located on the X chromosome, in which the morphogenetic changes included alterations of the head (eye, antenna, arista, palpus), wing, leg, haltere, abdomen, and genitalia (Simpson and Schneiderman, 1975). All of the mutants were autonomous for the cell death effect, but one of them [*l*(1)*ts*-1126], also reversibly affected the cell division rate and growth of both imaginal disc and larval tissues. The detailed studies with the genetic mosaics induced by X-ray and with gynanders on the effect of the $l(1)ts$-504 mutant $(1-6.0\pm 2)$ in combination with the *scalloped temperature sensitive (sd^{ts})* gene revealed that temperature treatment ($29°$ C) between the 24- and 72-h age period resulted in a high frequency of lethality and abnormality; deficiencies or duplications of imaginal disc derivatives occurred, as well as alteration of tarsal segmentation. Temperature treatment in early larval stages delayed pupation, but produced normal flies; the treatment in second and early third instar larva also delayed pupation but led a high frequency of deficiencies. (Similar duplication and deficiencies after heat pulses were observed in other *ts* mutants, such as $l(1)ts$-340, -445, -958, and -982.) Thus, the time dependence for deficiency and duplication was observed. These results were explained by the occurrence of a cell death area, which was directly observed by staining with trypan blue. The duplication of female nonmutant tissue of the gynanders also supported the conclusion.

Arking et al. (1975) found neurologic defects in the brain of both homo- and heterozygous ey^D mutants by histologic and transplantation studies. They found a biphasic lethal period; the first period is in the 1st or 2nd instar larval stage (55% homozyogtes die) and the second is in the pupal period (45% homozygotes die). Since a direct correlation between the extent of neurologic damage and the time of developmental arrest was found, they suggested that the ey^D gene acts through cell death both in homozygotes and heterozygotes.

These accumulated data suggest that the cell death caused by the ey^D gene at a specific time and at a specific area may lead to the duplication of a specific leg area, including $+$ and ey^D cells together in mosaics and thereby to multiple sex comb differentiation as was predicted earlier by J. Fristrom (1970). The pleiotropic effects of ey^D on eye degeneration, duplication of antennae and ocelli might be initiated by the prime effect, cell death, as these works with *ts* lethal mutants suggest. It should be pointed out that the time of the pleiotropic effects for each organ affected is not necessarily the same, a point suggested by the studies of some temperature-sensitive mutants (see review by Suzuki et al., 1976).

Just before Fristrom (1968, 1969) found that cell death caused the scalloping of the wing margin in mutants such as *ct*, Williams (1968) found that the scalloping of the wing margin in gynandric wings of mutants ct^6 and *sd* acts autonomously: the phenotype in any given homogeneous section of the margin was not influenced by the genotype of adjacent sections of the margin or wing blade. However, he noticed some interaction (nonautonomy) between opposed dorsal and ventral surfaces of the margin having different genotypes. A similar study with the Bx^3 mutant did not provide conclusive evidence due to the low expressivity of the Bx^3 phenotype in gynanders. In this early study, he avoided working with X-ray-induced mosaics, because it was known that the direct X-ray effects result in the phenocopy of these mutant effects on the wing margin, thereby obscuring the interpretation of the mosaic data. From this work the phenotypic expression of the scalloping of the wing margin by these mutants was suggested as probably having some relation to the evagination process of wing morphogenesis.

Recently, Santamaria and Garcia-Bellido (1975) studied the effect of mutants ct^6 and *Beadex of Jollos (Bx^J)* on wing margin formation, with gynanders and also with X-ray-induced mosaics provided with additional marker genes for higher resolution in clonal analyses. They found the behavior of homozygous ct^6

cells in mosaics to be largely autonomous as far as cell death, as Williams had observed, and this autonomy was only suppressed when the mutant clone became smaller. Mosaics in the margin resulted either in no gap or in a gap in the chaetal elements of both surfaces of the wing (nonautonomy of ct^6 or its + allele). There were no cases of gaps affecting one wing surface only, even in the X-ray-induced mosaics. This epigenetic effect is independent of the wing surface in which the mutant clone appeared, and occurred in the anterior as well as in the posterior margin. Therefore, they suggested that the differentiation of the marginal cuticular elements requires the interaction of the cells of the ventral and dorsal surfaces of the wing. In Bx^J experiments, similar cell behavior occurred. They also studied the X-ray-induced phenocopy, which has a sensitive period of from 72–104 h larval age in controls, and 64–120 h in ct^6. That the change of clone frequency and size relative to control wing discs sets in at about the age when cell death can be recognized histologically in these mutant discs during the late 3rd instar was found by Fristrom (1969).

In summary, numerous mutants are known to cause cell death and are distinguishable by the specific time, location, and intensity of their action. These genes act in various ts lethal mutants, and probably the ey^D mutant as well, on the morphogenesis of eye, ocelli, antenna, leg, wing margin, and other structures. The nonautonomy observed in mosaics of ct^6, sd, or Bx^J suggests that the wing margin might have an important role in pattern formation, perhaps controlling bristle formation and wing shape in these cases, similar to the role of the tarsal segment boundary found in the mosaics of fj and d mutants.

2. Mitotic Activities

a) Normal Development. Besides the direct histologic observation of cell numbers and mitotic index, the clonal analysis of X-ray-induced genetic mosaics by mitotic recombination at different developmental stages allows one to estimate the average cell division cycle. One cycle ranges from ca. 2.4 h for histoblasts during pupal mitosis (Garcia-Bellido and Merriam, 1971c), ca. 8.5 h (Garcia-Bellido and Merriam, 1971b) or 10 h for wing, and 12.5 h for mesonotum (Bryant, 1970) to ca. 15 h of leg disc dvelopment (Bryant and Schneiderman, 1969). However, these estimates as averages do not give information about regional differences in growth rates or about differences between the molting and intermolting periods in the cell division rate or the occurrence of mitotic waves (Becker, 1957; Foster and Suzuki, 1970; Garcia-Bellido and Merriam, 1971b). Nevertheless, differences of the average cell cycle between the discs studied have been shown.

The presence of regional differences of cell division rates in the disc itself was neatly demonstrated by Becker (1957) with the X-ray-induced eye mosaics. He utilized twin spots, both occupying the same sector of the eye, but in alternative positions, and has shown that the differences of the clone size between the partner of the twin spot in the sector is dependent on the location in the eye, not on the genotype of the cells. Similar observations followed later, and the presence of regional differences in mitotic activity has been found in the leg (ventral prothoracic) disc (Bryant and Schneiderman, 1969) and in the wing disc (Bryant, 1970;

Garcia-Bellido and Merriam, 1971 b). Bryant and Schneiderman (1969) also noted that twin spots induced earlier (48–72 h) are predominantly transverse to the main axis of the leg (tangential direction in the disc), and twin spots induced later are arranged predominantly in a longitudinal direction (radial orientation in the disc). A similar mitotic orientation change was observed in the presumptive wing area, although the mitotic orientation pattern in the presumptive mesonotum area of the same dorsal mesothoracic disc (wing disc) was different. They also speculated that cell movement occurring during eversion of the leg disc at meta-morphosis may play a part in producing the characteristic elongated shape of the mosaic patches. Bryant (1970) further suggested that the basic development of some imaginal discs can be understood in terms of the specification of a few factors, principally the rate of cell divisions and direction of cell divisions; what are specified are the probabilities of mitoses occurring in various directions.

As one of the possibilities, the *fj* and *d* gene discussed earlier might act by suppressing the longitudinal division phase of leg disc development at a specific region, but further studies, including twin spot are required to support this assumption.

b) Homoeotic Mutant Development. Over 30 years ago, a relationship between homoeotic changes and growth rate was suggested by studies of homoeotic mutants such as *tetraltera (tet)*, ss^a, and *proboscipedia* (Villee, 1942, 1943, 1945; Vogt, 1946). However, only recently has it become possible to analyze the relationship in more detail through the clonal analysis of genetic mosaics provided with proper marker genes.

The intensive clonal analysis of the antenna discs of wild-type and the homo-eotic mutant *Antennapedia of Rappaport ($Antp^R$)* by Postlethwait and Schneider-man (1969, 1971a, b) revealed a relationship between the growth pattern con-trolled by the mutant and its + allele. They observed that the cell proliferation dynamics of the $Antp^R$ antennae are the same as wild-type antennae until after 72 h, but between 72 and 96 h the growth rate of $Antp^R$ disc increases, while the rate of the wild-type antenna remains constant. When genetically marked clones were induced by mitotic recombination at or before 72 h in antenna discs of $Antp^R$, the marked clones included both antenna and homoeotic leg structures, whereas irradiation at 96 h resulted in no clones which included both antenna and leg structures. These results indicate that the time of final determination to be-come leg or antenna in an antenna disc must be between 72 and 96 h of age under normal conditions.

A classic morphologic study of a similar homoeotic mutant ss^a by Vogt (1946) indicates that at 72 h the antenna disc of the ss^a mutant is only different from that of the wild-type disc in that the end knob is about 50% larger. At the time of pupation, the mutant end knob became twice the size of that of the wild-type disc. Then segmentation occurred within the thickened end knob of the ss^a disc, whereas in the wild-type, the smaller end knob grew into a normal arista.

Transplantation experiments (Braun, 1940) and genetic mosaic studies with mitotic recombination mosaics induced during the first instar by X-ray (Table 1, Roberts, 1964) have shown that the ss^a gene acts autonomously for the homoeotic transformation of antenna to tarsus. However, it was noticed that in some mosaics, the proximal homoeotic tarsal segmentations in the ss^a spot were weak. Since considerable wild-type tissue was found to lie within these regions, the author suggested as one interpretation for this nonautonomy, that owing to the diffusion of substances from one genotype tissue to another, autonomy is not complete. Referring to the observation by Vogt (1946), he further suggested that the successful segmentation of the homoeotic tarsal regions seems to depend on their overall thickness, and furthermore, that the thickness in the mosaic ss^a spot seems to be affected by nonmutant cells of the surrounding tissue. Since the two alleles (ss^a and +) control, among other things, the growth rate of antenna cells, it is possible that the primary effect of dilution of ss^a tissue by wild-type tissue is to limit the thickness of the antenna (Roberts, 1964).

Ginter et al. (1974) also studied the mosaic spots of ss^a using the same genetic method for the production of genetic mosaics as Roberts (1964), and found autonomous gene action only in ss^a spots induced earlier than 75 h. Among the mosaic samples induced at 75 h, several spots were nonautonomous, that is, not showing the leg phenotype. Therefore, they suggested that the time interval between 75 and 95 h larval age is the most probable period for irreversible determination to develop into tarsal structures instead of the arista, and that the ss^a gene may be responsible for the regional determination. Only a month later, Postlethwait and Girton (1974) reported exactly the same experiment with ss^a mosaics with the similar results. Thus, they argued that this late nonexpression of ss^a may be due either to an influence of surrounding nonmutant tissues on the small ss^a clones, or to a persistence (perdurance) of the effect of a nonmutant genotype for one or two cell generations after it is removed from a cell line.

At this point, it is relevant to introduce the interesting temperature shift experiment with ss^{a40a} (an allele of ss^a), which was found to have a temperature-sensitive transformation of the arista segment to the tarsal segments (Grigliatti and Suzuki, 1971). The homoeotic transformation is complete (100%) at 17° C and 0% at 28° C. The temperature-sensitive transformation occurs during the second quarter of the 3rd larval instar (which corresponds to 84–96 h at 25° C). When cultured at 17° C until the TSP (temperature-sensitive period), and then shifted up to 28° C, the transformation to tarsal structures progressed from the proximal antenna part to the distal; the later the shift-up time, the more complete the transformation, proceeding from the proximal to the distal direction. In contrast to this, the shift-down experiment—that is, rear at 28° C until the TSP and then shift-down to 17° C during the TSP (about 6 h at 28° C, 18 h at 17° C)—has shown that the transformation starts at the proximal part, e.g., the shift-down in the later portion of the TSP produced only the proximal transformation; the distal part of the antenna is not transformed into the distal part of the tarsus. This situation suggests a relation between a growth pattern change and later transformation from arista to tarsal segments. That is, the TSP seems to reflect a period during which there is a change of determination induced by the temperature shift such that, afterward, segment formation follows as a result of the earlier change in growth pattern. It seems that the amount of growth depends on the length of 17° C treatment during the TSP. It does not seem to matter whether growth occurs by the treatment at the begining or end of the TSP, in determining the number of tarsal segments, and the segment formation starts at the proximal end of the antenna disc, as Vogt (1946) observed. These results of shift-down experiments with ss^{a40a} were unexpected after the similar study of N^{60g11}, which is a temperature-sensitive mutation that disrupts eye facet arrangement. In this case, it was observed that a wave of facet orientation proceeded from the posterior rim of the eye anteriorly, the later TSP treatment affecting only the anterior area (Foster and Suzuki, 1970). Thus, for the ss^{a40a} results, Grigliatti and Suzuki (1971) suspected that either there was significant cell movement within the late third instar discs, or alternatively, there could be a site within the antenna disc which, in response to cold temperature, produced a substance that alters the final determination of the existing predetermined arista cells. Based on the transplantation experiments with ss^a and ss^{a-UC1}, Gehring and Schubiger (1975) assumed that the transcription and translation of ss^a alleles occurs at least prior to and/or during the TSP. They also speculated that ss^a is a controlling gene involved in the alternative differentiation of antenna or leg structures.

These studies of homoeotic mutants all suggest that during the early 3rd instar, a critical determinative event occurs, at a time when extensive cell division and proliferation are taking place in the imaginal disc. Thus the time of this critical determination in normal development is long after the early determinative events are recognized at the blastoderm stage (Chan and Gehring, 1971). Transdetermination is also found to be dependent on the proliferation rate in studies of leg and wing discs (Tobler, 1966; see also review by Hadorn, 1967). In the studies of environmental effects on a homoeotic mutant, *loboid-opthalmoptera*, Ouweneel (1969) pointed out that since the homoeotic mutations often interact with agents (such as maternal age, temperature, UV irradiation, starvation and farnesol medium, acetamide, uracil, sodium tetraborate on *loboid-opthalmoptera* mutant), causing changes in growth rates, it is possible that the basic action of all homoeotic mutation has to do with such changes.

A relationship between mitotic activity—represented by division rate, orientation (clonal shape), and clonal restriction—and homoeotic transformations has been reported in *bx* pseudoalleles, *bx*, *pbx*, and *Ubx* (Morata, 1975; cited by Garcia-Bellido, 1975). The exception to the relationship was found by Morata (1975) in the study of the transformation of the wing blade by *Contrabithorax* (*Cbx*), which is a *bx* pseudoallele. Homo- or heterozygous *Cbx* partially transforms the posterior mesothorax into a structure resembling the posterior metathorax and often the transformation extends to the anterior mesothorax (Lewis, 1963; Garcia-Bellido and Lewis, 1976). The study of *mwh* clones induced by mitotic recombination at all ages of $mwh^+ Cbx/mwh Cbx^+$ larvae revealed that the clones forming exclusively a wing or haltere structure on the wing blade occurred in flies irradiated at 96–120 h, whereas earlier irradiation-induced *mwh* clones included both wing and haltere structures. Since the induced spot size (cell number) of clones producing haltere or wing structure is more or less similar, the author suggested that the division rate was not influencing the final homoeotic differentiation. In this work, the characteristic structural differences between the wing and halter studied on the wing blade were not clearly stated, but they seem to be the size of the trichome and of cell size or cell density. Thus, the importance of the data presented concerns morphogenetic aspects of the final determination of cell size and density, which occurs after 96 h.

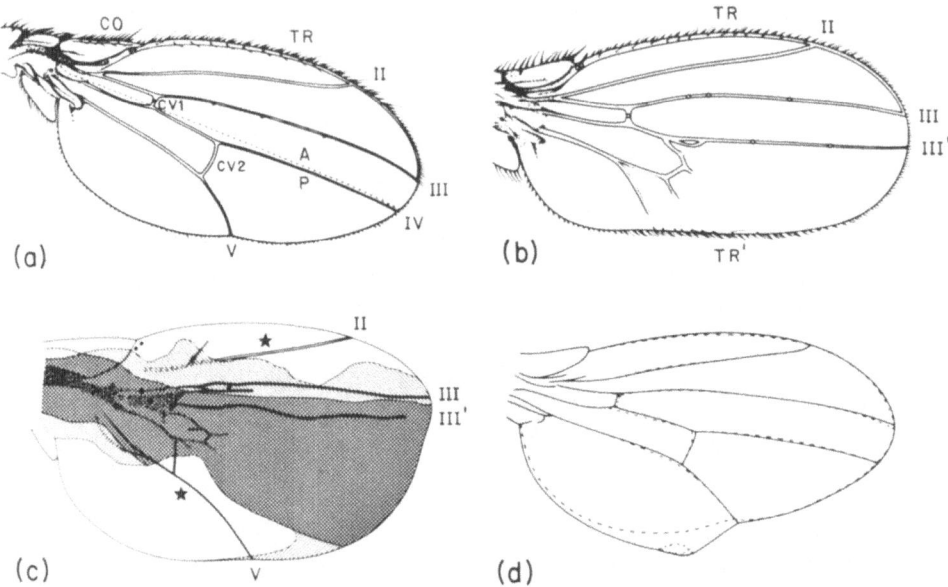

Fig. 17a–d. Diagrammactic illustrations of wings of *en/+* (a), *en/en* (b) and *en* mosaics (c, d). (a) *en/+* wing. *Dotted line* represents the clonal boundary between anterior *(A)* and posterior *(P)* compartments. (b) *en* wing. The posterior margin shows the presence of triple row bristles *(TR′)* typical of the anterior margin *(TR)*. A vein *(III′)* in a similar position to vein IV in wild-type wings bears sensilla. (c) *pwn en M(2)c⁺* clone of posterior origin induced in a fly of genotype *pwn en/M(2)c³³ᵃ*. Whereas, the *en⁺* cells form normal vein pattern (*), the *en* cells do not make normal anterior vein structures; the veins are deformed, and vein II is almost eliminated by the ventral clone that extends beyond it *(arrow)*. ////dorsal clone; \\\\ ventral clone. (d) Nonautonomy of wing shape; due to a small dorsal *en* clone *(shaded)* affecting the shape of a *pwn en/+ ; M(3)i⁵⁵/mwh jv* wing. The control nonmosaic wing on the other side is drawn and superimposed *(dotted line)*. Sensilla on vein are shown by a circle. *CO*, costa; *CV1*, first cross-vein; *CV2*, second cross-vein; *DR*, double row; *TR*, triple row; *II, III, IV, V*, longitudinal vein; *III′*, transformed vein by *en* corresponding to the vein *IV* in wild-type wing. (After Lawrence and Morata, 1976)

Another growth pattern change based on mitotic activity was found in the studies of the *en* gene effects on dorsal mesothoracic structures (Garcia-Bellido and Santamaria, 1972). As a pleiotropic effect, *en* leads to the transformation of the posterior structures of the wing disc into those characteristic of the anterior region of the same disc in mirror image fashion (similar to that observed in the *en* effect on secondary sex comb differentiation in the male first leg). Accompanied by a change of wing shape, the posterior wing margin differentiates a triple row of bristles instead of the normal double row of long hairs. The alula is generally reduced in size, and its socketless hair structures are replaced by long bristles similar to those found in the anterior costal region. Characteristics of longitudinal veins I to VI on the wing surfaces also indicate the mirror image changes of IV, V, and VI similar to III, II, and I, respectively. The homology aspects of veins are supported by the distribution pattern of extra bristles on the *en; h* wing. These transformation by *en* are not complete because of the low expressivity of the gene.

A clonal analysis of the *en* wing by induced mitotic recombination also indicated the transformation from posterior to anterior wing area revealed by the change of the size and shape of the clonal pattern on the wing blade from the normal posterior type to that of the anterior. In the studies of gene combinations with bx^3, or with *pbx*, which affect anterior or posterior wing regions, respectively, it was also found that the *en* gene acts preferentially on the posterior part of meso- and metathoracic discs. The transformation effects indicated the autonomous gene action of *en*, taking place several cell divisions prior to differentiation when the criterion was bristle differentiation. This can be expected, as discussed earlier. The homoeotic transformation was identified by its regional specificity; thus, the pattern of *en* was recognized as the partial substitution of the posterior part by anterior wing structures (see Fig. 17a, b).

Morata and Lawrence (1975) and Lawrence and Morata (1976) have confirmed the general autonomy of *en* gene action by detailed studies of *en* characteristics, including the medial marginal bristles, cell density, anteroposterior border (clonal boundary), the size and shape of the wing, and veins with or without sensilla in genetic mosaic wings. However, they also found some cases of nonautonomy.

One of the cases of nonautonomy observed concerned the wing shape. When *en* is combined with *M*, enlargement of the posterior wing area (clonal compartment) is greatly enhanced. By use of the clone marked with *pwn en; M (pwn en/ + ; M(3)i^{55}/mwh jv)* wing, the studies of cell density of the clones indicated that there were no differences in the anterior area between the dorsal and ventral wing surfaces in non-*en* and in *en* wings. Only the posterodorsal clones of *en* were high in their cell density compared with the posteroventral clones, and yet the wing shape was not changed, except when a small dorsal *en* spot (but induced before 96 h old) of a *pwn en; M* was located at the posterior margin of the wing. The latter situation causes *a change in the shape* and *an increase in the size* of the posterior marginal wing area, including the neighboring non-*en* tissue in mosaics (Fig. 17d). This nonautonomy and the fact that the increased density of cells on the dorsal surface did not change the wing shape (nonautonomy of wing shape in relation to the cell density) were interpreted as suggesting that probably the posterior dorsal wing margin is primarily responsible for the control of shape. A smaller spot induced at the margin after 96 h autonomously differentiated typical marginal bristles of the posterior *en* wing. Lawrence and Morata further suggested that the ventral surface is to some extent modeled on the dorsal (as far as size and the shape). This suggestion does not apply to the nonautonomy data from the studies of the mosaic wing margin with ct^6, *sd*, and Bx^J mutants. According to the illustrated gynandric mosaic data of Williams (1968) and the description by Santamaria and Garcia-Bellido (1975), the nonautonomy of these mutants and their wild-type alleles in wing margin formation mentioned earlier occurred regardless of the side of the marginal wing surface where these tissues were located. This suggests a different effect on the shape of the marginal wing formation by these cell death mutants and the *en* mutant.

Another case of apparent nonautonomy was found in vein formation. In the *pwn en* mosaics, Lawrence and Morata occasionally observed that pwn^+ en^+ cells become incorporated with an extra vein (which is typical of the *en* phenotype in the posterodorsal wing area) mostly made by the *pwn en* clone. The *en* gene increases the mitotic rate in the posterior wing area, as will be mentioned later, and *pwn* itself appeared to produce an increase in cell density on the wing blade (Lawrence and Morata, 1976). Whether the synergistic growth effect of the combined *pwn en* on the posterodorsal wing surface is related to this type of nonautonomy of $pwn^+ en^+$ cells found in the dorsal extra vein formation is an open question.

Thus, there are two nonautonomous expressions of the *en* gene in the posterior wing area: one in wing margin formation which relates to the wing shape, and another in wing vein formation, seemingly related to the change of mitotic activity.

Earlier, it had been suggested that the *en* wing imaginal disc grows differently from the normal wing disc from early stages of development (Garcia-Bellido and Santamaria, 1972). Lawrence and Morata (1976) found no significant effect of the *en* gene on early first larval clone size on the wing. This conclusion was derived from the method of estimating the mean area of clones expressed as a fraction of entire posterior wing area (Stern, 1940; Bryant and Schneiderman, 1969). They found a difference in clone size only when induced between 36 to 84 h in the posterior wing, suggesting that mitotic rate is greater in *en* than in +cells in that area. A temperature-shift experiment was used to demonstrate the late effect of *en* gene, which showed that the *en* phenotype in the medial triple bristle row at the posterior wing margin is affected by temperature shift during the period from 48 h larva to 12 h after pupation. According to the earlier work (Garcia-Bellido and Merriam, 1971b), the growth of the wing disc is nearly complete at 12 h after pupation, but these temperature-shift experiments show that the phenotype of the posterior medial marginal bristles is still sensitive at this stage. This suggests some morphogenetic difference between the bristles on the wing controlled by *Hw, h,* or *ac* mutants and the bristles of the posterior medial triple row controlled by *en*. Thus, they concluded that the pattern made by the clones of the posterior wing is dependent on the *en* gene until very late in development, and *en* begins to function during larval growth.

In summary, the rate and direction of cell division are important factors for basic imaginal disc development as suggested by Bryant. It also seems apparent that the homoeotic changes caused by mutants are somehow related (either as cause or effect) to a change of determination accompanied by a change of mitotic activity at a specific time and location, depending on the specific structure transformed during morphogenesis.

The nonautonomy found in these studies of homoeosis, as well as in the cell death mutants discussed earlier, was interpreted as an interaction between the cells of the mutant spot and the neighboring nonmutant cells. It is noteworthy that all of these cases also involved differences of mitotic rate or cell death (the latter causes not only deficiency, but also duplication of the structure) between the two different genotype cells.

The size and shape of mosaic wings has been discussed in an earlier section as well as here. The fact that cell death in the scalloped wing mutants occurs during the late 3rd instar, and that final cell density seems to be determined later than mid-3rd instar larval stages by *Cbx* mutant, suggest that the final determination of the size and shape of the wing occurs rather late in the wing disc. Difference in mitotic activity before the final determination between these scalloping mutants and *en* might be reflected in the nonautonomy of small *en* clones at the posterior wing margin induced before 96 h, and in the nonautonomy of cell death mutant spots in which cell death occurs in late 3rd instar.

B. Clonal Boundary Control by Pattern Mutants

As described earlier, the study of the time of clonal restriction to the tarsus or antenna structures of the antenna disc of the $Antp^R$ mutant led to the conclusion that the alternative determination occurs in a group of cells at a specific developmental stage between 72 to 96 h (Postlethwait and Schneiderman, 1969, 1971 a, b). It was noticed then that the clones induced in the leg disc after 84 h (3rd instar) were confined to a single leg segment (Bryant and Schneiderman, 1969), and that the clones induced after the 1st instar never crossed from the dorsal to the ventral side of the wing, whereas clones produced earlier could cross the border (Bryant, 1970; Garcia-Bellido and Merriam, 1971 b).

These clonal restrictions during disc cell proliferation were further studied by Garcia-Bellido et al. (1973, 1976). They showed that the clone of marked induced cells is restricted to an area, the developmental compartment, and the compartment to which the clone is confined is subsequently subdivided into smaller compartments at later developmental stages. In these experiments, M and non-M mosaics were used. [The proliferation rate of M tissue is much slower than that of the wild-type allele, and yet the compartment borders were observed by the mosaic clones (Morata and Ripoll, 1975)]. The same method applied to the cell lineage study of the leg revealed the presence of the anterior and posterior compartments parallel to the main axis of the leg segments (Steiner, 1976). When the induced clone are exclusively in the wing disc, the first clonal demarcation line separates the anterior and posterior compartments. At this stage, about twice as many induced clones appear in the anterior compartment as in the posterior one, based on the comparison of the frequency of induced clones. This was confirmed by Lawrence and Morata (1976); their estimation of the number of founder cells in the anterior versus posterior compartments was about 10 versus 5 cells for wild-type and also for the *en* wing (the notum was excluded). Each compartment was then subdivided into four by demarcation lines, separating dorsal from ventral (which occurred about 60 ± 12 h, Lawrence and Morata, 1976) and also wing from thorax area (proximal from distal).

Thus, they demonstrated that there is a successive clonal restriction taking place during normal development. It confines a group of cells, a polyclone (Crick and Lawrence, 1975), to regional compartments, which later subdivide into smaller compartments. By what mechanism the group of cells is split into subgroups within different developmental pathways is not yet known, but they (1976) suggested that the homoeotic mutants such as bx^3, *pbx*, *tet*, or *en* may be controlling the specific pathways of each compartment of the wing disc by defining morphogenetic cell properties. These include region-specific mitotic rates, mitotic waves, preferential orientation of the mitotic spindle, and cell recognition properties. [The latter have been revealed mainly by transplantation experiments with dissociated and reaggregated disc cells of mature 3rd instar larva (see Garcia-Bellido and Lewis, 1976 for references).] It was further suggested that each compartment is uniquely specified by the combination of a few controlling "selector genes" that are active within one or the other of the compartments (Garcia-Bellido, 1975).

This clonal compartment hypothesis was further supported by the study of *en* by Lawrence and Morata (1976). They found that the duplication theory, according to which the *en* wing has two anterior compartments instead of the normal anterior and posterior compartments, was not completely correct as regards vein formation. The extra pieces of vein found—other than vein IV (equivalent to vein III by mirror image symmetry, III' in Fig. 17b)—in the *en* wing or in an *en* clone in *en*/+ wing all have a bulge on the dorsal surface instead of on the ventral side, which is the characteristic of normal vein II. Also they frequently have sensillae that are not found in the normal vein II. Thus, these extra veins are not equivalent to vein II. Furthermore, the most posterior part of the anterior compartment—the regions of the 1st cross vein and those parts of vein III that are near it—are sometimes a little abnormal; thus the effect of *en* is not restricted exclusively to the posterior compartment, rather it also has a slight effect on the anterior. It was further found that $en; M^+$ clones in the posterior compartment crossed over the anteroposterior compartment border in M wing, although such clones arising in the anterior compartment do not cross over the compartment border. Lawrence and Morata concluded that the presence of the straight and precisely positioned compartment border is dependent on the activity of the *en* gene until late in development, and suggested that this is due to the gene effects on cell affinities on the posterior compartment, making these cells different from anterior compartments.

When *pwn en* M^+ clones [*pwn en* $M(2)c^+$] of posterior compartment origin in M wing [*pwn en*+/+ + $M(2)c^{33a}$] extend far into the anterior compartment, a replacement, not displacement, of the vein characteristics by these clones occurred (Fig. 17c). The veins formed by these clones are not typical of the anterior compartment; instead they show the characteristics of the veins mady by *en* cells in the posterior part of the *en* wing, e.g., they do not form vein II at the proper place but almost eliminate it. (At face value, this is a case of nonautonomy of *en* cells in vein formation at the anterior wing area, but as an interpretation, it is a case of autonomy of the posteriorly specified *en* cells.) Thus, any extension of posterior *en* cells into the anterior compartment is associated with the differentiation characteristic of the posterior compartment, as shown by the characteristic posterior veins of *en*. On the other hand, the anteriorly originated *en* clones of the same genotype never cross the anteroposterior demarcation line.

This result has revealed a difference between the same *en* genotype cells specified differently in the early stages of disc development; one as anterior, other as posterior compartment founder cells. As a working hypothesis to explain the phenomena, Lawrence and Morata adopted the "selector gene" hypothesis, which was originally proposed by Garcia-Bellido as mentioned earlier, and expanded it to incorporate the "bistable control circuit" hypothesis by Kauffman (1973). In normal development, cells of the founder group are related to each other by position but not by ancestry. When the founder cells are subdivided into two founder cell subgroups of the anterior and posterior compartments, a specific selector gene (A) in the anterior compartment acts to change the specification of that group of founder cells, making them different from the posterior group of founder cells. A similar function is proposed for the second selector gene (B), which acts subsequently to specify the founder cells of the dorsal and of the ventral compartments differently, and so on. They assume that in *en*, one of the early selector genes lacks normal function, at least of the selector gene A, which failure resulted in the incomplete specification of founder cells of the posterior compartment, leaving them as anteriorly specified cells.

Thus, according to this hypothesis, the step-by-step changes of cell specification during early developmental stages proceeds by the action of selector genes that control the groups of founder cells of specific compartments at specific positions.

According to the prepattern or positional information hypothesis, it is the two different singularities or specified positional information to which the same genotype cells respond via *en* gene action and specify the cells differently as the two founder cell groups for the anterior and posterior compartments. The difference in the competence of the cells to respond to the posterior singularity or specified positional information between the *en* and its + allele leads to the differences in posterior wing development between the *en* and wild-type wing.

Whether the clonally restricted compartment boundaries are equivalent to the reference point of the positional information hypothesis, or have some connection in defining the singularity in the prepattern hypothesis, is not yet clear. However, a clue to the relationship may be found in the observations by Lawrence and Morata on the anteroposterior demarcation line of the wing disc: posteriorly specified *en* cells crossed the demarcation line and formed a posterior *en* type vein at the anterior compartment site. That is, epigenetically differently specified posterior compartment clones differentiated autonomously the posterior compartment type of veins in the anterior compartment site. This suggests a change of singularity or specified positional information of the anterior compartment in normal development by these epigenetically changed clones. Thus, the mosaic prepattern singularities of an *en* mosaic wing became different from the non-*en* or normal prepattern singularities because of the anteroposterior compartment border's failure to perform its role in separating differently specified cell groups. The same explanation may also be applied to the slight abnormal vein formation in the proximal posterior part of the anterior compartment of an *en* wing as described. This may be caused by the posteriorly specified cells that crossed over the anteroposterior compartment border into the anterior area.

The data concerning the relationship between cell density and wing shape and size, and also nonautonomy in wing margin formation by the scalloping mutants, and by the posterodorsal *en* spot on the posterior wing margin are suggestive of the role of the dorsoventral demarcation line in controlling the wing shape and size, as described earlier.

The accumulated data suggests that the growth pattern change seems always to be involved in the aspects of *en* action in the posterior compartment discussed. That is, the first sign of *en* gene action is found in the change of mitotic activities at the middle of the 1st instar, as detected by clonal size differences. The nonautonomy found in vein formation and wing shape and size is always associated with the growth pattern change. Even the differentiation of the medial triple row of bristles seems to be associated with the extra growth of *en* cells at the posterior compartment. These characteristics, including the differentiation of the costal type of bristles, and typical sensillae on specific vein, are generally autonomously differentiated when *en* clones are induced in the posterior compartment of the non-*en* wing (at least before 96 h except for the medial triple row bristles). Thus, it is tempting to assume that *en* gene action might be primarily associated with the change of mitotic activity at specific sites, acting from mid-1st instar until as late as 132 h. The action of *fj* or *d* gene may also belong to a similar category.

VI. Conclusions

The classical prepattern hypothesis stimulated numerous genetic mosaic studies with pattern mutants of *Drosophila*. The results of these studies established that in general, the genes of pattern mutants act autonomously by specifying the response of individual developing cells to a hypothetical singularity (or specific positional information). With different singularities to provide epigenetic spatial differentness and with autonomous responses of genes, cells at one region of the fly could arrive step by step at a developmantel fate different from that of cells of the same genotype at other regions as development progresses. The original idea that mutants which change the prepattern of a large area could be found easily among the available pattern mutants had to be relinquished: since, in surviving flies, most pattern genes were found to act regionally, that is, not to change the overall organization, even in the case of homoeotic mutants studied so far. Still, the possibility exists that such prepattern mutants are lethal and thus could not be detected. The proof of this prediction had to await future conceptual and technical advancements in this field.

However, we were able to detect some mutants which, under some conditions, change the singularities involved in bristle pattern differentiation. This was possible by paying special attention to the following points: (1) the infrequent occurrences of nonautonomy among mosaics of generally autonomous mutants, in addition to the few cases of nonautonomy; (2) the time of mosaic induction in relation to the time of the final determination of the pattern element used for the identification of the unit structure of a pattern; (3) the border area where the two different genotype tissues meet in a mosaic.

All of the apparent singularity changes were associated with a change of growth pattern due to the studied pattern gene. The quantitative change of the sex comb singularity caused by the ey^D mutant, as suggested in the early genetic mosaic studies, can now be interpreted as a secondary response by $+$ and ey^D cells to the primary autonomous cell death of ey^D tissue at a specific time and region during development, as seems likely from recent studies by several authors using histologic as well as genetic mosaic studies on temperature-sensitive mutants. A growth pattern change, especially in mitotic activity, has also been reported for several homoeotic mutants. The well-analyzed cases have shown that the transformation of each specific structure occurs at a specific time independent of other characteristic structures transformed by the same homoeotic mutant. Nonautonomy phenomena in pattern formation in mosaics all appear to be associated with a changed growth pattern in one of the two genotypes of the mosaic studied.

The another factor that changes the singularity relates to tarsal intersegmental boundary formation, the failure of which changes the bristle pattern singularities within the segment, as found in the mosaic studies of *fj* and *d*.

The recent findings of the restriction of clones to compartments that subdivide during development led to the hypotheses that the clonal compartment is a unit for controlling a developmental pathway, and a unit for controlling the shape and size of a region as well as its pattern (Crick and Lawrence, 1975; Garcia-Bellido, 1975). These hypotheses are supported by the results of Lawrence and Morata (1976) with genetic mosaics of *en* wings. They have shown the change of specification of the founder cells of the posterior wing compartment by the *en* gene, which first expresses itself in terms of a change in mitotic activity in the mid-first instar larva, and by the nonautonomy in anterior vein formation by the posteriorly specified clones, which crossed the anteroposterior demarcation line. This effect suggests that the anteroposterior compartment boundary controls pattern formation. A similar role for the dorsoventral compartment boundary has been supported by the fact that the nonautonomy of cell density in the wing does not change the wing shape in *M* or *en* mosaics, and the nonautonomy of size and shape at the wing margin as affected by *en* in the posterior wing and in the anterior and posterior wing margins as affected by scalloping mutants.

In studies of insects other than *Drosophila*, namely *Oncopeltus* (Hemiptera), Lawrence (1973b) and Lawrence and Green (1975) supported the role of the segmental boundary as a clonal boundary, by clonal analysis and histologic observation. In *Drosophila*, the tarsal intersegmental boundary is known to affect bristle pattern formation in mutants, but its relation to the clonal compartment boundary has not yet been known in detail.

The common aspect in these genetic mosaic studies of *Drosophila* is that the nonautonomy of gene action has played a key role in finding factors that control developmental specification, such as the clonal compartment boundary of the wing and the intersegmental tarsal boundary. Furthermore, the change of growth pattern was also a key factor in finding the epigenetic step-by-step change in pattern formation processes through the use of genetic mosaics.

Thus, since the time when Sturtevant recognized the importance of genetic mosaics in developmental biology, and Stern first approached the pattern formation problem using genetic mosaics as "the issue of a late marriage between classical experimental embryology and classical genetics" (1968), the role of genetic mosaics has been fruitful in producing new insights into *Drosophila* development. However, we are still a long way from understanding the pattern formation process at the molecular level of gene action. (August, 1976).

References

Arking, R.: Temperature-sensitive cell-lethal mutants of Drosophila: isolation and character-ization. Genetics **80**, 519—537 (1975)

Arking, R., Putnam, R. L., Schubiger, M.: Phenogenetics of the *eyeless-Dominant* mutant of *Drosophila melanogaster*. J. Exp. Zool. **193**, 301—312 (1975)

Arnheim, N., Jr.: The regional effects of two mutants in Drosophila analysed by means of mosaics. Genetics **56**, 253—263 (1967)

Baker, B. S., Hall, J. C.: Meiotic mutants: Genetic control of meiotic recombination and chro-mosome segregation. In: Ashburner, M., Novitski, E. (Eds.): Genetics and Biology of Dro-sophila, Vol. Ia, pp. 351—434. London: Academic Press 1976

Becker, H. J.: On X-ray-induced somatic crossing over. Drosophila Inform. Serv. **30**, 101—102 (1956)

Becker, H. J.: Über Röntgenmosaikflecken und Defektmutationen am Auge von Drosophila und die Entwicklungsphysiologie des Auges. Z. Ind. Abst. Vererb. **88**, 333—373 (1957)

Becker, H. J.: The influence of heterochromatin, inversion-heterozygosity and somatic pairing on X-ray induced mitotic recombination in *Drosophila melanogaster*. Molec. Gen. Genet. **105**, 203—218 (1969)

Becker, H. J.: Mitotic recombination. In: Ashburner, M., Novitski, E. (Eds.): Genetics and Biology of Drosophila, Vol. Ic, pp. 1019—1088. London: Academic Press 1976

Braun, W.: Experimental evidence on the production of the mutant *"aristapedia"* by a change of developmental velocities. Genetics **25**, 143—149 (1940)

Bryant, P. J.: Cell lineage relationships in the imaginal wing disc of *Drosophila melanogaster*. Dev. Biol. **22**, 389—411 (1970)

Bryant, P. J.: Regeneration and duplication following operations in situ on the imaginal discs of *Drosophila melanogaster*. Dev. Biol. **26**, 637—651 (1971)

Bryant, P. J., Schneiderman, H. A.: Cell lineage, growth, and determination in the imaginal leg discs of *Drosophila melanogaster*. Dev. Biol. **20**, 263—290 (1969)

Chan, L. N., Gehring, W.: Determination of blastoderm cells in *Drosophila melanogaster*. Proc. Natl. Acad. Sci. USA **68**, 2217—2221 (1971)

Claxton, J. H.: Mosaic analysis of bristle displacement in *Drosophila*. Genetics **63**, 883—896 (1969)

Claxton, J. H., Kongsuwan, K.: Non-autonomy in *achaete* mosaics of Drosophila. Genet. Res., Camb. **27**, 11—22 (1976)

Crick, F. H. C., Lawrence, P. A.: Compartments and polyclones in insect development. Science **189**, 340—347 (1975)

Foster, G. G., Suzuki, D. T.: Temperature-sensitive mutations in *Drosophila melanogaster*. IV. A mutation affecting eye facet arrangement in a polarized manner. Proc. Natl. Acad. Sci. USA **67**, 738—745 (1970)

Fristrom, D.: Cellular degeneration in wing development of the mutant *vestigial* of *Drosophila melanogaster*. J. Cell Biol. **39**, 488—491 (1968)

Fristrom, D.: Cellular degeneration in the production of some mutant phenotypes in *Droso-phila melanogaster*. Molec. Gen. Genet. **103**, 363—379 (1969)

Fristrom, J. W.: The developmental biology of Drosophila. Ann. Rev. Genetics **4**, 325—346 (1970)

Garcia-Bellido, A.: Some parameters of mitotic recombination in *Drosophila* melanogaster. Molec. Gen. Genet. **115**, 54—72 (1972)

Garcia-Bellido, A.: Genetic control of wing disc development in Drosophila. In: Cell Pattern-ing. Ciba Foundation Symposium **29**, 161—182 (1975)

Garcia-Bellido, A., Lewis, E. B.: Autonomous cellular differentiation of homoeotic *bithorax* mutants of *Drosophila melanogaster*. Dev. Biol. **48**, 400—410 (1976)

Garcia-Bellido, A., Merriam, J. R.: Genetic analysis of cell heredity in imaginal discs of *Droso-phila melanogaster*. Proc. Natl. Acad. Sci. USA **68**, 2222—2226 (1971 a)

Garcia-Bellido, A., Merriam, J. R.: Parameters of the wing imaginal disc development of *Dro-sophila melanogaster*. Dev. Biol. **24**, 61—87 (1971 b)

Garcia-Bellido, A., Merriam, J. R.: Clonal parameters of tergite development in *Drosophila*. Dev. Biol. **26**, 264—276 (1971 c)

Garcia-Bellido, A., Ripoll, P., Morata, G.: Developmental compartmentalization of the wing disk of *Drosophila*. Nature (Lond.) **245**, 147, 251—253 (1973)

Garcia-Bellido, A., Ripoll, P., Morata, G.: Developmental compartmentalization in the dorsal mesothoracic disk of *Drosophila*. Dev. Biol. **48**, 132—147 (1976)

Garcia-Bellido, A., Santamaria, P.: Developmental analysis of the wing disk in the mutant *engrailed* of *Drosophila melanogaster*. Genetics **72**, 87—104 (1972)

Gehring, W. J., Schubiger, G.: Expression of homeotic mutations in duplicated and regenerated antennae of *Drosophila melanogaster*. J. Embryol. Exp. Morph. **33**, (2), 459—469 (1975)

Ginter, E. K., Ivanov, V. I., Mglinets, V. A.: Morphogenetic mosaicism for homoeotic mutation *aristapedia* in *Drosophila melanogaster*. Genetika (GNKAA) **10**, 67—75 (1974)

Goldschmidt, R.: Gen und Außencharakter. III. (Untersuchungen an Drosophila). Biol. Zentralbl. **55**, 535—554 (1935)

Gottlieb, F. J.: Genetic control of pattern determination in Drosophila. The action of *Hairy-wing*. Genetics **49**, 739—760 (1964)

Grigliatti, T., Suzuki, D. T.: Temperature-sensitive mutations in *Drosophila melanogaster*. VIII. The homeotic mutant, ss^{a40a}. Proc. Natl. Acad. Sci. USA **68**, 1307—1311 (1971)

Guerra, M., Postlethwait, J. H., Schneiderman, H. A.: The development of the imaginal abdomen of *Drosophila melanogaster*. Dev. Biol. **32**, 361—372 (1973)

Hadorn, E.: Dynamics of determination. In: Locke, M. (Ed.): Major Problem in Developmental Biology, pp. 85—104. London: Academic Press 1967

Hannah, A.: Non-autonomy of *yellow* in gynandromorphs of *Drosophila melanogaster*. J. Exp. Zool. **123**, 523—555 (1953)

Hannah, A.: Environmental factors affecting elimination of the ring-X chromosome in *Drosophila melanogaster*. Z. Vererbungslehre **86**, 600—621 (1955)

Hannah-Alava, A.: Morphology and chaetotaxy of the legs of *Drosophila melanogaster*. J. Morphol. **103**, 281—310 (1958 a)

Hannah-Alava, A.: Developmental genetics of the posterior legs in *Drosophila melanogaster*. Genetics **43**, 878—905 (1958 b)

Hinton, C. W.: The behaviour of an unstable ring chromosome of *Drosophila melanogaster*. Genetics **40**, 951—961 (1955)

Hollingsworth, M. J.: Sex-combs of intersexes and the arrangement of the chaetae on the legs of Drosophila. J. Morphol. **115**, 35—51 (1964)

Janning, W.: Aldehyde oxidase as a cell marker for internal organs in *Drosophila melanogaster*. Naturwissenschaften **59**, 516—517 (1972)

Janning, W.: Developmental studies on gynandromorphs of *Drosophila melanogaster*. III. Some observations on larval gynandromorphs. Verh. Dtsch. Zool. Ges. **67**, 134—138 (1975)

Kauffman, S. A.: Control circuits for determination and transdetermination. Science **181**, 310—317 (1973)

King, J. L.: The formation of *dumpy* vortices in mosaics of *Drosophila melanogaster*. Genetics **49**, 425—438 (1964)

Lawrence, P. A.: The development of spatial pattern in the integument of insects. In: Counce, S. J., Waddington, C. H. (Eds.): Developmental Systems: Insects, Vol. 2, pp. 157—209, London-New York: Academic Press 1973 a

Lawrence, P. A.: A clonal analysis of segment development in Oncopeltus (Hemiptera). J. Embryol. Exp. Morph. **30**, 681—699 (1973 b)

Lawrence, P. A., Green, S. M.: The autonomy of a compartment border. J. Cell Biol. **65**, 373—382 (1975)

Lawrence, P. A., Morata, G.: Compartments in the wing of Drosophila: A study of the *engrailed* gene. Dev. Biol. **50**, 321—337 (1976)

Lee, A. D., Waddington, C. H.: The development of the bristles in normal and some mutant types of *Drosophila melanogaster*. Proc. R. Soc. London Ser. B **131**, 87—110 (1942)

Lewis, E. B.: Genes and developmental pathways. Am. Zool. **3**, 33—56 (1963)

Lewis, E. B.: Genetic control and regulation of developmental pathways. In: Locke, M. (Ed.): Role of Chromosomes in Development, pp. 231—252. London-New York: Academic Press 1964

Lewis, E. B., Bacher, F.: Method of feeding ethyl methane sulfonate (EMS) to Drosophila. Drosophila Inform. Serv. **43**, 193 (1968)

Lindsley, D., Grell, E. H.: Genetic variations of Drosophila melanogaster. Carnegie Inst. Wash. Publ. No. 627 (1968)

Merriam, J. R.: Autonomy of mutant bristle frequency in gynandromorphs with different sc alleles in Drosophila melanogaster. Am. Nat. **103**, 672—675 (1969)

Morata, G.: Analysis of gene expression during development in the homeotic mutant Contrabithorax of Drosophila melanogaster. J. Embryol. Exp. Morph. **34**, 19—31 (1975)

Morata, G., Lawrence, P. A.: Control of compartment development by the engrailed gene in Drosophila. Nature (Lond.) **255**, 614—617 (1975)

Morata, G., Ripoll, P.: Minutes: Mutants of Drosophila autonomously affecting cell division rate. Dev. Biol. **42**, 211—221 (1975)

Morgan, L. V.: Correlation between shape and behaviour of a chromosome. Proc. Natl. Acad. Sci. USA **12**, 180—181 (1926)

Morgan, L. V.: A closed X chromosome in Drosophila melanogaster. Genetics **18**, 250—283 (1933)

Mukherjee, A. S.: The effects of sexcombless on the forelegs of Drosophila melanogaster. Genetics **51**, 285—304 (1965)

Mukherjee, A. S., Stern, C.: The effect of sexcombless in genetic mosaics of Drosophila melanogaster. Z. Vererb. **96**, 36—48 (1965)

Murphy, C. G., Tokunaga, C.: Cell lineage in the dorsal mesothoracic disc of Drosophila. J. Exp. Zool. **175**, 197—219 (1970)

Noujdin, N. I.: Genetic analysis of certain problems of the physiology of development of Drosophila melanogaster. (In Russian). Biol. Zhur. v. Armenii. **5**, 571—624 (1936)

Ouweneel, W. J.: Influence of environmental factors on the homeotic effect of loboid-opthalmoptera in Drosophila melanogaster. Wilhelm Roux' Arch. **164**, 15—36 (1969)

Postlethwait, J. H., Girton, J. R.: Development in genetic mosaics of aristapedia, a homeotic mutant of Drosophila melanogaster. Genetics **76**, 767—774 (1974)

Postlethwait, J. H., Schneiderman, H. A.: A clonal analysis of determination in Antennapedia. A homeotic mutant of Drosophila melanogaster. Proc. Natl. Acad. Sci. USA **64**, 176—183 (1969)

Postlethwait, J. H., Schneiderman, H. A.: A clonal analysis of development in Drosophila melanogaster. Morphogenesis, determination, and growth in the wild-type antenna. Dev. Biol. **24**, 477—519 (1971 a)

Postlethwait, J. H., Schneiderman, H. A.: Pattern formation and determination in the antenna of the homoeotic mutant Antennapedia of Drosophila melanogaster. Dev. Biol. **25**, 606—640 (1971 b)

Postlethwait, J. H., Schneiderman, H. A.: Pattern formation in imaginal discs of Drosophila melanogaster after irradiation of embryos and young larvae. Dev. Biol. **32**, 345—360 (1973)

Reed, C. T., Murphy, C. G., Fristrom, D.: The ultrastructure of the differentiating pupal leg of Drosophila melanogaster. Wilhelm Roux' Arch. **178**, 285—302 (1975)

Roberts, P.: Bristle formation controlled by the achaete locus in genetic mosaics of Drosophila melanogaster. Genetics **46**, 1241—1243 (1961)

Roberts, P.: Mosaics involving aristapedia, a homoeotic mutant of Drosophila melanogaster. Genetics **49**, 593—598 (1964)

Russell, M. A.: Pattern formation in the imaginal discs of a temperature sensitive cell-lethal mutant of Drosophila melanogaster. Dev. Biol. **40**, 24—39 (1974)

Santamaria, P., Garcia-Bellido, A.: Developmental analysis of two wing scalloping mutants ct^6 and Bx^J of Drosophila melanogaster. Wilhelm Roux' Arch. **178**, 233—245 (1975)

Schweizer, P.: Wirkung von Röntgenstrahlen auf die Entwicklung der männlichen Genitalprimordien von Drosophila melanogaster und Untersuchung von Erholungsvorgängen durch Zellklon-Analyse. Biophysik **8**, 158—188 (1972)

Simpson, P., Schneiderman, H. A.: Isolation of temperature sensitive mutations blocking clone development in *Drosophila melanogaster*, and the effects of a temperature sensitive cell lethal mutation on pattern formation in imaginal discs. Wilhelm Roux' Arch. **178**, 247—275 (1975)

Steiner, E.: Establishment of compartments in the developing leg imaginal discs of *Drosophila melanogaster*. Wilhelm Roux' Arch. **180**, 9—30 (1976)

Stern, C.: Somatic crossing over and segregation in *Drosophila melanogaster*. Genetics **21**, 625—730 (1936)

Stern, C.: The prospective significance of imaginal discs in Drosophila. J. Morphol. **67**, 107—122 (1940)

Stern, C.: Genes and developmental patterns. Caryologia, suppl. **VI**, 355—369 (1954)

Stern, C.: The genetic control of developmental competence and morphogenetic tissue interactions in genetic mosacis. Wilhelm Roux' Arch. **149**, 1—25 (1956a)

Stern, C.: Genetic mechanisms in the localized initiation of differentiation. Cold Spring Harbor Symp. Quant. Biol. **21**, 375—381 (1956b)

Stern, C.: Developmental genetics of pattern. In: Genetic Mosaics and Other Essays, pp. 135—173. Cambridge, Mass.: Harvard Univ. Press 1968

Stern, C., Hannah, A. M.: The sex comb in gynanders of *Drosophila melanogaster*. Portugaliae Acta Biol. Ser. A, R. B. Goldschmidt Vol., 798—812 (1950)

Stern, C., Swanson, D. L.: The control of the ocellar bristle by the *scute* locus in *Drosophila melanogaster*. J. Fac. Sci. Hokkaido Univ. Ser. VI, Zool. **13**, 303—347 (1957)

Stern, C., Tokunaga, C.: Nonautonomy in differentiation of pattern-determining genes in Drosophila. I. The sex comb of *eyeless-Dominant*. Proc. Natl. Acad. Sci. USA **57**, 658—664 (1967)

Sturtevant, A. H.: The *vermilion* gene and gynandromorphism. Proc. Soc. Exp. Biol. **17**, 70—71 (1920)

Sturtevant, A. H.: The effects of the *Bar* gene of *Drosophila* in mosaic eyes. J. Exp. Zool. **46**, 493—498 (1927)

Sturtevant, A. H.: The *claret* mutant type of *Drosophila simulans*: a study of chromosome elimination and of cell lineage. Z. Wiss. Zool. **135**, 323—356 (1929)

Sturtevant, A. H.: The use of mosaics in the study of the developmental effects of genes. Proc. 6th Int. Cong. Genetics. **1**, 304—307 (1932)

Suzuki, D. T.: Temperature-sensitive mutations in *Drosophila melanogaster*. Science **170**, 695—706 (1970)

Suzuki, D. T., Kaufman, T., Falk, D. et al.: Conditionally expressed mutations in *Drosophila melanogaster*. In: Ashburner, M., Novitski, E. (Eds.): Genetics and Biology of Drosophila, Vol. Ia, pp. 207—263. London: Academic Press 1976

Tobler, H.: Zellspezifische Determination und Beziehung zwischen Proliferation und Transdetermination in Bein- und Flügelprimodien von *Drosophila melanogaster*. J. Embryol. Exp. Morph. **16**, 609—633 (1966)

Tokunaga, C.: The differentiation of a secondary sex comb under the influence of the gene *engrailed* in *Drosophila melanogaster*. Genetics **46**, 157—176 (1961)

Tokunaga, C.: Cell lineage and differentiation on the male foreleg of *Drosophila melanogaster*. Dev. Biol. **4**, 489—516 (1962)

Tokunaga, C.: *Msc: Multiple sex comb*. Drosophila Inform. Serv. **41**, 57 (1966)

Tokunaga, C.: Nonautonomy in differentiation of pattern-determining genes in Drosophila. II. Transplantation of *eyeless-Dominant* leg disks. Dev. Biol. **18**, 401—413 (1968)

Tokunaga, C.: The effect on somatic crossing over of an ey^D insertion into chromosome 3. Drosophila Inform. Serv. **45**, 98 (1970)

Tokunaga, C.: Autonomy or nonautonomy of gene effects in mosaics. Proc. Natl. Acad. Sci. USA **69**, 3383—3386 (1972)

Tokunaga, C., Arnheim, N., Jr.: Age dependence of the locations of X-ray induced somatic crossing over in Drosophila. Genetics **54**, 267—276 (1966)

Tokunaga, C., Gerhart, J. C.: The effect of growth and joint formation on bristle pattern in *D. melanogaster*. J. Exp. Zool. **198**, 79—95 (1976)

Tokunaga,C., Stern,C.: The developmental autonomy of *extra sex combs* on *Drosophila melanogaster*. Dev. Biol. **11**, 50—81 (1965)

Tokunaga,C., Stern,C.: Determination of bristle direction in Drosophila. Dev. Biol. **20**, 411—425 (1969)

Villee,C.A.: A study of hereditary homoeosis: the mutant *tetraltera* in *Drosophila melanogaster*. Univ. Calif. Publ. in Zoology **49**, 125—183 (1942)

Villee,C.A.: Phenogenetic studies of the homoeotic mutants of *Drosophila melanogaster*. 1. The effects of temperature on the expression of *aristapedia*. J. Exp. Zool. **96**, 75—98 (1943)

Villee,C.A.: Developmental interactions of homoeotic and growth rate genes in *Drosophila melanogaster*. J. Morphol. **77**, 105—118 (1945)

Vogt,M.: Zur labilen Determination der Imaginalscheiben von *Drosophila*. II. Die Umwandlung präsumptiven Fühlergewebes in Beingewebe. Biol. Zentralbl. **65**, 238—254 (1946)

Waddington,C.H.: The genetic control of wing development in *Drosophila*. J. Genet. **41**, 75—139 (1940)

Waddington,C.H.: Some developmental effects of X-rays in *Drosophila*. J. Exp. Biol. **19**, 101—117 (1942)

Wigglesworth,V.B.: Local and general factors in the development of "pattern" in *Rhodnius prolizus* (Hemiptera). J. Exp. Biol. **17**, 180—200 (1940)

Wigglesworth,V.B.: The Control of Growth and Form. Ithaca: Cornell Univ. Press 1959

Williams,G.O.: Wing mosaics in *Drosophila melanogaster*. M.A. Thesis, Univ. California, Berkeley, Calif. (1968)

Wolpert,L.: Positional information and the spacial pattern of cellular differentiation. J. Theoret. Biol. **25**, 1—47 (1969)

Wolpert,L.: Positional information and pattern formation. In: Moscona,A.A., Monroy,A. (Eds.): Current Topics in Developmental Biology, Vol. 6, pp. 183—224. London-New York: Academic Press 1971

Young,S.S.Y., Lewontin,R.C.: Differences in bristle-making abilities in *scute* and wild-type *Drosophila melanogaster*. Genet. Res. Camb. **7**, 295—301 (1966)

The Relationship Between Cell Lineage and Differentiation in the Early Mouse Embryo

RICHARD L. GARDNER

Department of Zoology, University of Oxford, Oxford, Great Britain

I. Introduction

The notion that particular types of terminally differentiated cells originate from specific ancestral cells during ontogeny is one of wide currency and long standing. It arose principally from studies on lower chordate and invertebrate embryos carried out in the latter part of the last century and the beginning of the present one (see Davidson, 1976, for a recent review). The concept of stem cells, which may persist into adulthood in a variety of tissues, is clearly related to that of specific cell lineages (Cairnie et al., 1976). Thus neural crest cells, haematocytoblasts, and neuroblasts are believed to give rise to distinct constellations of specific mature cell types.

Early attempts to trace cell lineages during development depended either on exploiting intrinsic cytoplasmic markers such as pigment granules (e.g., Conklin, 1905), or applying extrinsic markers such as vital dyes or carbon particles to cells of early embryos (Vogt, 1929; Weston, 1967; Horstadius, 1973). However, rigorous analysis of cell lineage over longer intervals requires a means of indelibly marking cells that is not subject to dilution or modification through metabolism or mitosis. Hence, genetically determined phenotypic differences between constituent cells of the embryo afford the only unequivocal solution to the marker problem. The power of the genetic approach to cell lineage analysis is most clearly demonstrated by the elegant studies on insects discussed in other contributions to this volume.

With a few notable exceptions (e.g., Metcalf and Moore, 1971; Le Douarin, 1976) cell lineages have not been established with any precision in embryos of higher vertebrates. This is particularly so in mammals in which, until recently, understanding of the fate of cells in the early embryo depended entirely on histologic analysis of fixed and sectioned material. This situation has changed as a result of development of techniques to be described shortly, which enable production of genetic mosaicism in the preimplantation embryo. The mouse has been

chosen for most of this work for several reasons. First, the genetics of this eutherian mammal is understood in considerable detail, and many defined inbred
strains are available (Staats, 1972). Second, it has both a short gestation and
generation time. Finally, preimplantation embryos of this species have proved
very amenable to in vitro culture and manipulation (Daniel, 1971, 1978).

This review has two basic aims. The first is to discuss current knowledge of cell
lineage in the early mouse embryo that has been derived largely from examining
the fate of genetically labelled cells. The second is to attempt to explore the
relationship between cell lineage and differentiation in this system. Here, a central
issue is the extent to which phases in the differentiation of cells depend on their
ancestry as opposed to current environmental cues. The primacy of cell lineage
has been advocated by Holtzer and his colleagues (e.g., Holtzer et al., 1975;
Dienstman and Holtzer, 1975) who maintain that inducers, hormones, and other
means of intercellular communication play a purely elective rather than an instructive role in cyto-differentiation. They emphasise that even early embryonic
cells such as those of blastula and gastrula stages are differentiated, expressing
programmes of genetic activity that differ from those of their antecedents and
successors. Re-programming of cells is said to require special "quantal" mitoses,
as a result of which two options at most are made available to their daughters.
Holtzer et al. allow for a variable number of intervening proliferative mitoses in
regulative embryos, whereby the size of the cell population expressing a particular
programme can be varied. Finally, changes in gene activity that take place independently of DNA synthesis and cell division are attributed to modulation rather
than re-programming. However, absence of criteria by which one can distinguish
at present between the two types of changes preclude critical appraisal of this
aspect of their lineage hypothesis. In insects, which have proved most amenable to
fate mapping, development seems to be based broadly on cell lineage, while "fine-
tuned" differentiation depends on positional cues (Lawrence, 1975). Thus, while
polyclones respect compartment boundaries in *Drosophila*, lineages within compartments are indeterminate (e.g., Crick and Lawrence, 1975). The reader is referred to Lawrence (1975) for a critical review of the significance of cell lineage in
relation to differentiation in insects.

II. Sources of Genetic Mosaicism in Mammals

As emphasised earlier, genetic markers provide the only really reliable means
of tracing cell lineages throughout development. Genetic mosaicism may arise in
two basic ways. The first is by the occurrence of stable structural or functional
change in the genome of one or more cells derived from a single zygote. The
second is by integration of cells derived from more than one zygote in a single
organism. Animals belonging to the former category have been termed genetic
mosaics, and those belonging to the latter category chimaeras (Anderson et al.,

1961). Though not always practicable in cases of natural mosaicism (see Ford, 1969; McLaren, 1976a), the distinction is useful because, in general, the degree of genetic disparity between constituent cell populations in chimaeras will be greater than between those of genetic mosaics. While this terminology will be adopted in the following pages, it is relevant to point out that alternatives have been proposed (e.g., Mintz, 1974).

Several types of both genetic mosaics and chimaeras occur naturally in mammals (Ford, 1969; Benirschke, 1970; McLaren, 1976a), though usually too infrequently to enably their ready exploitation in developmental studies. An obvious exception is the functional genetic mosaicism attributable to random X-chromosome inactivation in female eutherian mammals (Lyon, 1961, 1974). However, while such mosaics have been employed in a variety of retrospective studies of development (see later), there are several aspects of X-inactivation that limit their value in cell lineage analysis. For example, mosaicism is restricted to polymorphic loci that are either carried on the X-chromosome or translocated to it. Such loci provide useful markers for only a limited number of tissues in the mouse. Also, representation of the two cell populations in tissues tends to be balanced; potentially informative instances of extremely unequal contributions are rare. Finally, it is still not clear when X-inactivation takes place during development, and indeed whether it does so at the same stage in all tissues (Gardner and Lyon, 1971; Nesbitt, 1971, 1974; Deol and Whitten, 1972a, b; Gardner, 1974; McLaren, 1976a). Radiation-induced genetic mosaics or mosaics produced by hyper-mutable genes have been studied to a limited extent in mammals (e.g., Fraser and Short, 1958; Russell, 1964; Melvold, 1971). However, neither these mosaics nor those that may arise as a result of somatic recombination (Gruneberg, 1966a) appear very promising for cell lineage work because only those affecting the coat or germ line are likely to be detected.

Viviparity is clearly responsible for the fact that chimaerism is a well-documented natural phenomenon in eutherian mammals. In general, the extent of chimaerism is greater the earlier in development cell mixing occurs. Anomalies of fertilisation or accidental union of pre-implantation embryos can lead to so-called "primary" chimaerism in which mosaicism extends to most if not all tissues of resulting offspring (McLaren, 1976a). The possibility of routine production of primary chimaeras in mammals was first realized by Tarkowski (1961) and by Mintz (1962). They independently showed that pairs of denuded mouse morulae could be aggregated in culture to form unitary embryos in which both cell populations contributed to the development of normal offspring. This important discovery has enabled cells carrying any prescribed genotypic differences to be incorporated into a single organism virtually at the outset of its development. Such "aggregation" chimaeras have been used to investigate diverse aspects of both normal and abnormal differentiation (see Mintz, 1974; McLaren, 1976a for recent reviews). The technique for producing them is very simple and is described in detail elsewhere (Mintz, 1971).

Subsequently, the author demonstrated that mouse chimaeras could also be produced experimentally by injecting dissociated cells or pieces of tissue into the more advanced blastocyst stage embryo (Gardner, 1968, 1971). The technique is

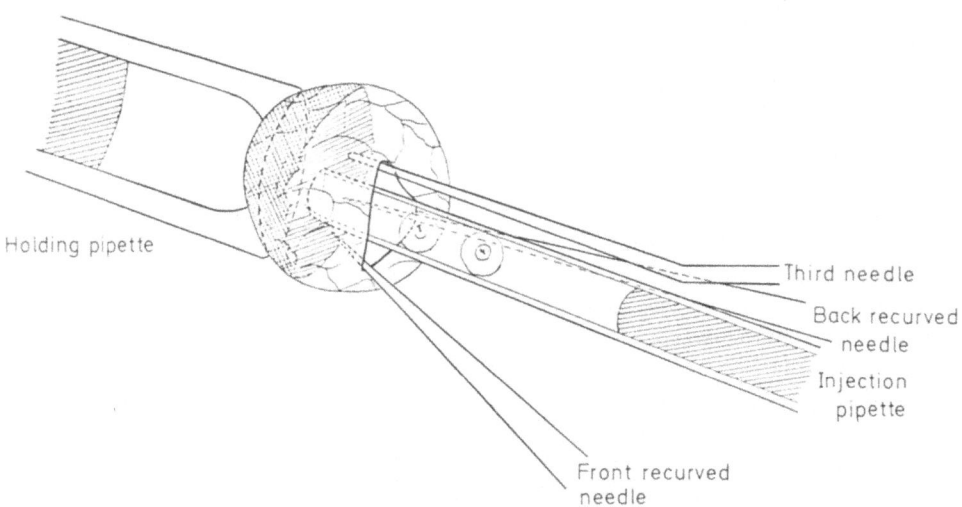

Holding pipette

Third needle

Back recurved needle

Injection pipette

Front recurved needle

Fig. 1. Diagram illustrating the technique of injecting cells into the mouse blastocyst. The blastocyst is immobilised on the tip of the holding pipette by suction applied to the embryonic pole. The two sharp-tipped recurved glass needles are used to make a slit in the abembryonic trophectoderm wall into which the third, blunt-tipped needle is introduced. By raising the latter needle and lowering the recurved pair the slit is converted into a triangular hole through which the cell injection pipette may be inserted into the blastocoel. (From Gardner, 1978, with permission of Academic Press)

described fully elsewhere (Gardner, 1978). Briefly, three glass micro-needles are used to form a triangular hole in the abembryonic trophectoderm wall of the blastocyst while it is held by gentle suction on the tip of a micro-pipette. The cells or tissue to be injected is then introduced into the blastocoelic cavity through this aperture by means of a second micro-pipette (Fig. 1). Though obviously a more difficult technique than embryo aggregation, blastocyst injection offers several advantages that are vital for cell lineage studies. For example, chimaerism can be obtained with cells that have already undergone restriction in potency (Gardner and Papaioannou, 1975; Rossant, unpublished data). Furthermore, relatively high frequencies of chimaerism can be attained following injection of single embryo- or teratoma-derived cells (Gardner and Lyon, 1971; Gardner and Papaioannou, 1975; Illmensee and Mintz, 1976), thus enabling strict clonal analysis in vivo. Finally, development of inter-specific chimaeras has been more successful employing injection than aggregation (Gardner and Johnson, 1973, 1975; Johnson and Gardner, 1975; Mystkowska, 1975; Rossant, 1976), possibly because one can thereby ensure that the entire trophectoderm of the composite embryos is of the same species origin as the uterine foster-mother. The importance of interspecies chimaeras is that they provide at present the only way of identifying all cells of either origin in sectioned material so that their spatial arrangement is conserved (Gardner and Johnson, 1973, 1975). A more detailed comparison of embryo aggregation versus blastocyst injection can be found elsewhere (Gardner, 1978).

III. Early Development of the Mouse Embryo

Cleavage is characteristically a very slow process in mammals. Thus the mouse embryo takes approximately $3^1/_2$ days from the estimated time of fertilisation to reach the stage of an expanded blastocyst consisting of some 60 cells. Two other interesting points about cleavage in the mouse embryo have been investigated recently. One is the stage at which products of specific paternally inherited alleles can first be detected. Results of these studies attest to very early expression of parts of the embryonic genome in relation to cell number (Brinster, 1973; Chapman et al., 1976). Second, the surface properties of cells clearly change during this initial phase of development. Thus, beginning at the late 8-cell stage blastomeres assume very extensive mutual contact. This phenomenon, termed compaction, seems to be a prerequisite for forming focal tight junctions and later complete *zonulae occludentes* between outer cells of the morula (Dulcibella and Anderson, 1975). The latter barrier takes approximately 8 h to develop, and is completed by about 70 h post coitum (p.c.) judging by exclusion of lanthanum tracer. Changes in cell surface glycoproteins have been detected during this period (Pinsker and Mintz, 1973), as has also the agglutinability of embryos by concanavalin A (Rowinski et al., 1976). Scanning electron-microscopic observations have revealed that from the 8-cell stage, blastomeres extend thin processes over their neighbours (Calarco and Epstein, 1973). These processes may be responsible for compaction since the latter is inhibited reversibly by absence of external calcium or presence of cytochalasin B (Dulcibella and Anderson, 1975). Between the 16- and 32-cell stage, blastomeres show intracellular accumulation of fluid (Calarco and Brown, 1969), release of which leads to formation of the characteristic structure of the blastocyst (Fig. 2).

The expanded $3^1/_2$-day p.c. blastocyst is thus a spherical fluid-filled vesicle whose surface is composed of approximately 45 flattened polygonal trophectoderm cells united by *zonulae occludentes*. A further 15 or so cells are packed together as a plano-convex disc, the inner cell mass (ICM), which is applied to part of the inner surface of the trophectoderm wall. Both the outer trophectoderm and enclosed ICM consist of two morphologically distinct cell populations by $4^1/_2$ days p.c., when the blastocyst has attached to the uterine epithelium (Fig. 3a). The mural trophectoderm cells embracing the blastocoelic cavity have ceased dividing and have started to enlarge. This process of giant transformation begins opposite the ICM at approximately midnight on day 4, and reaches the edge of the ICM 12 h or so later (Dickson, 1966). Blastocysts have been found to contain cells with DNA values in excess of 4 C at the latter stage (Barlow and Sherman, 1972). The polar trophectoderm cells overlying the ICM remain diploid and mitotically active. During the same period a distinct monolayer of primitive endoderm cells forms on the blastocoelic surface of the ICM, and begins to extend over the inner surface of adjacent mural trophectoderm cells (Fig. 3a). This tissue is separated from the underlying primitive ectoderm by a partial basement membrane by $4^1/_2$ days p.c. (Enders, 1971).

Implantation heralds the beginning of a period of very rapid growth and complex morphogenetic transformation of the embryo, stages of which are illustrated in Figure 3. The peripheral mantle of the primitive endoderm eventually

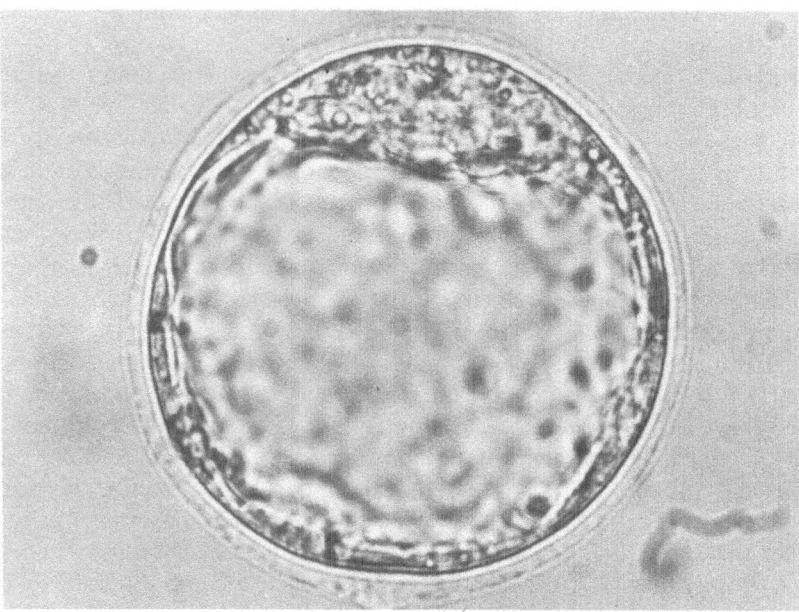

Fig. 2. A well-expanded 3.5 day p.c. mouse blastocyst within an intact zona pellucida. (From Gardner, 1971, with permission of Vieweg and Sohn)

extends over the entire inner surface of the so-called primary giant cells of the mural trophectoderm. The primitive ectoderm, together with the adjacent part of the polar trophectoderm (the extra-embryonic ectoderm), enlarges within a stocking of primitive endoderm so as to occupy most of the former blastocoelic cavity (Fig. 3b). The outer zone of the polar trophectoderm projects externally above this cylinder as the ectoplacental cone. Peripheral cells in this cone also undergo giant transformation and move laterally and ventrally. This process continues over a period of several days so that the entire conceptus is eventually surrounded by several layers of these so-called secondary giant cells, which are morphologically indistinguishable from the primary ones of the mural trophectoderm. Biochemical studies on trophoblastic giant cells have established that they are formed by repeated endoreduplication of the entire genome rather than by cell fusion (Chapman et al., 1972; Gearhart and Mintz, 1972; Sherman et al., 1972;

Fig. 3a–f. Tentative revised fate map of the trophectoderm, primitive endoderm, and primitive ▷ ectoderm of the late blastocyst projected onto to successive developmental stages redrawn and modified from Snell and Stevens (1966); (a) 4 days 5 h, (b) 5- or 6-day egg cylinder, (c) 5 days 12 h, (d) 7 days 1 h, (e) 7 days 6 h, (f) 8 days 11 h. The parts of the figure are drawn to different scales. *al*, allantois; *am*, amnion; *am. cav*, amniotic cavity; *bc*, blastocoel; *ch*, chorion; *d. en*, distal endoderm; *em ec*, embryonic ectoderm; *em. en*, *definitive* embryonic endoderm; *epc*, ectoplacental cone; *ep. cav*, ectoplacental cavity; *ex. ec*, extra-embryonic ectoderm; *exo*, exocoelom; *f*, fetus; *mes*, mesoderm; *m.t.* mural trophoblast; *m. tr*, mural trophectoderm; *p. em. en*, proximal embryonic endoderm; *p. en*, proximal endoderm; *p. ex. en*, proximal extra-embryonic endoderm; *pr*, proamnion; *p. tr*, polar trophectoderm; *y.s.s.*, yolk sac splanchnopleure. (From Gardner and Papaioannou, 1975, with permission of Cambridge University Press)

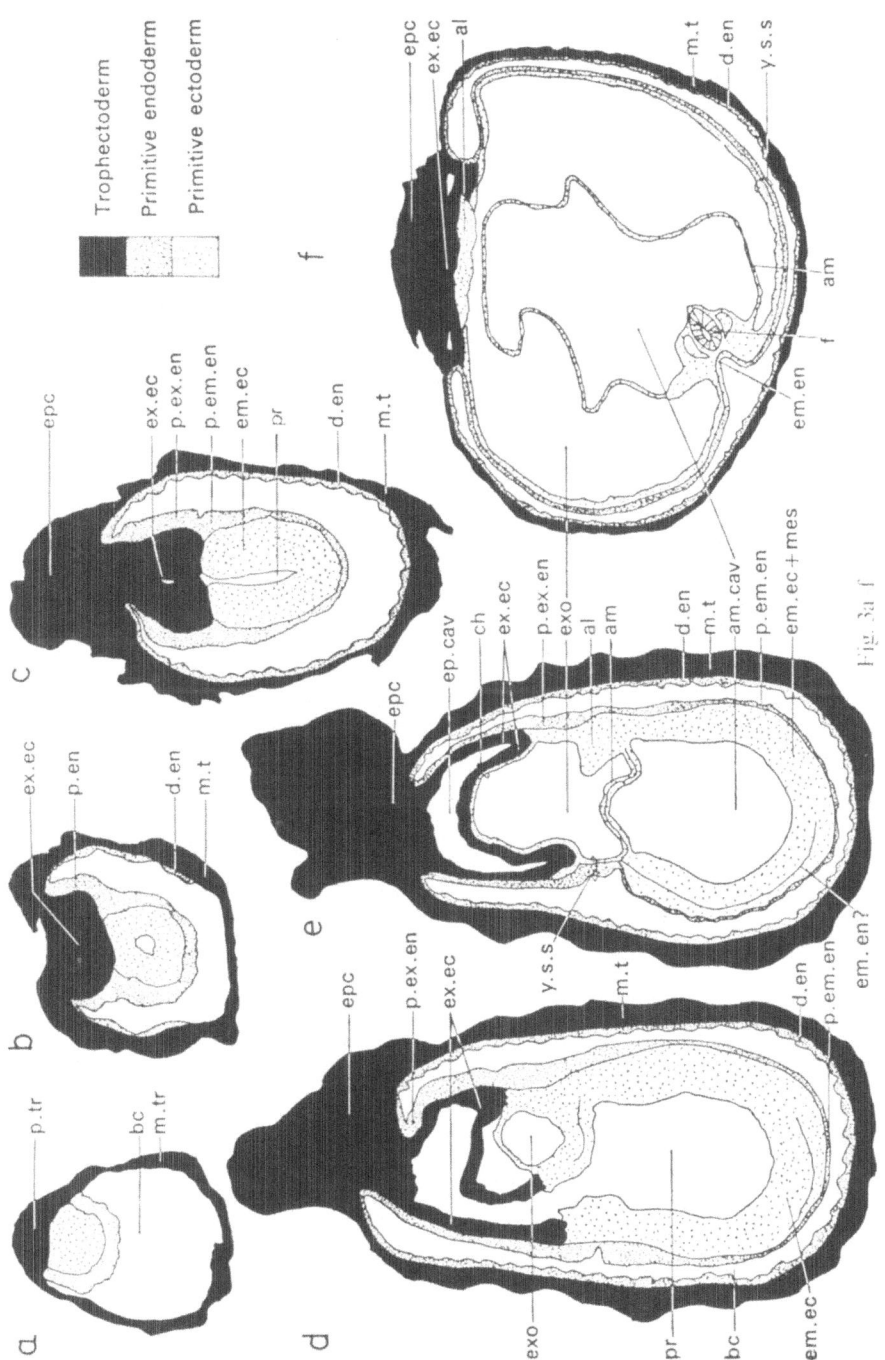

Fig. 3a–f

Gardner et al., 1973). Furthermore, they appear to be polytene rather than polyploid (Snow and Ansell, 1974).

Further development of the egg cylinder begins with formation of the primitive streak along one side of the primitive ectoderm (Snell and Stevens, 1966). The process of gastrulation in which this region engages is poorly understood, as indeed are most of the subsequent steps in the morphogenesis of the definitive embryo and membranes. Nevertheless, the foregoing brief description of early development, in conjunction with the stages illustrated in Figure 3, will hopefully provide an adequate background for discussion of the following studies. More detailed descriptions of mouse development can be found elsewhere (Snell and Stevens, 1966; Rugh, 1968; Theiler, 1972).

IV. Prospective Studies on Early Development

A. Determination in the Early Embryo

Pure mural trophectoderm tissue can be isolated by cutting $3^1/_2$-day-p.c. blastocysts in two, close to the ICM (Gardner, 1971; Gardner and Johnson, 1972). ICMs can also be isolated either microsurgically or by destruction of the investing trophectoderm with anti-mouse serum and complement (Gardner, 1971; Gardner and Johnson, 1972; Solter and Knowles, 1976). While trophectoderm tissue has acquired several specialised properties by $3^1/_2$ days p.c., this does not appear to be so for the ICM (Gardner and Johnson, 1972; Gardner, 1975a, b). Nevertheless, various isolation, recombination, and transplantation experiments have provided strong evidence that both types of cells are committed to mutually exclusive pathways of development at this stage (Gardner, 1975a, b; Rossant, 1975a, b, 1976; Gardner and Rossant, 1976). Particularly compelling results were obtained from experiments in which rat ICMs of equivalent stages were aggregated with morulae or injected into blastocysts of the mouse because of the single-cell resolution attainable with immunofluorescence (Gardner and Johnson, 1973, 1975; Rossant, 1976).

The main conclusion from the preceding studies is that all cells of the blastocyst are determined as either ICM or trophectoderm cells by the 60-cell stage. This prompts two questions. When does determination actually take place? Also, more fundamentally, what is its causal basis? It is perhaps appropriate to deal with the latter question first, particularly since it has received most attention.

The early development of a variety of metazoan species seems to depend to a greater or lesser extent on regional cytoplasmic differentiation that is established within the egg prior to fertilisation (Raven, 1961; Reverberi, 1971; Anderson, 1973; Davidson, 1976). Such stable organisation is discernible in living eggs of some species as a result of inhomogeneity in the distribution of certain cytoplasmic inclusions (e.g., Conklin, 1905). Cytoplasmic differentiation in mammalian eggs has been detected consistently only after application of various cytochemical procedures to fixed material. So-called "dorsal" and "ventral" cytoplasmic territories have thus been revealed, and appear to confer a bilaterally symmetrical organisation on the egg (see Dalcq, 1957, for a review). Furthermore, it was claimed that this cytoplasmic differentiation persisted throughout cleavage, "dorsal" and "ventral" cytoplasm being segregated to different quartets of blastomeres at the 8-cell stage, and to ICM and trophectoderm respectively in the blastocyst. These observations led Dalcq and his colleagues to postulate that blastomeres were determined to form ICM or trophectoderm cells by the type of cytoplasm that they inherited during cleavage (Dalcq, 1957; Mulnard, 1965).

Critical evaluation of the importance of egg organisation in early mammalian development is complicated by the absence of stable reference points in living material. Nonetheless, the foregoing hypothesis has been disproved by the recent demonstration of the totipotency of blastomeres at the 8-cell stage. Thus Kelly (1975) showed that each of two daughter cells produced by division of isolated 4-cell stage blastomeres was capable of both ICM and trophectodermal differentiation when combined with separate groups of "carrier" blastomeres of a different genotype. While not altogether excluding cytoplasmic factors from playing a role in early differentiation (see Denker, 1976), these results combined with other experimental data discussed below suggest that they are unlikely to be of major importance.

Though tending to remain together, [^{3}H]-thymidine-labelled cells showed no consistent pattern of distribution between trophectoderm and ICM of composite blastocysts produced by aggregating pairs of labelled and unlabelled morulae (Mintz, 1964, 1965; Garner and McLaren, 1974). Mintz (1965) also found that larger numbers of unlabelled morulae could form morphologically normal blastocysts following arrangement in a variety of spatial configurations likely to militate against extensive cell movement. She concluded from such experiments that blastomeres retain developmental lability at least until the late morula stage, and that attention should be focussed on disparate micro-environments produced by cleavage rather than "pre-patterning" in the egg for causal factors in early differentiation (Mintz, 1965). Subsequent studies on the in vitro development of blastomeres isolated from 4- and 8-cell embryos led Tarkowski and Wroblewska (1967) to suggest more specifically that their fate depended on whether they occupied an inside or outside location during cleavage. Cells that became wholly surrounded by others were thereby determined to become ICM cells, whereas those with part of their surface exposed were determined to form trophectoderm cells. The evidence on which this "inside–outside" hypothesis is based has been discussed in detail in several recent reviews (e.g., Herbert and Graham, 1974; Gardner, 1975a; Wilson and Stern, 1975; Rossant and Papaioannou, 1977), and appears to be consistent with the predictions that it makes (Gardner and Rossant, 1976). The single most compelling study is by Hillman et al. (1972), who demonstrated that they could predictably bias the fate of blastomeres or even entire cleaving embryos by placing them in either an exposed or enclosed position in artificial aggregates.

The main obstacle to rigorous appraisal of the inside-outside hypothesis is the persisting uncertainty as to when determination takes place (Gardner and Rossant, 1976). Snow (1973a, b) found that culture of cleaving eggs in medium containing [^{3}H]-thymidine led to reduction or absence of an ICM following blastulation. While the radio-isotope appears to exert its effect at the 16-cell stage, it is not clear whether it does so by acting specifically on presumptive ICM cells or by causing a non-specific reduction in the number of viable blastomeres. The conclusion drawn from a variety of embryo aggregation experiments is that determination takes place later, possibly in the early blastocyst (Mintz, 1964, 1965; Stern, 1972; Stern and Wilson, 1972). Observations on the development of isolated blastomeres and aggregated embryos in vitro suggest that the timing of blastulation depends on the number of cell divisions that have taken place since fertilisation rather than on cell number per se (Mintz, 1965; Tarkowski and Wroblewska, 1967). Thus isolated blastomeres appear to blastulate after completing the same number of divisions that they would if left in the intact embryo (Tarkowski and Wroblewska, 1967). Hence, the fact that single blastomeres isolated at the 8-cell stage can form miniature blastocysts in culture also favours late determination if the inside–outside hypothesis is correct (Gardner and Rossant, 1976).

However, none of the above experiments and observations provides compelling evidence concerning time of determination of trophectoderm versus ICM. Indeed, it is not even clear if both types of cells are determined simultaneously. Critical examination of this problem requires investigation of the potency of outside and inside cells isolated from progressively more advanced embryos. This obviously depends on finding efficient ways of isolating pure populations of the two categories of cells. The immunosurgical technique of Solter and Knowles (1976) has made possible the isolation of inside cells by lysis of those on the outside. Handyside and Johnson (1977) claim in a preliminary report that, using this approach, they have found that ICMs isolated from early blastocysts are still capable of producing trophectoderm cells. Unfortunately, since no cell counts were undertaken, it is not clear up to what stage ICM cells retain this capacity. Nevertheless, providing it can be established that all outside cells are destroyed by immunosurgery, these results suggest that determination in the ICM takes place close to the 60-cell stage.

Dalcq (1957) also drew attention to polarity in the mammalian egg, the animal pole being defined as the site of extrusion of the polar bodies. Polar organisation of the mouse egg is also evident from such surface properties as the distribution of microvilli and accessibility of concanavalin A binding sites (Johnson et al., 1975; Eager et al., 1976). However, it is not clear whether this cortical polarisation is produced as a consequence of the peripheral migration of the meiotic apparatus or reflects pre-existing asymmetry in egg organisation. That earlier asymmetry exists is indicated by the eccentric location of the germinal vesicle in pre-maturation oocytes (Dalcq, 1957). In addition, if germinal vesicle breakdown is inhibited by treating such oocytes with dibutyryl cyclic AMP, exposure to cytochalasin B induces "pseudo-cleavage," in which nucleated and non-nucleated fragments differ strikingly in both surface morphology and mitochondrial distribution (Wassarman et al., 1976). However, the relationship between this earlier polarity and that of the mature oocyte has yet to be determined.

Direct investigation of the significance of egg polarity in early development is precluded by the fact that the polar bodies have been found by time-lapse cinematography to move about on the surface of the embryo during cleavage (Borghese and Cassini, 1963; Mulnard, 1967). However, the blastocyst is a clearly polarised

stage as a consequence of attachment of the ICM locally to part of the trophectoderm wall (Fig. 2). Earlier claims that the ICM moves relative to overlying trophectoderm cells when the blastocyst orientates during early implantation (Kirby
et al., 1967) have not been substantiated by marking trophectoderm cells (Gardner, 1975b). Results of both these experiments and others involving transplantation of second ICMs to $3^1/_2$-day-p.c. blastocysts demonstrate that the position of
the ICM is fixed by this stage of development (Gardner, 1977b). The significance
of this polarity in the blastocyst is two-fold. First it serves to define (though
probably not determine) the dorso-ventral axis of the future embryo. Second, as
will be discussed later, it dictates the boundary of regional differentiation within
the trophectoderm.

If polarity of the blastocyst was determined by that of the egg, one might
anticipate regularities in cleavage similar to those found in many sub-mammalian
species (e.g., Davidson, 1976; Horstadius, 1973). This aspect of early mammalian
development has been almost totally neglected, probably because of the lack of
stable reference points in living material. Nonetheless, Gulyas (1975) recently
reported consistent features in the early cleavage of rabbit embryos. Thus, either
before or during second cleavage one blastomere appears to rotate so that its
polar axis comes to lie at a right angle to that of the other blastomere. This results
in a cross-wise configuration of blastomeres at the 4-cell stage. This predictable
pattern of cleavage seems to persist until at least the 8-cell stage. The author
claims to observe a similar pattern of cleavage in the majority of mouse eggs
examined, but in only a small percentage of those of the rat (Gulyas, 1975).
Clearly this merits further study. Absence of extensive cytoplasmic rearrangement
during cleavage is evident from experiments in which drops of silicone oil were
injected either peripherally or centrally into early blastomeres (Wilson et al.,
1972). All peripheral drops ended up in trophectoderm cells following blastulation; central ones tended to be enclosed in ICM cells. However, the fact that
multi-embryo aggregates consistently formed unitary blastocysts capable of normal post-implantation morphogenesis (Mintz, 1965; Hillman et al., 1972) argues
against the polarity of the egg dictating that of the blastocyst.

Hence, while one cannot exclude altogether the possibility that egg organisation plays a part in undisturbed development (Denker, 1976), cellular relationships during cleavage appear to be of over-riding importance. How relative cell
position controls determination is a problem for which there are at present no
data and little speculation. The phenomenon of compaction may well be a necessary precondition for producing disparate inside and outside cellular micro-environments by enabling formation of *zonulae occludentes* between outside cells
(Dulcibella and Anderson, 1975). This raises further questions concerning factors
responsible for altering cell surface properties during cleavage and for terminating
this phase of development by initiating blastulation (Izquierdo and Ortiz, 1975).

B. Differentiation Within the Trophectoderm

In addition to forming the actual exchange surface of the chorio-allantoic
placenta, trophoblast cells provide a complete cellular barrier between conceptus

and mother during the greater part of pregnancy. All trophoblast cells are believed to originate from the trophectoderm of the blastocyst and the extra-embryonic ectoderm of the later egg-cylinder stage of development (Fig. 3 b). The
origin of the latter tissue has long been disputed (see review by Rossant and
Papaioannou, 1977), though an ICM derivation has been generally accepted in
recent years (Snell and Stevens, 1966; Rugh, 1968). However, since histologic
studies cannot provide a definitive answer, the problem has been re-examined by
reconstituting mouse blastocysts from genetically distinct trophectoderm and
ICM tissue, and by injecting rat ICMs into mouse blastocysts (Gardner et al.,
1973; Gardner and Johnson, 1973, 1975). Post-implantation analysis of such chimaeras has provided compelling evidence for a trophectodermal origin of the
extra-embryonic ectoderm. This conclusion has been reinforced by comparison of
the development of this tissue with that of a known trophectodermal derivative,
the ectoplacental cone (Gardner et al., 1973), in ectopic grafts and in culture
(Gardner and Papaioannou, 1975; Rossant and Ofer, 1977). Hence available data
are compatible with unitary origin of all mature trophoblast cells from the trophectoderm of the blastocyst. However, confirmation of this hypothesis awaits
discovery of genetic markers that would enable critical analysis of the derivation
of cell populations in the mature placenta.

Whether produced by microsurgery on the $3^1/_2$-day-p.c. blastocyst (Gardner,
1971; Gardner and Johnson, 1972) or by exposing cleaving embryos to [^{3}H]-
thymidine (Snow, 1973 a, b), trophectodermal vesicles exhibit the pattern of differentiation characterising mural rather than polar cells. Thus they show no further
mitotic activity and produce only giant cells in utero, in ectopic sites, and in
culture (Gardner, 1971; Gardner and Johnson, 1972; Snow, 1973 a, b; Ansell and
Snow, 1975). However, if ICMs are injected into such vesicles produced micro-
surpically prior to transplantation, normal conceptuses develop in utero (Gardner, 1971). Furthermore, the ectoplacental cones and secondary trophoblastic
giant cells of these conceptuses express the genetic marker of mural trophectoderm rather than ICM donor blastocysts (Gardner et al., 1973). These results
demonstrate that, in the presence of ICM tissue, mural cells can give rise to polar
trophectoderm cells at $3^1/_2$ days p.c. Initiation of giant transformation by mural
cells in intact blastocysts does not depend on implantation in utero, or on attachment and spreading of the trophectoderm in vitro (Barlow and Sherman, 1972).
Indeed, endoreduplication of DNA can begin in blastocysts that fail to hatch from
the investing zona pellucida (Barlow and Sherman, 1972), as well as in vesicles in
which ICM cells have not developed (Sherman, 1975). Thus it seems likely that
uncoupling of cytokinesis from DNA synthesis is a consequence of commitment
of cells to a trophectodermal path of differentiation which is prevented or delayed
in polar cells by contact with the ICM (Gardner et al., 1973; Gardner, 1975a).
This hypothesis affords an explanation for the regional differentiation of the
trophectoderm observed in the late blastocyst, as also for the later production of
numerous secondary giant cells from the regions of the ectoplacental cone that
are most remote from ICM derivatives. Indications that trophoblast proliferation
continued to depend on ICM derivatives beyond implantation came from histologic studies on abnormal conceptuses (Gardner and Johnson, 1972; Gardner,
1975 b). This impression has been fully substantiated by recent experiments.

Giant cell formation is a conspicuous feature of the ectoplacental cone, but is not observed in the extra-embryonic ectoderm of normal embryos. However, preliminary experiments demonstrated that when isolated from $5\frac{1}{2}$-day-p.c. embryos, this tissue formed such giant cells both in ectopic grafts and in culture (Gardner and Papaioannou, 1975; Gardner and Ofer, unpublished data). These findings have been confirmed and extended recently by Rossant and Ofer (1977). They compared the development of ectoplacental cones, extra-embryonic ectoderm, and embryonic ectoderm (primitive ectoderm) isolated from $5\frac{1}{2}$- to $8\frac{1}{2}$-day-p.c. embryos after one week in adult testes or in culture. Testis grafts of embryonic ectoderm formed compact teratomas composed exclusively of small cells, while extra-embryonic ectoderm yielded haemorrhagic grafts containing giant cells, as did those of ectoplacental cone tissue. Densitometry measurements on feulgen-stained nuclei established that extra-embryonic ectoderm was composed of diploid cells at explantation. Furthermore, its mitotic index was found to be at least as high as that of the embryonic ectoderm. However, the frequency of mitoses fell dramatically during the first day of culture and was negligible by 48 h. Virtually no diploid cells remained after a week, and the cultures resembled those of ectoplacental tissue in containing conspicuous giant cells. In contrast, embryonic ectoderm cells continued to proliferate in culture, the explants often attaining considerable size. No giant cells were seen after a week, and most nuclei exhibited DNA values around 2C (Rossant and Ofer, 1977).

The above study strongly suggests that maintenance of diploid proliferating trophoblast cells depends on the presence of ICM derivatives at least until day 9 of gestation. Extension of this work to even later stages of development is complicated by the fact that ICM and trophectoderm-derived cells become intimately admixed in the placental region shortly thereafter. Nothing is known about the nature of the interaction between ICM and trophectoderm cells. As pointed out by Rossant and Ofer (1977) the ICM may exert its effect by maintaining a particular organisation of polar trophectoderm and its derivatives rather than by transmitting a specific inductive signal. Either way the subject merits further study, especially in relation to elucidating control mechanisms in the eukaryotic cell cycle.

C. Subsequent Differentiation Within the Inner Cell Mass (ICM)

Evidently ICM cells cannot be re-directed along a trophectodermal path of differentiation at $3\frac{1}{2}$ days p.c., when they number approximately 15. ICM tissue has been found to exhibit several qualitative differences from trophectoderm at this stage, in terms of polypeptides newly incorporating ^{35}S-methionine (Van Blerkom et al., 1976). While this might represent synthesis of ICM-specific gene products, chemical or enzymatic modification of common proteins has yet to be excluded. Alternatively, the apparently faster cycling of inside cells from the time of their first occurrence in the morula (Barlow et al., 1972) could lead to differential requirement for resumption of synthesis of certain "household" proteins.

A day later at $4\frac{1}{2}$ days p.c. the ICM contains some 45 cells, of which a little over half are arranged as an epithelial sheet of primitive endoderm coating its blastocoelic surface. This endodermal monolayer is eventually separated by a basement membrane (Enders, 1971) from the apparently less ordered primitive ectoderm cells that are sandwiched between it and the polar trophectoderm. Dalcq (1957) suggested on the basis of similarity in their cytochemical properties in rat blastocysts that the primitive endoderm was formed by inward movement of trophectoderm cells in the junctional region between mural and polar tissue. Since ICMs routinely develop an external coat of endoderm cells after isolation from $3^1/_2$-days-p.c. blastocysts (Rossant, 1975a; Hogan and Tilly, 1977), such migration of cells would have to take place in the early blastocyst. Neither numerous light and electron-microscopic studies on fixed blastocysts, nor several time-lapse cinematography studies on living ones (e.g., Borghese and Cassini, 1963; Cole and Paul, 1965; Snell and Stevens, 1966; Mulnard, 1967; Nadijcka and Hillman, 1974; Dulcibella et al., 1975) provide evidence of such a process. Thus, while confirmation depends on investigation of the potency of early ICM cells, the primitive endoderm probably originates from this tissue.

Fairly clean microsurgical separation of primitive endoderm from ectoderm has been obtained at $4^1/_2$ days p.c. and led to the finding that cells of the two tissues could be readily distinguished following dissociation (Gardner and Papaioannou, 1975). Those of the endoderm layer have a "rough-surfaced" appearance compared to ectodermal cells. Preliminary experiments revealed that the two types of cells gave different patterns of chimaerism following injection into genetically dissimilar host blastocysts (Gardner and Papaioannou, 1975). While progeny of the primitive ectoderm cells contributed widely to organs formed from all three foetal germ layers as well as certain extra-embryonic structures, those of the primitive endoderm appeared to be restricted in distribution to extra-embryonic endoderm.

These findings have been consolidated and extended by more recent experiments in which the yolk sacs of presumptive chimaeras were separated enzymatically into endodermal and mesodermal components prior to analysis (Gardner and Rossant, 1978). Isozymal variants of glucose phosphate isomerase *(Gpi-1)* were employed as genetic markers. The electrophoretic assay that was used to resolve the isozymes enables routine detection of a 3% or greater donor cell contribution. At this level of sensitivity the two types of ICM cells exhibited mutually exclusive patterns of colonisation of host conceptuses (see Fig. 4 for details). Several points are relevant in attempting to assess the significance of the results. First there was an exact correspondence between the morphology of the donor cells and their subsequent distribution, regardless of the number of one or other type injected into each host blastocyst. Second, the donor cell contribution to chimaeric tissues often reached 50%, sometimes more. Finally, detailed cytogenetic analysis of adult chimaeras produced by injecting 2–5 dissociated $3^1/_2$-day-p.c. ICM cells or single $4^1/_2$-day-p.c. embryonic ectoderm cells into blastocysts demonstrates that their progeny become widely disseminated (Ford et al., 1975; Ford et al., unpublished data). These considerations seem to exclude trivial explanations for the results, such as chance partitioning of basically equipotential cells whose clonal expansion is insufficient to embrace both ectodermal and primitive endodermal derivatives. Instead, they favour highly selective colonisation of host embryos by each type of donor cell. This in turn implies that primitive

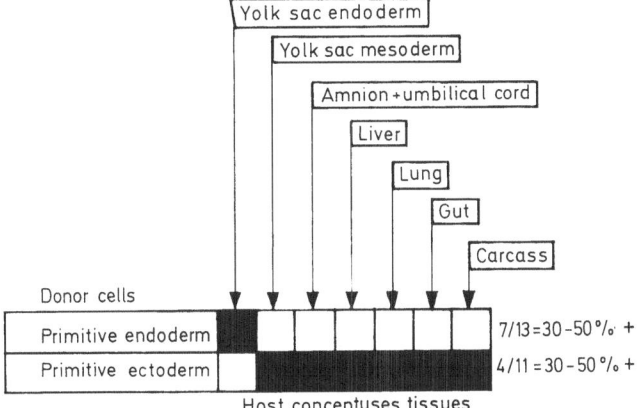

Fig. 4. Diagram illustrating the distribution of chimaerism in embryonic and extra-embryonic tissues of advanced (mainly $15\frac{1}{2}$ days p.c.) conceptuses developing from blastocysts injected with disaggregated $4\frac{1}{2}$-day-p.c. primitive endoderm or primitive ectoderm cells of different *Gpi-1* genotype. The carcass samples represent entire foetuses, excluding liver, lungs, and gut. The proportion of chimaeras in which the donor contribution reached 30–50% or more in one or more host tissues is indicated at the right of the diagram. Analysis of the distal extra-embryonic endoderm could not be undertaken because it degenerates prior to this stage of development. Both categories of donor cells typically yielded chimaerism in the chorio-allantoic placenta, as expected from the histologic composition of this organ. However, in the absence of a means of separating it into its constituent tissues prior to electrophoetic analysis (as was achieved with the yolk sac), the placenta is uninformative

ectoderm and endoderm cells have enduring surface differences at $4\frac{1}{2}$ days p.c. Sorting out of cells in mixed aggregates has been used as an indication of determination in studies on imaginal disc cells in *Drosophila* (e.g., Nöthiger, 1972). Recent in vitro studies on isolated ICMs further support the view that primitive ectoderm cells are determined by the stage at which primitive endoderm has delaminated. Thus Hogan and Tilly (1977) found that the ectoderm did not produce further endoderm after the latter had been removed immunosurgically. Further attempts to elucidate the developmental status of the two types of cells are in progress. These include injection of rat ectodermal and primitive endodermal cells into mouse blastocysts in order to examine their distribution more precisely in sectioned material by immunofluorescence.

It is evident from these experiments that the primitive endoderm does not make a cellular contribution to endodermal or, indeed, any other organs of the foetus. The entire foetus is in fact formed by progeny of cells present in the $4\frac{1}{2}$-day-p.c. primitive ectoderm. This conclusion accords with work of Škreb and his colleagues (reviewed in Škreb et al., 1976) on grafting germ layers isolated from post-implantation rat embryos under the kidney capsules of syngeneic adult hosts. Primitive ectoderm isolated at the two-layered embryonic shield stage or ectoderm plus mesoderm at the slightly more advanced primitive streak stage formed teratomas containing tissues representative of all three foetal germ layers. However, by the head-fold stage (Snell and Stevens, 1966), ectoderm had lost the capacity to form definitive endodermal derivatives while endo-mesodermal grafts had acquired it. Grafts of endoderm alone showed no differentiation, regardless of the stage of isolation. These findings have been confirmed recently by similar studies in the mouse (Diwan and Stevens, 1976). Hence the primitive endoderm of the rodent embryo resembles the hypoblast of the avian embryo in contributing to only extra-embryonic tissues (Nicolet, 1967; Škreb et al., 1976).

Little is known at present about differentiation within the primitive endoderm and ectoderm, though the latter in particular embarks on a phase of very rapid proliferation following implantation (Snow, 1976). Three morphologically distinct endodermal regions are discernible in the early post-implantation embryo, for two of which some evidence of biochemical differentiation has been obtained very recently. The first of these is the distal or parietal endoderm lining the mural trophectoderm (Fig. 3). This tissue engages in a phase of plasminogen activator production (Strickland et al., 1976). The latter is probably a distinct species of molecule from that produced later by ectoplacental trophoblast (S. Strickland and E. Reich, personal communication of unpublished data). The proximal or visceral endoderm is divisible morphologically into regions surrounding the embryonic and extra-embryonic ectoderm. The cells of the former have been found by im-muno-peroxidase labelling to contain α foeto-protein (M. Dziadek, personal communication of unpublished data). The fate of these cells is not known, though, as mentioned earlier, they do not contribute to definitive foetal endoderm. The remainder of the proximal endoderm embracing extra-embryonic ectoderm is believed to form the endoderm of the yolk sac (Snell and Stevens, 1966). However, it is important to bear in mind that while all three endodermal regions probably take origin from the primitive endoderm of the late blastocyst, this has yet to be established by means of unequivocal genetic markers.

Investigations on pre- and post-natal chimaeras produced by injecting single $4\frac{1}{2}$-day-p.c. primitive ectoderm cells into blastocysts indicate that, apart from inability to form trophectoderm and extra-embryonic endoderm, these cells are unrestricted in potency (Gardner and Papaioannou, 1975; Gardner and Rossant, 1977; Ford et al., unpublished data). Interestingly, individual chimaeras can exhibit both somatic and germ cell mosaicism, proving that at least some primitive ectoderm cells are ancestral in the germ lineage (see Gardner and Rossant, 1976). This conflicts with the widely held view that primordial germ cells are of yolk sac endodermal origin (Brambell, 1956; Hamilton and Mossman, 1972), and accords with the work of Ozdzenski (1967), who claimed to detect these cells first in the primitive streak or allantoic mesoderm at $7\frac{1}{2}$ days p.c. using alkaline phosphatase staining as a cytochemical marker. Presumably, therefore, segregation of germ-line cells occurs in the primitive ectoderm some time between $4\frac{1}{2}$ and $7\frac{1}{2}$ days p.c.

Attempts to investigate onset of differentiation within the primitive ectoderm by transplanting cells from later donor embryos have been thwarted by failure to obtain chimaerism. The reason for lack of success is not clear, especially since extra-embryonic ectoderm and endoderm cell injections have succeeded (Rossant and Ofer, 1977; Gardner et al., unpublished observations). The only approach that has yielded insight into this problem so far is the ectopic grafting of Škreb et al., mentioned earlier. Before the primitive streak stage they found no evidence of regional differentiation when different areas of the primitive ectoderm were placed under the kidney capsule. They concluded that it is the stage of formation of mesoderm from primitive ectoderm that heralds onset of differentiation within the latter (Škreb et al., 1976).

Having briefly discussed the fate of primitive endoderm and ectoderm, it is appropriate to examine observations relevant to initial determination of these cells. These are as follows:

1. Rat ICMs isolated prior to endoderm formation appear to exhibit position-dependent differentiation following injection into $3^1/_2$-day-p.c. mouse blastocysts. The majority spread over the blastocoelic surface of host mouse ICMs and contribute mainly to the extra-embryonic endoderm of later conceptuses. However, in a few instances they remain separate and form complete secondary embryos attached to host mural trophectoderm (Gardner and Johnson, 1975).

2. Aggregated pairs of isolated $3^1/_2$-day-p.c. mouse ICMs transplanted to the oviduct inside empty zonae delaminate a complete surface monolayer of cells within 24 h (Rossant, 1975b). These not only have the morphologic characteristics of primitive endoderm cells (Fig. 5), but give the appropriate pattern of chimaerism following injection into host blastocysts (Rossant, 1977).

3. Histologic analysis of abnormal conceptuses developed from blastocysts whose ICM cell number had been reduced experimentally revealed the presence of extra-embryonic endoderm in all cases in which ICM derivatives were detected. Primitive ectoderm derivatives were apparent only in cases where ICM development was more extensive (Gardner, 1975b).

4. Endoderm also differentiates on the exposed surfaces of clumps of embryonal carcinoma cells undergoing "embryoid" development in vitro (Martin and Evans, 1975). The fact that embryonal carcinoma cells can participate in normal development when transplanted into blastocysts (Mintz et al., 1975; Mintz and Illmensee, 1975; Papaioannou et al., 1975; Illmensee and Mintz, 1976) argues that their differentiation is governed by the same mechanisms as cells of the embryo.

Collectively, the findings suggest that differentiation of primitive ectoderm versus endoderm may depend on cell position, as is evidently so for trophectoderm versus ICM (Gardner and Johnson, 1975; Gardner and Papaioannou, 1975; Gardner and Rossant, 1976). According to this hypothesis ICM cells that are totally enclosed between other ICM and polar trophectoderm cells are thereby determined to form primitive ectoderm, while those on the blastocoelic surface are determined to form endoderm. This obviously presupposes that at an earlier stage individual ICM cells should be able to contribute to all derivatives of this tissue. The considerable regulative capacity of the $3^1/_2$-day-p.c. ICM is compatible with this notion (Lin, 1969; Gardner, 1971, 1975b). However, it has yet to be demonstrated by the rigorous test of injecting single early ICM cells into genetically marked blastocysts (Gardner, 1975b; Gardner and Rossant, 1977).

Clearly, prospective studies afford as yet a very incomplete picture of early murine development. Nevertheless, several conclusions may be drawn tentatively from the data that have been amassed so far. First, determination appears to depend primarily on the position that cells occupy within initially homogeneous populations rather than on the parcelling out during cleavage of information already present in the egg. Second, determination for a particular type of cyto-differentiation takes place in relatively small populations of cells (Fig. 6). Third,

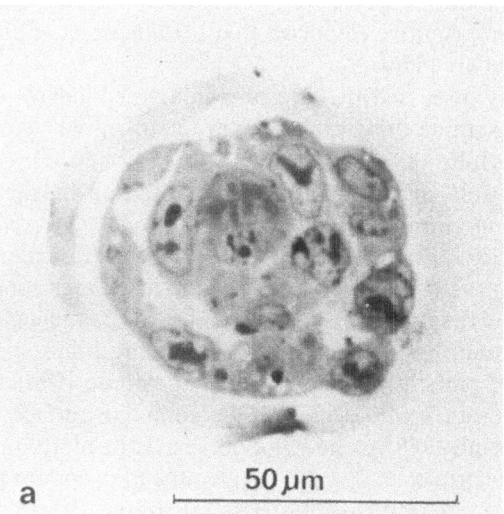

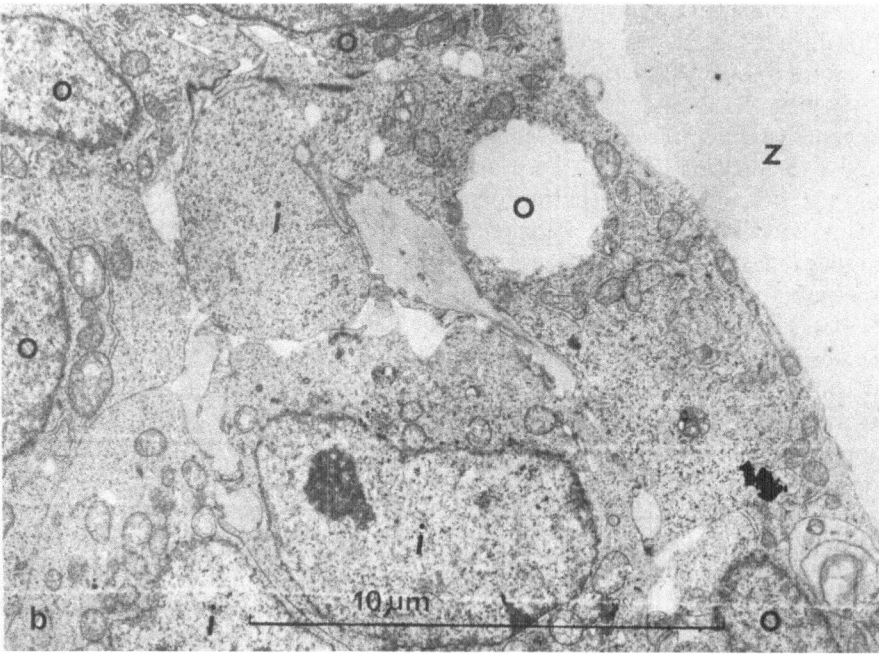

Fig. 5a and b. The structure of aggregated pairs of isolated $3\frac{1}{2}$-day-p.c. ICMs injected into empty zonae, after 1 day in the oviduct: (a) light micrograph of a sectioned aggregate showing the outer monolayer of primitive endoderm cells and inner cluster of primitive ectoderm cells; (b) electron micrograph of a sectioned aggregate showing that endoplasmic reticulum, peculiar to primitive endoderm cells at this stage of development, is apparent in all outside (o) cells but absent from inside (i) cells. Z, zona pellucida. (From Rossant, 1976b, with permission of the author and the Company of Biologists)

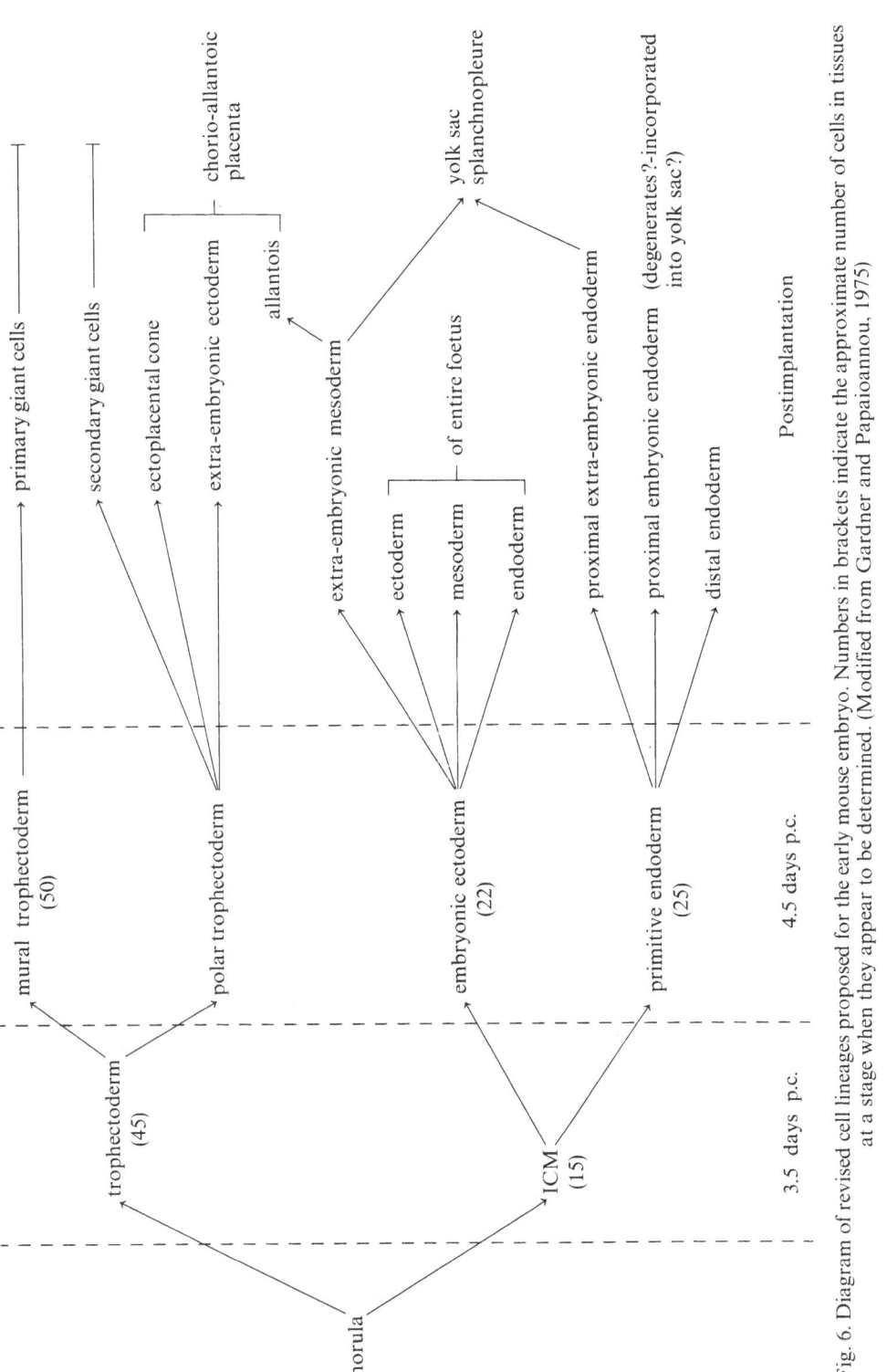

Fig. 6. Diagram of revised cell lineages proposed for the early mouse embryo. Numbers in brackets indicate the approximate number of cells in tissues at a stage when they appear to be determined. (Modified from Gardner and Papaioannou, 1975)

with the exception of the polar trophectoderm, which clearly retains the capacity to form giant cells, each determinative step results in commitment of cells within a defined population to one of two mutually exclusive paths of subsequent differentiation. Fourth, while there is a very close temporal relationship between determination and "overt" physiologic and/or morphologic differentiation in the trophectoderm and primitive endoderm that are destined to form exclusively extra-embryonic tissues, this does not seem to be so in the primitive ectoderm. Finally, these studies emphasise the importance of the use of unequivocal genetic markers in tracing cell lineages. Thus, in addition to establishing ICM control of trophectodermal proliferation, experiments employing such markers have necessitated revision of ideas on the fate of primitive endoderm and on the origin of extra-embryonic ectoderm and germ cells.

V. Retrospective Studies

A. Adult Chimaeras and X-Inactivation Mosaics

Studies discussed in the last section depend largely on isolating defined populations of cells and examining their developmental potential either collectively or individually in genetically dissimilar host embryos, ectopic grafts, or in culture. While such prospective studies have provided valuable data on cell lineage and determination during early development, later differentiation of the primitive ectoderm has so far proved refractory to detailed analysis by these means. This tissue is of central interest to the mammalian embryologist because it gives rise to the definitive embryo.

An alternative way of attempting to gain information about development of the embryo proper is by the study of genetically mosaic adult organisms. Insects are eminently suitable for this type of approach because very little cell mixing occurs during ontogeny. In consequence, much has been learnt about insect development thus, as will be apparent from other contributions to this volume. Similar studies have been undertaken in mammals on post-natal chimaeras and X-inactivation mosaics; but here interpretation of the data is often more difficult. This is due in part to the paucity of suitable marker genes (Gardner and Johnson, 1975; McLaren, 1976a) and also to the pattern of growth of the embryo. While available data indicate that the early rodent embryo exhibits coherent clonal growth (Garner and McLaren, 1974; Gardner and Johnson, 1975), it is clear from analyses on adult chimaeras and mosaics that this is succeeded by very extensive cell mixing (e.g., Ford et al., 1975; P. Iannacone, personal communication). How and when the transition takes place during development is at present obscure.

Investigations on later development using chimaeras and X-inactivation mosaics have been reviewed comprehensively in several recent publications (Nesbitt and Gartler, 1971; Mintz, 1974; McLaren, 1976a). Hence the following discussion will be restricted to a general appraisal of those relevant to elucidation of the relationship between cell lineage and differentiation. In this context there are basically two ways in which genetically mosaic mammals have been employed. The first is to determine the number of precursor cells from which the embryo and its constituent organs and tissues originate. The second is to establish lineage relationships between tissues. In most cases available marker genes enable only estimation of the proportion of cells of either genotype and afford no clue as to their spatial arrangement. The variance in proportions for a given tissue in a series of individuals, or the frequency of individuals in which it does not show mosaicism, has been used to estimate the number of its precursor cells (e.g., Mintz, 1970; Nesbitt, 1971; Fialkow, 1973). Similarly, correspondence in mosaic composition between different tissues (Gandini et al., 1968; Mintz, 1971) has been taken to imply their common origin from the same pool of precursor cells.

B. Precursor Cell Numbers

These numerical studies are based on two major assumptions. The first is that neither cellular genotype contributing to a chimaera or mosaic has a selective advantage in terms of rate of proliferation or survival. The second assumption is that they are distributed randomly at the stage when the precursor cells are "set aside." The validity of both these assumptions is questionable. There is evidence for cell selection in both X-inactivation mosaics and chimaeras (reviewed by McLaren, 1976a). Furthermore, as indicated earlier, growth of the embryo is evidently coherent, at least initially. This means that both the accuracy and significance of such numerical estimates are debatable. Most discussion has been concerned with the stage in the developmental history of the embryo and its constituent organs and tissues to which the derived numbers relate. This has often been couched in rather vague terms (see Lewis et al., 1972; McLaren, 1972, 1976a), though the stage at which restriction of cellular potency takes place has usually been implied (Nesbitt, 1971; Tettenborn et al., 1971). Mintz (1970, 1971, 1972a, b) has developed this notion most explicitly on the basis of pre- and postnatal studies on aggregation chimaeras. She postulates that the number derived for an organ or tissue represents its "clonal-initiator cells" and defines a clone as "the mitotic progeny of one cell in which a specific constellation of gene loci first become active (or derepressed) and has remained either active or mobilizable as a cell heredity" (Mintz, 1970). This appears, in effect, to be identifying determination as the stage in development of a structure to which the numerical estimates relate. She regards development as a progressive series of clonal determinations in which constant small numbers of cells initiate differential genetic activity, starting with the embryo and then proceeding to its constituent organs and tissues. The hypothesis thus resembles that of polyclones in the development of compartments in insects (Crick and Lawrence, 1975). However, it is worth emphasising that determination can only be established operationally by studying the developmental potential of cells in altered circumstances (e.g., Gehring, 1972; Gardner and Papaioannou, 1975). The fact that polyclones have been shown to respect compartment boundaries in *Drosophila* suggests that they may indeed be determined (Morata and Lawrence, 1975), but there is as yet no evidence that numbers extrapolated from mammalian mosaics and chimaeras apply to the stage of restriction in cellular potency. It is the assertion that they do, rather than Mintz's hypothesis per se, that has been challenged both by McLaren (1972, 1976a) and by Lewis et al. (1972). Essentially, they argue that the statistical approach is only meaningful in systems in which an indefinite period of random cell mixing is succeeded by a period of coherent clonal growth during which the structure in question is formed. McLaren (1976a) applies the term *allocation* to the stage in development of a tissue or organ primordium at which movement of cells between it and neighbouring primordia effectively ceases. In such systems binomially derived initial cell numbers relate either to the stage of *allocation* or to that of initiation of mosaicism, whichever occurs later in development. The final point emphasised by McLaren (1972, 1976a) is that there is no a priori reason why allocation should coincide with determination rather than any one of several other events in the ontogenesis of an adult structure.

In the one case in which the retrospective statistical approach can be related to direct study of cellular potency an obvious numerical discrepancy is indicated. Mintz (1970) tentatively inferred a 3-cell origin for the definitive mouse embryo from an estimated 75% frequency of chimaerism in 129 mice produced by C3H—C57/BL morula aggregations. However, as noted earlier, single primitive ectoderm cells isolated from $4\frac{1}{2}$-day-p.c. blastocysts contribute progeny to both embryonic and extra-embryonic tissues when injected into host blastocysts. The primitive ectoderm is composed of more than 20 cells at this stage (McLaren, 1976b). Hence, in this case at least, the derived number of 3 cells is unlikely to represent the number of determined embryo progenitor cells. It is more likely to represent the average number of cells present at the time of morula aggregation which contribute a proportion of their mitotic descendents to the region of primitive ectoderm from which the primordium of the definitive embryo becomes determined some time after $4\frac{1}{2}$ days p.c. This conclusion is consistent with the lack of cell mixing observed in chimaeric blastocysts by Garner and McLaren (1974).

The two genetically distinct populations of cells in aggregation chimaeras coexist prior to the onset of cellular determination. The fact that all tissues of such chimaeras that have been analysed can contain cells of both genotypes has led to the conclusion that each must originate from at least two cells. This seems reasonable on the grounds that development would be remarkably precarious were it otherwise. However, the assumption that numerical estimates derived from chimaeras apply to standard mice is more questionable. Aggregation chimaeras are composed initially of twice as many cells as standard embryos, and contain more than double the normal number of ICM cells at the blastocyst stage (Buehr and McLaren, 1974). Cell number evidently remains elevated until $5\frac{1}{2}$ days p.c. when it is rapidly restored to normal by as yet elusive means (Buehr and McLaren, 1974). Hence, if *allocation* of primordia occurs prior to this stage, as is almost certainly the case with the ICM itself, derived numbers would be larger in chimaeras. This problem obviously does not arise in the case of X-inactivation mosaics in which it is replaced by one of uncertainty as to when mosaicism arises.

One obvious way of attempting to reduce some of the uncertainties inherent in retrospective analysis is to study tissues in which genetic mosaicism yields phenotypic differences that are manifest at the level of individual cells in situ. Only thus can the spatial arrangement of cells of both genotypes be visualised in chimaeras and mosaics. Unfortunately, there is a dearth of such cell autonomous marker genes in mammals (Gardner and Johnson, 1975; McLaren, 1976a). Melanocytes are undoubtedly the most amenable cell type for this approach since allelic differences exist at several loci which control the structure and arrangement of melanosomes (Wolfe and Coleman, 1966). Pigmentation patterns can be visualised readily in the coat, iris, choroid, retina, and inner ear. However, they are strictly cell autonomous only in the latter two situations; melanosomes are secreted by melanocytes in hair follicles and choroid (McLaren, 1976a). Considerable attention has been devoted to patterns of coat pigmentation in chimaeras and X-inactivation mosaics because they are so readily studied in living mice. The fact that pigmentation mosaicism can be observed in individual hairs argues that secretory activity of more than one melanocyte per hair follicle is involved (McLaren and Bowman, 1969; Cattanach et al., 1972).

The definitive experimental investigation on the origin and migration of melanoblasts in the mouse was undertaken by Rawles (1947). She grafted pieces of 8- to 12-day embryos of a pigmented strain into the coelom of White Leghorn chick embryos. The capacity to produce melanocytes was found thus to be associated initially with grafts that included the neural tube, and later with those taken from progressively more lateral locations. Furthermore, this transverse migration of melanoblasts appeared to begin earlier at anterior than more posterior levels of the neuraxis. The experimental findings correlated well with morphologic observations on the development and dispersal of neural crest tissue (Rawles, 1947).

The patterns of coat pigmentation displayed by post-natal X-inactivation mosaics and chimaeras are consistent with the above findings. They are most striking in mice in which the constituent populations of melanocytes differ in genetic activity at the albino locus (e.g., Mintz, 1967). Completely random admixture of pigmented, albino, and mixed hairs is never seen. Rather, more or less discrete transverse stripes or bands are the rule, in which intensity of pigmentation often trails off in a ventral direction. Furthermore, patterning on the two sides of the body seems to be autonomous, as evidenced by the frequent discontinuity of bands both mid-dorsally and mid-ventrally (Fig. 7). Mintz (1967, 1970, 1974) argues that the occurrence of mid-dorsal discontinuity requires that melanoblasts are determined by 7 days p.c. before closure of the neural folds. However, since migration of neural crest cells does not begin until after closure (Rawles, 1947), it would seem necessary to postulate additionally or alternatively that cells on each side of the neural tube exhibit polarised movement away from it. The discontinuity mid-ventrally suggests that migration is completed prior to closure of the ventral body wall.

The simplest explanation for the observed patterns is that melanoblasts proliferate extensively during migration so as to colonise transverse areas of the body surface. Hence, some of the smaller stripes would be expected to represent single melanoblast clones. This notion receives support from work of Russell and Major (1957) who X-irradiated $10\frac{1}{2}$-day-p.c. pregnant mice carrying foetuses that were heterozygous for recessive alleles at several loci expressed in melanocytes. Approximately 11% of the resulting offspring showed areas of mutant coat pigmentation. Single coherent patches were found in all except two mice. Though very

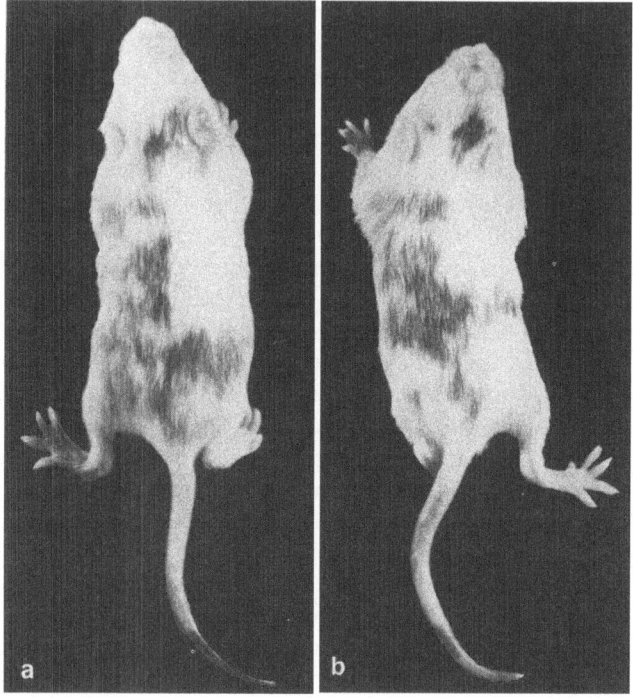

Fig. 7a and b. Two 11-day post-partum chimaeras produced by transplanting ICMs isolated from 4½ day p.c. Balb/c blastocysts of albino genotype to 3½ day p.c. C57/BL blastocysts of a pigmented genotype. Note the transverse arrangement of the pigmented areas of their coats, and the often striking discontinuity along the dorsal mid-line

variable in both size and location, they exhibited a mainly transverse orientation and respected both dorsal and ventral mid-lines.

Attempts have been made to estimate the number of primordial melanoblast clones by studying the banding in mice displaying melanocyte mosaicism. This task is complicated by various factors, such as differences in the proportion of the two cell types, cell selection, and mutual encroachment between adjacent patches (see Mintz, 1970; McLaren, 1976a). For example, melanocytes of one genotype consistently predominated in the anterior coat while those of the other genotype predominated posteriorly in one series of chimaeras (West and McLaren, 1976).

Mintz (1967) claimed that, despite these sources of developmental "noise," she could discern an "archetypal" pattern in chimaeras consisting of three pairs of stripes on the head, six pairs on the trunk, and a further eight pairs along the tail. This was attributed to clonal proliferation of a chain of 17 primordial melano-blasts extending along each side of the neuraxis. Following examination of a larger series of aggregation chimaeras she later withdrew the additional postulate that melanoblasts within each chain alternate in genotype (Mintz, 1970). While this model has been accepted by others (e.g., Cattanach et al., 1972; Cattanach, 1974), retraction of the latter, biologically improbable postulate, accentuates the problem highlighted by Wolpert and Gingell (1970). These authors pointed out that if adjacent clones are as likely to be of the same as of opposite genotype, chimaeras displaying all 34 bands would be expected to be exceedingly rare on statistical grounds. They argued that the observed frequency of maximally patterned chimaeras would require approximately 32 clones on each side of the body, nearly twice the total number postulated by Mintz (1967). Schaible (1969) also tackled the problem of clonal numbers by introducing various spotting genes into mouse stocks in order to reduce the extent of coat pigmentation. He identified approximately 14 discrete centres by this method (6 pairs of lateral centres and 2 medial). However, as pointed out by Mintz (1974), this model is less readily reconciled with the embryologic data of Rawles (1947). Furthermore, since both the site and mode of action of spotting genes are controversial (Searle, 1968; Mayer, 1970; Deol, 1971, 1973) the technique may be identifying centres of survival rather than origin of melanoblasts. Thus, while there is general agreement that coat colour patterns of mice exhibiting melanocyte mosaicism have a clonal basis, the number of such clones is still uncertain.

It has been known for some time that not all genes affecting coat colour do so via melanocytes that populate the hair follicles. The most familiar exception is the *agouti* locus. In phenotypically agouti mice carrying the *A* allele overhairs are black with yellow subterminal bands (Wolfe and Coleman, 1966). Transplantation studies by Silvers and Russell (1955) suggested that this transitory switch from eumelanin to phaeomelanin production by melanocytes depended on the genotype of other cells in the hair follicle. More recent dermo-epidermal recombination experiments have established that, with the exception of the dominant *yellow lethal (A^y)* allele (Poole, 1974), it is the genotype of the mesodermal component of the follicle that determines this aspect of hair phenotype (Mayer and Fishbane, 1972). Studies on aggregation chimaeras have indeed confirmed that the agouti locus is expressed in cells other than melanocytes. Thus mice obtained by combining embryos that differed in genotype at both the *agouti* locus and loci expressed in melanocytes showed two new "recombinant" hair phenotypes in addition to the parental ones (McLaren and Bowman, 1969; Mintz and Silvers, 1970). Chimaeras obtained by combining *agouti (A/A)* with *non-agouti (a/a)* morulae also show transversely arranged stripes of one or other phenotype which can be out of register on the two sides of the body (Mintz, 1970, 1971). However, these stripes are much narrower than those attributable to melanocyte mosaicism, and seem to be more regular in width (McLaren, 1976a). This is particularly clear in mice exhibiting both agouti and melanocyte mosaicism (e.g., Mintz and Silvers, 1970). Very similar narrow stripes are seen in mice that are heterozygous for the X-linked mutant gene *tabby* (Gruneberg, 1966b) and in tabby↔wild-type chimaeras (Cattanach et al., 1972), as well as in chimaeras produced by aggregating embryos that are homozygous for the autosomal recessive *fuzzy* with their wild-type counterparts (Mintz, 1970, 1971).

Although hairs develop from the ectodermal cells of follicles, their morphogenesis requires an inductive stimulus from the mesodermally derived dermal papillae (see Sengel, 1976 for a review). Mintz (1970, 1971) therefore postulated that the coat patterns produced as a consequence of mosaicism at all three of the above loci have a common basis, and reflect clonal development of the mesodermal component of the hair follicles. She finds that in the trunk region, where comparisons can be made, the number of stripes corresponds well to the somite number, and argues that unit stripes represent clones formed by single cells of somitic dermatome origin. The developmental "archetype" for coat mesoderm is thus attributed to medio-lateral growth of approximately 166 progenitor cells. However, several aspects of the observed patterns seem to be at variance with this model. One is that, although no measurements have been published, the width of these stripes appears to be rather regular compared with those attributable to melanocyte mosaicism (McLaren, 1976 a). Also, certain areas of the coat seem to be occupied consistently by follicles of one phenotype rather than the other in agouti chimaeras (McLaren, 1976 a). Finally, while mutant hairs predominate in dark bands of tabby heterozygotes and wild-type hairs in light bands, the reverse was found in one tabby↔wild-type aggregation chimaera (Cattanach et al., 1972). Hence, the colour of a band is not dependent simply on its composition of hairs. Alternative explanations for these non-melanocytic coat patterns are discussed elsewhere (Gruneberg, 1966 b; McLaren, 1976 a).

Melanocytes of the retinal pigment epithelium and inner ear are morphologically different from those of the coat, iris, and choroid and are believed to originate directly from the central nervous system rather than the neural crest (see McLaren, 1976 a). As noted earlier, their pigmentation is strictly cell autonomous. Patterns of mosaicism in the retina and inner ear have been studied both from the viewpoint of determining progenitor cell numbers (Mintz and Sanyal, 1970) and of attempting to establish the time at which the cells undergo X-chromosome inactivation (Deol and Whitten, 1972 a, b). The former authors concluded that the pigment epithelium of each retina develops from 10 precursor cells, each of which proliferates clonally to form a radiating sector (Mintz and Sanyal, 1970; Mintz, 1971). Deol and Whitten (1972 a, b) compared patch size in this tissue and the inner ear in albino↔pigmented chimaeras versus heterozygotes exhibiting X-autosome-induced mosaicism at the same locus. They inferred that X-chromosome inactivation was a late event (probably taking place after establishment of these primordia), from the fact that patches were smaller and hence more numerous in the latter.

However, West (1975) has pointed out that the preceding studies failed to take into account the relative proportion of the two types of melanocytes as a factor affecting the relationship between patch and clone size. Put in crudest terms, the greater the departure from equal representation, the more clones of the numerically dominant genotype will be adjacent to like ones, resulting in increase in average patch size. When allowance is made for this, the estimated average number of clones per patch in the $12\frac{1}{2}$-day-p.c. retinal pigment epithelium was similar in mosaics and chimaeras. The number of clones increased by a factor of three between this stage and adulthood, and the average number of cells per clone to between five and six (West, 1976 a). These findings argue for the occurrence of extensive cell mixing during early development of this tissue, and slight mixing later. West's failure to detect any coherent clonal patterns at $12\frac{1}{2}$ days p.c. suggests that the radiating sectors noted by Mintz and Sanyal (1970) arise later in development of this epithelium [1].

C. Cell Lineage Analysis

Few retrospective studies have been undertaken on chimaeras and X-inactivation mosaics that relate specifically to this aspect of development. Correlation in mosaicism between adult tissues cannot necessarily be taken to imply that they share a common ancestry from the same pool of precursor cells. It could be the result of selection operating through a mutant allele carried in one cellular component of a mosaic or chimaera that is expressed in cells belonging to several distinct lineages. Mutant alleles at the W-locus, for example, affect the proliferation of melanoblasts, primordial germ cells, and erythroid stem cells (Green, 1966; Metcalf and Moore, 1971).

In addition, one might anticipate fortuitous correlations between remote tissues if cells of the two constituent genotypes were very thoroughly mixed throughout the embryo, and relatively large numbers of cells were *allocated* to each (e.g., Fialkow, 1973). Detailed marker chromosome analysis of a series of blastocyst injection chimaeras indeed points to the occurrence of a marked degree of intermingling of cells during embryogenesis (Ford et al., 1975; Ford et al., unpublished observations). Such considerations may account for the fact that mosaicism in several internal cell populations of chimaeras, including the lypho-myeloid and skeletal systems, seems to be positively correlated with the melanocyte populations of the coat (Ford et al., 1974; Gruneberg and McLaren, 1972).

Correlations have been found between the relative proportions of the two haemoglobin and gamma-globin types in aggregation chimaeras (Mintz and Palm, 1969; Wegmann and Gilman, 1970) as also between red and white blood cells (B. Mintz and R. Niece, unpublished data quoted in Mintz, 1971). The significance of these findings is questionable in the light of the foregoing discussion, since no attempt was made to examine the extent of correlation between these and other cell types. However, studies involving transplantation of haemopoietic cells carrying radiation-induced marker chromosomes to lethally irradiated adult rodents provide independent evidence that lymphocytes, granulocytes, and erythrocytes may share a common stem cell (Whang et al., 1963; Wu et al., 1968; Nowell et al., 1970).

---

[1] Examination of the neural retinas of chimaeras produced by aggregating embryos that were homozygous for the gene retinal degeneration *(rd/rd)* with wild-type embryos led Mintz and Sanyal (1970) to propose that they also developed by coherent radial growth from 10 initiator cells. These sectors did not correspond to those of the overlying pigment epithelium, thus implying a separate clonal origin. However, measurement of patch size in such chimaeric neural retinas is complicated by the occurrence of phenotypically intermediate areas (West, 1976b). These might be due to lack of cell autonomy of the mutant allele, or to compensatory migration of wild-type cells.

The need to exercise caution in interpreting correlations in mosaicism is also illustrated by studies on X-inactivation mosaics. Gandini and his colleagues found that women who were proven heterozygotes for the *deficient (mediterranean)* allele of X-linked *glucose-6-phosphate-dehydrogenase* (G-6-P-D) and showed the hemizygous deficient phenotype in red cells, also did so in granulocytes, but not in lymphocytes and other tissues sampled (Gandini et al., 1968). They therefore inferred a common lineage for these two types of haemopoietic cells. However, more recently, Fialkow (1973) undertook a similar study of women that were heterozygous for genetically determined electrophoretic variants of the same enzyme. He found that the correlation in mosaic composition was as high between granulocytes and lymphocytes on the one hand as between granulocytes and skin or muscle on the other. He concluded from analysis of the variance between the same cell types in a series of such heterozygotes that all tissues that were sampled had a common origin from approximately 16 cells present at the time of X-chromosome inactivation. Furthermore he attributed the high correlation between all the different cell types sampled in individual heterozygotes to their each being established from at least 80 cells at the time of *allocation*. Finally, he suggested that by choosing heterozygotes for a deficient allele, Gandini et al. (1968) may have been looking at cell selection rather than random sampling from precursor cell pools. Nesbitt (1971) drew similar conclusions to those of Fialkow (1973) from a study of X-inactivation mosaicism in mice.

No attempt has been made to provide a comprehensive review of retrospective studies that claim to illuminate cell lineage and determination during mammalian embryogenesis since this task has been undertaken recently by McLaren (1976a) in an excellent monograph. Theoretical treatment of various aspects of clonal analysis can also be found elsewhere (Wolpert and Gingell, 1970; Lewis et al., 1972; McLaren, 1972; Hutchison, 1973; Lewis, 1973; Nesbitt, 1974; West, 1975). Nevertheless, the above survey serves to illustrate some of the pitfalls and uncertainties that are encountered when trying to interpret these data. These appear to stem from two interrelated problems. First, there is at present a state of complete ignorance concerning the temporal relationship between *allocation* and determination of cells. Second, a satisfactory genetic marker has yet to be found which would enable direct visualisation of cellular deployment throughout embryogenesis. Use of interspecific chimaeras in this context is questionable because of the very real possibility of preferential adhesion of cells according to genotype (Gardner and Johnson, 1975). The same objection may apply to intra-specific chimaeras if the constituent cell populations carry genotypic differences at loci affecting surface properties. Evidence that patch size is indeed influenced by the degree of genetic disparity between cells in mouse chimaeras is discussed by McLaren (1976a). Furthermore, it is likely that the degree of cell mixing during development differs from one region to another. West (1976a) has deduced that there is very "fine-grained" mosaicism in the early retinal pigment epithelium. In contrast, Gearhart and Mintz (1972) found that in chimaeras in which the constituent cell populations differed in genotype at the *Gpi-1* locus, hetero-polymeric enzyme was much rarer in the extrinsic muscles of the eye than in other striated muscle, implying a much higher degree of coherent clonal growth of pre-myoblasts in the former case.

VI. Conclusions

The task of preparing this article has served to make the author acutely aware of the extent of our ignorance regarding early development of the mouse embryo. Much of the information needed to undertake a critical evaluation of the relationship between cell lineage and differentiation is not available. Nonetheless, it is perhaps instructive to attempt to relate what data are at hand to the lineage hypothesis of Holtzer and his colleagues (Dienstman and Holtzer, 1975; Holtzer et al., 1975), if only to pinpoint some of the vital gaps in knowledge. Cell commitment is, at least during the initial preimplantation phase of development, explicable on a binary basis. Late cleavage blastomeres can form only ICM or trophectoderm cells. Somewhat later, within the ICM itself, cells have the option of becoming either primitive endodermal or ectodermal cells. Furthermore, attempts to manipulate early development have provided no evidence that temporal compartments within lineages can be by-passed. However, the question whether such steps in cytodifferentiation are linked in an obligatory manner with DNA synthesis and cell division has yet to be examined.

The principal difficulty encountered in attempting to apply the lineage hypothesis to the mouse embryo concerns the role of extracellular factors in differentiation. Holtzer and colleagues attribute the sequential programming of cells to endogenous nuclear–cytoplasmic interactions, and relegate exogenous factors to a purely permissive status. There is little problem in envisaging how such a model might be applied to species with mosaic eggs in which developmentally significant regional cytoplasmic differentiation has been found to exist. However, it is evident from data discussed earlier that the mouse egg is not organised in this way, and that the spatial relationship between cells emerges as a primary factor in early differentiation in this species. Furthermore, results of experiments involving rearrangement of blastomeres cannot be explained simply by invoking selective survival and proliferation of one or other of two already distinct types of cells (Hillman et al., 1972). Instead, they seem to require that position is in some way instrumental in selecting which of two programmes individual cells will express rather than merely permitting realisation of that which has been determined endogenously.

It is important to bear in mind that conclusions about differentiation in early mouse development are based largely on detailed experimental analysis of the few steps that take place prior to implantation. One might question whether these are likely to be representative on the grounds that preimplantation development is adapted specifically to meet the demands of viviparity. Thus, while there is clearly a need to consolidate work on the preimplantation embryo, the principal challenge for the future is extension of experimental studies to early postimplantation stages. This will certainly require development of fresh approaches and new techniques. The limitations of some of the existing ones have been alluded to earlier. Retrospective studies on their own are likely to raise more questions than they answer. Recently, progress has been made in obtaining development of mouse embryos through the implantation period in culture (reviewed in McLaren and Hensleigh, 1975). So far, however, few attempts have been made to carry out

experiments on embryos under these conditions. Finally, the striking parallel that
has emerged concerning endoderm formation in the mouse and chick embryo
serves as a reminder that the latter is also an amniote, all phases of whose
development have been subjected to very extensive experimental investigation
(Bellairs, 1971). Greater familiarity with avian development might assist the mammalian embryologist in asking the most pertinent questions of his more elusive
material.

Acknowledgements. I wish to thank Dr. J.B.Gurdon, Dr. C.F.Graham, and Dr. M.F.Lyon
for valuable discussion, and Mrs. P.Little and Mrs. M.White for their help in preparation of
the manuscript. I am also indebted to the Medical Research Council for support.

References

Anderson,D.T.: Embryology and Phylogeny in Annelids and Arthropods. Oxford: Perga-
    mon Press 1973
Anderson,D., Billingham,R.E., Lampkin,G.H., Medawar,P.B.: The use of skin grafting to
    distinguish between monozygotic and dizygotic twins in cattle. Heredity **5**, 379—397
    (1961)
Ansell,J.D., Snow,M.H.L.: The development of trophoblast in vitro from blastocysts con-
    taining varying amounts of inner cell mass. J. Embryol Exp. Morph. **33**, 177—185 (1975)
Barlow,P.W., Owen,D., Graham,C.F.: DNA synthesis in the preimplantation mouse em-
    bryo. J. Embryol. Exp. Morph. **27**, 431—445 (1972)
Barlow,P.W., Sherman,M.I.: The biochemistry of differentiation of mouse trophoblast: stud-
    ies on polyploidy. J. Embryol. Exp. Morph. **27**, 447—465 (1975)
Bellairs,R.: Developmental Processes in Higher Vertebrates. London: Logos Press 1971
Benirschke,K.: Spontaneous chimerism in mammals, a critical review. In: Current Topics in
    Pathology, Vol. 51, pp. 1—61. Berlin-Heidelberg-New York: Springer 1970
Blerkom,J., Van, Barton,S.C., Johnson,M.H.: Molecular differentiation in the preimplanta-
    tion mouse embryo. Nature (Lond.) **259**, 319—321 (1976)
Borghese,E., Cassini,A.: Cleavage of mouse egg. In: Rose,G.C. (Ed.): Cinemicrography in
    Cell Biology, pp. 263—277. New York: Academic Press 1963
Brambell,F.W.R.: Ovarian changes. In: Parkes,A.S. (Ed.): Marshall's Physiology of Repro-
    duction, 3rd Ed., pp. 397—542. London: Longman's Green 1956
Brinster,R.L.: Parental glucose phosphate isomerase activity in three-day mouse embryos.
    Biochem. Genet. **9**, 187—191 (1973)
Buehr,M., McLaren,A.: Size regulation in chimaeric mouse embryos. J. Embryol. Exp.
    Morph. **31**, 229—234 (1974)
Cairnie,A.B., Lala,P.K., Osmond,D.G. (Eds.): Stem Cells of Renewing Cell Populations.
    New York: Academic Press 1976
Calarco,P.G., Brown,E.H.: An ultrastructural and cytological study of preimplantation de-
    velopment of the mouse. J. Exp. Zool. **171**, 253—284 (1969)
Calarco,P.G., Epstein,C.J.: Cell surface changes during preimplantation development in the
    mouse. Dev. Biol. **32**, 208—213 (1973)
Cattanach,B.M.: Position effect variegation in the mouse. Genet. Res. **23**, 291—306 (1974)
Cattanach,B.M., Wolfe,H.G., Lyon,M.F.: A comparative study of the coats of chimaeric
    mice and those of heterozygotes for X-linked genes. Genet. Res. **19**, 213—228 (1972)
Chapman,V.M., Adler,D., Labarca,C., Wudl,L.: Genetic variation of β-glucuronidase ex-
    pression during early embryogenesis. In: Embryogenesis in Mammals. Ciba Foundation
    Symposium 29 (new series), pp. 115—124. Amsterdam: Elsevier, Excerpta Medica 1976

Chapman,V.M., Ansell,J.D., McLaren,A.: Trophoblast giant cell differentiation in the mouse: expression of glucose phosphate isomerase (GP1-1) electrophoretic variants in transferred and chimeric embryos. Dev. Biol. **29**, 48—54 (1972)

Cole,R.J., Paul,J.: Properties of cultured preimplantation mouse and rabbit embryos, and cell strains derived from them. In: Wolstenholme,G.E.W., O'Connor,M. (Eds.): Preimplantation Stages of Pregnancy. Ciba Foundation Symposium, pp. 82—112. London: Churchill 1965

Conklin,E.G.: The organisation and cell-lineage of the ascidian egg. J. Acad. Natl. Sci. Philad. **13**, 5—119 (1905)

Crick,F.H.C., Lawrence,P.A.: Compartments and polyclones in insect development. Science **189**, 340—347 (1975)

Dalcq,A.M.: Introduction to General Embryology. London: Oxford University Press 1957

Daniel,J.C., Jr.: Methods in Mammalian Embryology. San Francisco: Freeman 1971

Daniel,J.C., Jr.: Methods in Mammalian Reproduction. New York: Academic Press 1978

Davidson,E.H.: Gene Activity in Early Development, 2nd ed. New York: Academic Press 1976

Denker,H.-W.: Formation of the blastocyst: determination of trophoblast and embryonic knot. In: Gropp,A. (Ed.): Current Topics in Pathology, Vol. 62, pp. 59—79. Berlin-Heidelberg-New York: Springer 1976

Deol,M.S.: Spotting genes and internal pigmentation patterns in the mouse. J. Embryol. Exp. Morph. **26**, 123—133 (1971)

Deol,M.S.: The role of the tissue environment in the expression of spotting genes in the mouse. J. Embryol. Exp. Morph. **30**, 483—489 (1973)

Deol,M.S., Whitten,W.K.: Time of X-chromosome inactivation in retinal melanocytes of the mouse. Nature (New Biol.) **238**, 159—160 (1972a)

Deol,M.S., Whitten,W.K.: X-chromosome inactivation: does it occur at the same time in all cells of the embryo? Nature (New Biol.) **240**, 277—279 (1972b)

Dickson,A.D.: The form of the mouse blastocyst. J. Anat. **100**, 335—348 (1966)

Dienstman,S.R., Holtzer,H.: Myogenesis: a cell lineage interpretation. In: Reinert,J., Holtzer,H. (Eds.): Results and Problems in Cell Differentiation, Vol. VII: Cycle and Cell Differentiation, pp. 1—25. Berlin-Heidelberg-New York: Springer 1975

Diwan,S.B., Stevens,L.C.: Development of teratomas from the ectoderm of mouse egg cylinders. J. Natl. Cancer Inst. **57**, 937—939 (1976)

Dulcibella,T., Albertini,D.F., Anderson,E., Biggers,J.D.: The preimplantation mammalian embryo: characterisation of intercellular junctions and their appearance during development. Dev. Biol. **45**, 231—250 (1975)

Dulcibella,T., Anderson,E.: Cell shape and membrane changes in the eight-cell mouse embryo: prerequisites for morphogenesis of the blastocyst. Dev. Biol. **47**, 45—58 (1975)

Eager,D.D., Johnson,M.H., Thurley,K.W.: Ultrastructural studies on the surface membrane of the mouse egg. J. Cell Sci. **22**, 345—353 (1976)

Enders,A.C.: The fine structure of the blastocyst. In: Blandau,R.J. (Ed.): Biology of the Blastocyst, pp. 71—94. Chicago: Univ. of Chicago Press 1971

Fialkow,P.J.: Primordial cell pool size and lineage relationships of five human cell types. Ann. Hum. Genet. (Lond.) **37**, 39—48 (1973)

Ford,C.E.: Mosaics and chimaeras. Br. Med. Bull. **25**, 104—109 (1969)

Ford,C.E., Evans,E.P., Burtenshaw,M.D., Clegg,H., Barnes,R.D., Tuffrey,M.: Marker chromosome analysis of tetraparental AKR—CBA-T6 mouse chimaeras. Differentiation **2**, 321—333 (1974)

Ford,C.E., Evans,E.P., Gardner,R.L.: Marker chromosome analysis of two mouse chimaeras. J. Embryol. Exp. Morph. **33**, 447—457 (1975)

Fraser,A.S., Short,B.F.: Studies of sheep mosaic for fleece type I. Patterns and origins of mosaicism. Aust. J. Biol. Sci. **11**, 200—208 (1958)

Gandini,E., Gartler,S.M., Angioni,G., Argiolas,N., Dell'Acqua,G.: Developmental implications of multiple tissue studies in glucose-6-phosphate dehydrogenase-deficient heterozygotes. Proc. Natl. Acad. Sci. USA **61**, 945—948 (1968)

Gardner,R.L.: Mouse chimaeras obtained by the injection of cells into the blastocyst. Nature (Lond.) **220**, 596—597 (1968)

Gardner,R.L.: Manipulations on the blastocyst. Adv. Biosci. **6**, 279—296 (1971)

Gardner,R.L.: Microsurgical approaches to the study of early mammalian development. In: Moghissi,K.S. (Ed.): Birth Defects and Fetal Development: Endocrine and Metabolic Factors, pp. 212—233. Illinois: Thomas 1974

Gardner,R.L.: Origin and properties of trophoblast. In: Edwards,R.G., Howe,C.W.S., Johnson,M.H. (Eds.): Immunobiology of Trophoblast, pp. 43—65. London: Cambridge Univ. Press 1975a

Gardner,R.L.: Analysis of determination and differentiation in the early mammalian embryo using intra- and inter-specific chimaeras. In: Markert,C.L. (Ed.): The Developmental Biology of Reproduction. 33rd Symposium of The Society for Developmental Biology, pp. 207—238. New York: Academic Press 1975b

Gardner,R.L.: The problem of intrauterine orientation of the implanting mouse blastocyst. J. Anat. (Lond.) **124**, 236 (1977)

Gardner,R.L.: Production of chimeras by injecting cells or tissue into the blastocyst. In: Daniel,J.C., Jr. (Ed.): Methods in Mammalian Reproduction. New York: Academic Press 1978

Gardner,R.L., Johnson,M.H.: An investigation of inner cell mass and trophoblast tissues following their isolation from the mouse blastocyst. J. Embryol. Exp. Morph. **28**, 279—312 (1972)

Gardner,R.L., Johnson,M.H.: Investigation of early mammalian development using interspecific chimaeras between rat and mouse. Nature (New Biol.) **246**, 86—89 (1973)

Gardner,R.L., Johnson,M.H.: Investigation of cellular interaction and deployment in the early mammalian embryo using inter-specific chimaeras between rat and mouse. In: Cell Patterning. Ciba Foundation Symposium 29 (new series), pp. 183—200. Amsterdam: Elsevier, Excerpta Medica 1975

Gardner,R.L., Lyon,M.F.: X-chromosome inactivation studied by injection of a single cell into the mouse blastocyst. Nature (Lond.) **231**, 385—386 (1971)

Gardner,R.L., Papaioannou,V.E.: Differentiation in the trophectoderm and inner cell mass. In: Balls,M., Wild,A.E. (Eds.): The Early Development of Mammals. 2nd Symposium of The British Society for Developmental Biology, pp. 107—132. London: Cambridge Univ. Press 1975

Gardner,R.L., Rossant,J.: Determination during embryogenesis. In: Embryogenesis in Mammals. Ciba Foundation Symposium 40 (new series), pp. 5—18. Amsterdam: Elsevier, Excerpta Medica 1976

Gardner,R.L., Rossant,J.: The potency of primitive endoderm and primitive ectoderm cells isolated from $4\frac{1}{2}$ day p.c. mouse blastocysts. (in preparation) (1977)

Gardner,R.L., Papaioannou,V.E., Barton,S.C.: Origin of the ectoplacental cone and secondary giant cells in mouse blastocysts reconstituted from isolated trophoblast and inner cell mass. J. Embryol. Exp. Morph. **30**, 561—572 (1973)

Garner,W., McLaren,A.: Cell distribution in chimaeric mouse embryos before implantation. J. Embryol. Exp. Morph. **32**, 495—503 (1974)

Gearhart,J.D., Mintz,B.: Glucosephosphate isomerase subunit-reassociation tests for maternal-fetal and fetal-fetal cell fusion in the mouse placenta. Dev. Biol. **29**, 55—64 (1972)

Gehring,W.J.: The stability of the determined state in cultures of imaginal disks in Drosophila. In: Ursprung,H., Nöthiger,R. (Eds.): Results and Problems in Cell Differentiation, Vol. V: The Biology of Imaginal Disks, pp. 35—58. Berlin-Heidelberg-New York: Springer 1972

Green,M.C.: Mutant genes and linkages. In: Green,E.L. (Ed.): Biology of the Laboratory Mouse, 2nd ed, pp. 87—150. New York: McGraw-Hill 1966

Gruneberg,H.: The case for somatic crossing over in the mouse. Genet. Res. **7**, 58—75 (1966a)

Gruneberg,H.: More about the tabby mouse and about the Lyon hypothesis. J. Embryol. Exp. Morph. **16**, 569—590 (1966b)

Gruneberg,H., McLaren,A.: The skeletal phenotype of some mouse chimaeras. Proc. R. Soc. Lond. B., **182**, 183—192 (1972)

Gulyas,B.J.: A reexamination of cleavage patterns in eutherian mammalian eggs: rotation of blastomere pairs during second cleavage in the rabbit. J. Exp. Zool. **193**, 235—248 (1975)

Hamilton, W. J., Mossman, H. W.: Hamilton, Boyd and Mossman's Human Embryology, 4th ed. Cambridge: Heffer 1972

Handyside, A. H., Johnson, M. H.: Control mechanisms in the differentiation of the mouse blastocyst. II. The timing of determination and differentiation. J. Anat. (Lond.) **124**, 236 (1977)

Herbert, M. C., Graham, C. F.: Cell determination and biochemical differentiation of the early mammalian embryo. In: Current Topics in Developmental Biology, Vol. VIII, pp. 152—178. New York: Academic Press 1974

Hillman, N., Sherman, M. I., Graham, C. F.: The effect of spatial arrangement on cell determination during mouse development. J. Embryol. Exp. Morph. **28**, 263—278 (1972)

Hogan, B., Tilly, R.: In vitro culture and differentiation of normal mouse blastocysts. Nature (Lond.) **265**, 626—629 (1977)

Holtzer, H., Rubinstein, N., Fellini, S., Yeoh, G., Chi, J., Birnbaum, J., Okayama, M.: Lineage, quantal cell cycles, and the generation of diversity. Quart. Rev. Biophys. **8**, 523—557 (1975)

Horstadius, S.: Experimental Embryology of Echinoderms. Oxford: Clarendon Press 1973

Hutchison, H. T.: A model for estimating the extent of variegation in mosaic tissues. J. Theor. Biol. **38**, 61—79 (1973)

Illmensee, K., Mintz, B.: Totipotency and normal differentiation of single teratocarcinoma cells cloned by injection into blastocysts. Proc. Natl. Acad. Sci. USA **73**, 549—553 (1976)

Izquierdo, L., Ortiz, M. E.: Differentiation in the mouse morulae. Wilhelm Roux' Arch. **177**, 67—74 (1975)

Johnson, M. H., Eager, D., Muggleton-Harris, A., Grave, H. M.: Mosaicism in organisation of concanavalin A receptors on surface membrane of mouse eggs. Nature (Lond.) **257**, 321—322 (1975)

Johnson, M. H., Gardner, R. L.: Analysis of rat: mouse chimaeras by immunofluorescence; a preliminary report. In: Centaro, A., Caretti, N. (Eds.): Immunology in Obstetrics and Gynaecology. Proc. 1st Int. Cong. Immunology of Obstetrics and Gynaecology, Padua, 1973, pp. 312—314. Amsterdam: Elsevier, Excerpta Medica 1975

Kelly, S. J.: Studies of the potency of the early cleavage blastomeres of the mouse. In: Balls, M., Wild, A. E. (Eds.): The Early Development of Mammals. 2nd Symposium of The British Society for Developmental Biology, pp. 97—105. Cambridge: Cambridge Univ. Press 1975

Kirby, D. R. S., Potts, D. M., Wilson, I. B.: On the orientation of the implanting blastocyst. J. Embryol. Exp. Morph. **17**, 527—532 (1967)

Lawrence, P. A.: The cell cycle and cellular differentiation in insects. In: Reinert, J., Holtzer, H. (Eds.): Results and Problems in Cell Differentiation, Vol. VII: Cell Cycle and Cell Differentiation, pp. 111—121. Berlin-Heidelberg-New York: Springer 1975

Le Douarin, N.: Cell migration in early vertebrate development studied in inter-specific chimaeras. Embryogenesis in Mammals. Ciba Foundation Symposium 40 (new series), pp. 71—97. Amsterdam: Elsevier, Excerpta Medica 1976

Lewis, J.: The theory of clonal mixing during growth. J. Theor. Biol. **39**, 47—54 (1973)

Lewis, J. H., Summerbell, D., Wolpert, L.: Chimaeras and cell lineage in development. Nature (Lond.) **239**, 276—278 (1972)

Lin, T. P.: Microsurgery of inner cell mass of mouse blastocysts. Nature (Lond.) **222**, 480—481 (1969)

Lyon, M. F.: Gene action in the X-chromosome of the mouse *(Mus musculus L.).* Nature (Lond.) **190**, 372—373 (1961)

Lyon, M. F.: Mechanisms and evolutionary origins of variable X-chromosome activity in mammals. Proc. R. Soc. Lond. B. **187**, 243—268 (1974)

Martin, G., Evans, M. H.: Differentiation of clonal lines of teratocarcinoma cells: formation of embryoid bodies in vitro. Proc. Natl. Acad. Sci. USA **72**, 1441—1445 (1975)

Mayer, T. C.: A comparison of pigment cell development in albino, steel, and dominant-spotting mutant mouse embryos. Dev. Biol. **23**, 297—309 (1970)

Mayer, T. C., Fishbane, J. L.: Mesoderm-ectoderm interactions in the production of agouti pigmentation pattern in mice. Genetics **71**, 297—303 (1972)

McLaren, A.: Numerology of development. Nature (Lond.) **239**, 274—276 (1972)

McLaren, A.: Mammalian Chimaeras. Cambridge: Cambridge University Press 1976 a

McLaren, A.: Growth from fertilisation to birth in the mouse. In: Embryogenesis in Mammals. Ciba Foundation Symposium 40 (new series), pp. 47—51. Amsterdam: Elsevier, Excerpta Medica 1976 b

McLaren, A., Bowman, P.: Mouse chimaeras derived from fusion of embryos differing by nine genetic factors. Nature (Lond.) **224**, 236—240 (1969)

McLaren, A., Hensleigh, H. C.: Culture of mammalian embryos over the implantation period. In: Balls, M., Wild, A. E. (Eds.): The Early Development of Mammals. 2nd Symposium of the British Society for Developmental Biology, pp. 45—60. Cambridge: Cambridge University Press 1975

Melvold, R. W.: Spontaneous somatic reversion in mice. Effects of parental genotype on stability at the p-locus. Mutat. Res. **12**, 171—174 (1971)

Metcalf, D., Moore, M. A. S.: Haemopoietic Cells. Amsterdam: North-Holland 1971

Mintz, B.: Formation of genotypically mosaic mouse embryos. Am. Zool. **2**, 432 (Abstr. 310) (1962)

Mintz, B.: Formation of genotypically mosaic mouse embryos and early development of "lethal (t^{12}/t^{12})-Normal" mosaics. J. Exp. Zool. **157**, 273—292 (1964)

Mintz, B.: Experimental genetic mosaicism in the mouse. In: Wolstenholme, G. E. W., O'Connor, M. (Eds.): Preimplantation Stages of Pregnancy. Ciba Foundation Symposium, pp. 194—207. London: Churchill 1965

Mintz, B.: Gene control of mammalian pigmentary differentiation. I. Clonal origin of melanocytes. Proc. Natl. Acad. Sci. USA **58**, 344—351 (1967)

Mintz, B.: Gene expression in allophenic mice. In: Padykula, H. A. (Ed.): Control Mechanisms in the Expression of Cellular Phenotypes, pp. 15—42. New York: Academic Press 1970

Mintz, B.: Clonal basis of mammalian differentiation. In: Davies, D. D., Balls, M. (Eds.): Control Mechanisms of Growth and Differentiation. Symposia of the Society for Experimental Biology XXV, pp. 345—370. Cambridge: Cambridge University Press 1971

Mintz, B.: Clonal differentiation in early mammalian development. In: Sussman, M. (Ed.): Molecular Genetics and Developmental Biology, pp. 455—474. New Jersey: Prentice-Hall 1972 a

Mintz, B.: Clonal units of gene control in mammalian differentiation. In: Harris, R., Allin, P., Viza, D. (Eds.): Cell Differentiation, pp. 267—271. Copenhagen: Munksgaard 1972 b

Mintz, B.: Gene control of mammalian differentiation. Ann. Rev. Genet. **8**, 411—470 (1974)

Mintz, B., Illmensee, K.: Normal genetically mosaic mice produced from malignant teratocarcinoma cells. Proc. Natl. Acad. Sci. USA **72**, 3585—3589 (1975)

Mintz, B., Illmensee, K., Gearhart, J. D.: Developmental and experimental potentialities of mouse teratocarcinoma cells from embryoid body cores. In: Sherman, M. I., Solter, D. (Eds.): Teratomas and Differentiation, pp. 59—82. New York: Academic Press 1975

Mintz, B., Palm, J.: Gene control of hematopoiesis. I. Erythrocyte mosaicism and permanent immunological tolerance in allophenic mice. J. Exp. Med. **129**, 1013—1027 (1969)

Mintz, B., Sanyal, S.: Clonal origin of the mouse visual retina mapped from genetically mosaic eyes. Genetics **64**, (suppl.), 43—44 (1970)

Mintz, B., Silvers, W. K.: Histocompatibility antigens on melanoblasts and hair follicle cells. Transplantation **9**, 497—505 (1970)

Morata, G., Lawrence, P. A.: Control of compartment development by the engrailed gene in Drosophila. Nature (Lond.) **255**, 614—617 (1975)

Mulnard, J. G.: Studies of regulation of mouse ova in vitro. In: Wolstenholme, G. E. W., O'Connor, M. (Eds.): Preimplantation Stages of Pregnancy. Ciba Foundation Symposium, pp. 123—138. London: Churchill 1965

Mulnard, J. G.: Analyse microcinématographique du développement de l'oeuf de souris du stade II au blastocyste. Arch. Biol. (Liege) **78**, 107—138 (1967)

Mystkowska, E. T.: Development of mouse-bank vole interspecific chimaeric embryos. J. Embryol. Exp. Morph. **33**, 731—744 (1975)

Nadijcka, M., Hillman, N.: Ultrastructural studies of mouse blastocyst substages. J. Embryol. Exp. Morph. **32**, 675—695 (1974)

Nesbitt, M. N.: X-chromosome inactivation mosaicism in the mouse. Dev. Biol. **26**, 252—263 (1971)

Nesbitt, M. N.: Chimaeras vs. X inactivation mosaics: significance of differences in pigment distribution. Dev. Biol. **38**, 202—207 (1974)

Nesbitt, M. N., Gartler, S. M.: The application of genetic mosaicism to developmental problems. Ann. Rev. Genet. **5**, 143—162 (1971)

Nicolet, G.: La chronologie d'invagination chez le poulet: etude à l'aide de la thymidine tritiée. Experientia **23**, 576—577 (1967)

Nöthiger, R.: The larval development of imaginal disks. In: Ursprung, H., Nöthiger, R. (Eds.): Results and Problems in Cell Differentiation, Vol. V: The Biology of Imaginal Disks, pp. 1—34. Berlin-Heidelberg-New York: Springer 1972

Nowell, P. C., Hirsch, B. E., Fox, D. H., Wilson, D. B.: Evidence for the existence of multipotential lymphohaemopoietic stem cells in the adult rat. J. Cell. Physiol. **75**, 151—158 (1970)

Ożdżeński, W.: Observations on the origin of primordial germ cells in the mouse. Zool. Pol. **17**, 367—379 (1967)

Papaioannou, V. E., McBurney, M. W., Gardner, R. L., Evans, M. H.: Fate of teratocarcinoma cells injected into early mouse embryos. Nature (Lond.) **258**, 70—73 (1975)

Pinsker, M. C., Mintz, B.: Change in cell surface glycoproteins of mouse embryos before implantation. Proc. Natl. Acad. Sci. USA **70**, 1645—1648 (1973)

Poole, T. W.: Dermal-epidermal interactions and the site of action of the yellow (A^y) and nonagouti (a) coat colour genes in the mouse. Dev. Biol. **36**, 208—211 (1974)

Raven, Chr. P.: Oogenesis. Oxford: Pergamon Press 1961

Rawles, M. E.: Origin of pigment cells from the neural crest in the mouse embryo. Physiol. Zool. **20**, 248—266 (1947)

Reverberi, G.: Experimental Embryology of Marine and Fresh-water Invertebrates. Amsterdam: North-Holland 1971

Rossant, J.: Investigation of the determinative state of the mouse inner cell mass. I. Aggregation of isolated inner cell masses with morulae. J. Embryol. Exp. Morph. **33**, 979—990 (1975a)

Rossant, J.: Investigation of the determinative state of the mouse inner cell mass. II. The fate of isolated inner cell masses transferred to the oviduct. J. Embryol. Exp. Morph. **33**, 991—1001 (1975b)

Rossant, J.: Investigation of inner cell mass determination by aggregation of isolated rat inner cell masses with mouse morulae. J. Embryol. Exp. Morph. **36**, 163—174 (1976)

Rossant, J.: Cell commitment in early rodent development. In: Johnson, M. H. (Ed.): Development in Mammals, Vol. II, pp. 119—150. Amsterdam: Elsevier, North-Holland 1977

Rossant, J., Ofer, L.: Properties of extraembryonic ectoderm isolated from postimplantation mouse embryos. J. Embryol. Exp. Morph. (in press) (1977)

Rossant, J., Papaioannou, V. E.: The Biology of embryogenesis. In: Sherman, M. I. (Ed.): Concepts in Mammalian Embryogenesis, pp. 1—36. Cambridge/Mass.: MIT Press 1977

Rowinski, J., Solter, D., Koprowski, H.: Change of concanavalin A induced agglutinability during preimplantation mouse development. Exp. Cell Res. **100**, 404—408 (1976)

Rugh, R.: The Mouse: Its Reproduction and Development. Minneapolis: Burgess 1968

Russell, L. B.: Genetic and functional mosaicism in the mouse. In: Locke, M. (Ed.): The Role of Chromosomes in Development. 23rd Symposium of the Society for the Study of Development and Growth, pp. 153—181. New York: Academic Press 1964

Russell, L. B., Major, M. H.: Radiation-induced presumed somatic mutations in the house mouse. Genetics **42**, 161—175 (1957)

Schaible, R. H.: Clonal distribution of melanocytes in piebald-spotted and variegated mice. J. Exp. Zool. **172**, 181—200 (1969)

Searle, A. G.: Comparative Genetics of Coat Colour in Mammals. London: Logos Press 1968

Sengel, P.: Morphogenesis of Skin. Cambridge: Cambridge Univ. Press 1976

Sherman, M. I.: The role of cell-cell interactions during early mouse embryogenesis. In: Balls, M., Wild, A. E. (Eds.): The Early Development of Mammals: 2nd Symposium of The British Society for Developmental Biology, pp. 145—165. Cambridge: Cambridge University Press 1975

Sherman, M. I., McLaren, A., Walker, P. M. B.: Mechanism of accumulation of DNA in giant cells of mouse trophoblast. Nature (New Biol.) **238**, 175—176 (1972)

Silvers, W. K., Russell, E. S.: An experimental approach to action of genes at the agouti locus in the mouse. J. Exp. Zool. **130**, 199—220 (1955)

Škreb, N., Švajger, A., Levak-Švajger, B.: Developmental potentialities of the germ layers in mammals. In: Embryogenesis in Mammals. Ciba Foundation Symposium 40 (new series), pp. 27—39. Amsterdam: Elsevier, Excerpta Medica 1976

Snell, G. D., Stevens, L. C.: Early embryology. In: Green, E. L. (Ed.): Biology of the Laboratory Mouse, pp. 205—245. New York: McGraw-Hill 1966

Snow, M. H. L.: The differential effect of [^{3}H]-thymidine upon two populations of cells in pre-implantation mouse embryos. In: Balls, M., Billett, F. S. (Eds.): The Cell Cycle in Development and Differentiation. 1st Symposium of The British Society for Developmental Biology, pp. 311—324. London: Cambridge University Press 1973a

Snow, M. H. L.: Abnormal development of pre-implantation mouse embryos grown in vitro with [^{3}H]-thymidine. J. Embryol. Exp. Morph. **29**, 601—615 (1973b)

Snow, M. H. L.: Embryo growth during the immediate postimplantation period. In: Embryogenesis in Mammals. Ciba Foundation Symposium 40 (new series), pp. 53—66. Amsterdam: Elsevier, Excerpta Medica 1976

Snow, M. H. L., Ansell, J. D.: The chromosomes of giant trophoblast cells of the mouse. Proc. R. Soc. (Lond.) B. **187**, 93—98 (1974)

Solter, D., Knowles, B. B.: Immunosurgery of mouse blastocysts. Proc. Natl. Acad. Sci. USA **72**, 5099—5102 (1976)

Staats, J.: Standardized nomenclature for inbred strains of mice: fifth listing. Cancer Res. **32**, 1609—1646 (1972)

Stern, M. S.: Experimental studies on the organisation of the early rodent embryo. II. Reaggregation of disaggregated embryos. J. Embryol. Exp. Morph. **28**, 255—261 (1972)

Stern, M. S., Wilson, I. B.: Experimental studies on the organisation of the preimplantation mouse embryo. I. Fusion of asynchronously cleaving eggs. J. Embryol. Exp. Morph. **28**, 247—254 (1972)

Strickland, S., Reich, E., Sherman, M. I.: Plasminogen activator in early embryogenesis: enzyme production by trophoblast and parietal endoderm. Cell **9**, 231—240 (1976)

Tarkowski, A. K.: Mouse chimaeras developed from fused eggs. Nature (Lond.) **190**, 857—860 (1961)

Tarkowski, A. K., Wroblewska, J.: Development of blastomeres of mouse eggs isolated at the four- and eight-cell stage. J. Embryol. Exp. Morph. **18**, 155—180 (1967)

Tettenborn, U., Dofuku, R., Ohno, S.: Noninducible phenotype exhibited by a proportion of female mice heterozygous for X-linked testicular feminization mutation. Nature (Lond.) **234**, 37—40 (1971)

Theiler, K.: The House Mouse. Berlin-Heidelberg-New York: Springer 1972

Vogt, W.: Gestaltungsanalyse am Amphibienkeim mit örtlicher Vitalfärbung. II. Teil: Gastrulation und Mesodermbildung bei Urodelen und Anuren. Wilhelm Roux' Arch. **120**, 385—706 (1929)

Wassarman, P. M., Albertini, D. F., Josefowicz, W. J., Letourneau, G. E.: Cytochalasin B-induced pseudo-cleavage of mouse oocytes in vitro: asymmetric localization of mitochondria and microvilli associated with a stage-specific response. J. Cell Sci. **21**, 523—535 (1976)

Wegmann, T. G., Gilman, J. G.: Chimerism for three genetic systems in tetraparental mice. Dev. Biol. **21**, 281—291 (1970)

West, J. D.: A theoretical approach to the relation between patch size and clone size in chimaeric tissues. J. Theor. Biol. **50**, 153—160 (1975)

West, J. D.: Clonal development of the retinal epithelium in mouse chimaeras and X-inactivation mosaics. J. Embryol. Exp. Morph. **35**, 445—461 (1976a)

West, J. D.: Distortion of patches of retinal degeneration in chimaeric mice. J. Embryol. Exp. Morph. **36**, 145—149 (1976b)

West, J. D., McLaren, A.: The distribution of melanocytes in the dorsal coats of a series of chimaeric mice. J. Embryol. Exp. Morph. **35**, 87—93 (1976)

Weston, J. A.: Cell marking. In: Wilt, F. H., Wessells, N. K. (Eds.): Methods in Developmental Biology, pp. 723—736. New York: Crowell 1967

Whang, J., Frei, E., Tjio, J. H., Carbone, P. P., Brecher, G.: The distribution of the philadelphia chromosome in patients with chronic myelogenous leukaemia. Blood **22**, 664—673 (1963)

Wilson, I. B., Bolton, E., Cuttler, R. H.: Preimplantation differentiation in the mouse egg as revealed by microinjection of vital markers. J. Embryol. Exp. Morph. **27**, 467—479 (1972)

Wilson, I. B., Stern, M. S.: Organisation in the preimplantation embryo. In: Balls, M., Wild, A. E. (Eds.): The Early Development of Mammals. 2nd Symposium of The British Society for Developmental Biology, pp. 81—95. Cambridge: Cambridge University Press 1975

Wolfe, H. G., Coleman, D. L.: Pigmentation. In: Green, E. L. (Ed.): Biology of the Laboratory Mouse, pp. 405—425. New York: McGraw-Hill 1966

Wolpert, L., Gingell, D.: Striping and the pattern of melanocyte cells in chimaeric mice. J. Theor. Biol. **29**, 147—150 (1970)

Wu, A. M., Till, J. E., Siminovitch, L., McCulloch, E. A.: Cytological evidence for a relationship between normal haemopoietic colony-forming cells and cells of the lymphoid system. J. Exp. Med. **127**, 455—464 (1968)

Sexual Differentiation in Mammalian Chimaeras and Mosaics

Anne McLaren

MRC Mammalian Development Unit, London, Great Britain

I. Definitions and Explanations

Animals in which two genetically distinct populations of cells coexist are termed mosaics if both populations stem from a single fertilized egg, the product of two gametes. If more than one fertilized egg or more than two gametes are involved, the term chimaera is used.

Mosaics can originate at any point in development from the 2-cell stage onwards, by somatic mutation, or chromosome loss, or chromosome non-disjunction. Loss of a Y chromosome, for example, from one cell in an embryo, could lead to an XY/XO mosaic; non-disjunction of the same chromosome could give an XYY/XO mosaic. In female placental mammals, XX in chromosome constitution, each somatic cell has one or other X chromosome genetically active but not both: In this way every woman is genetically composite for any X-linked locus at which she happens to be heterozygous. Thus all females except the most uniformly homozygous are mosaics (X-inactivation mosaics). Female marsupials, however, are not mosaics, since it seems always to be the paternally derived X chromosome that is inactive.

Chimaeras may be designated "primary" if they originate at fertilization or during very early embryonic development, before implantation, and "secondary" if an additional cell population is introduced after implantation.

Primary chimaerism may arise spontaneously either during the fertilization process, for example if one spermatozoon fertilizes the egg and another the second polar body, or if two mitotic products of the egg nucleus are independently fertilized, or perhaps during cleavage, by spontaneous aggregation of two preimplantation embryos. A few cases of primary chimaerism in man have been identified, usually as a result of one of the two genetically distinct cell populations being XX in chromosome constitution, and the other XY. Primary chimaeras can also be produced experimentally in animals, either by aggregation (sometimes termed fusion) of cleaving embryos (aggregation chimaeras) or by injection of embryonic cells of contrasting genetic constitution into the preimplantation blastocyst (injection chimaeras). The genetically composite individuals created in this way are sometimes termed tetraparental or allophenic. Experimental chimaerism has been achieved by embryonic aggregation or injection in mice, rats, rabbits, and sheep, as well as between mouse and rat and mouse and bank-vole embryos, but most of the experimental analysis so far has been carried out on mouse chimaeras. A comprehensive review of the literature in this field is given by McLaren (1976).

Secondary chimaerism most often comes about when anastomosis of placental blood vessels makes possible the exchange of cells between neighbouring foetuses. The condition has been reported only rarely in man, but in cattle it is common, since twin placentas usually share a common blood supply. When the twins are of opposite sex, the freemartin syndrome develops. In marmosets, twin births are the rule, and vascular anastomosis is usual, so almost every individual is a chimaera, though none shows freemartinism. Spontaneous chimaerism may also arise during pregnancy by passage of cells across the placenta from foetus to mother or mother to foetus, but such chimaerism is usually transient. Because of the ways in which it originates, spontaneous secondary chimaerism usually involves blood cells only. Passage of primordial germ cells from one foetus to another has been claimed in both cattle and marmosets, but not yet conclusively established. Blood chimaerism can also be produced experimentally, by transfusion of blood cells (usually postnatally) into, e.g., neonatally thymectomized or heavily irradiated animals (radiation chimaeras). Other examples of the experimental production of secondary chimaerism would include individuals bearing allografts of skin, gonad, or other tissues.

Sexual differentiation, including differentiation of the germ cells and the rest of the reproductive system, may encounter special problems in genetically composite animals. These will be largely independent of whether the individual is a chimaera or a mosaic, but may depend critically on when in development the composite originated, and what tissues it involves. Blood chimaerism induced in the adult animal would hardly be expected to have an influence on the differentiation of the reproductive tract. The other crucial factor is whether the components of the composite are of the same chromosomal sex, i.e., single-sex composites (both XX, or both XY) or of opposite chromosomal sex (XX/XY). In this chapter we shall discuss the development of XX/XY individuals first, and then consider single-sex composites, including X-inactivation mosaics.

II. XX/XY Composites

A. Primary Chimaeras and Mosaics

1. Determination of Phenotypic Sex in Mice

Since the primary sex ratio in mice does not appear to differ significantly from unity, and since one cannot tell by inspection the sex of an embryo before implantation, one would expect that half of all the aggregation and injection chimaeras that have been produced would be of the XX/XY type. Limited evidence, based on four independent bodies of data, agrees with this expectation: Out of a total of 49 randomly selected mouse aggregation chimaeras in which the chromosomal sex of both components has been determined, 25 XX/XY individuals have been found (McLaren, 1976). Seventeen of these were males, five were females, and three were hermaphrodites.

No a priori reasons exist for expecting XX/XY animals to be male, or female, or intermediate. In the initial series of aggregation chimaeras described by Tarkowski (1961), a striking preponderance of males was seen (11/13); Tarkowski tentatively concluded that XX/XY individuals tended to develop as phenotypic males, an interpretation upheld by later evidence (Mystkowska and Tarkowski, 1968; McLaren, 1972, 1975a). The strength of the tendency varies from one strain combination to another: Of nine strain combinations listed by McLaren (1976), the sex ratio of overt chimaeras (in which both cell populations are known to be present) in five combinations did not differ significantly from the 75% expected if all XX/XY mice developed as males; in two the proportion of males was 50%, and the remaining two (from Mintz, 1969) showed an intermediate sex ratio, significantly different from both 50 and 75%, suggesting that most but not all of the XX/XY individuals developed as males. The two strain combinations giving equal numbers of male and female chimaeras were those referred to by Mullen and Whitten (1971) as "unbalanced," in which one component population is represented to only a very limited extent in the coat, and perhaps also in the gonads. In such "unbalanced" XX/XY chimaeras, the gonads may therefore be made up predominantly either of XX or XY cells, and will develop in the female or the male direction accordingly.

The evidence thus indicates that the presence of XY cells in an XX/XY gonad exerts a strong masculinizing influence on development. A possible mechanism for this masculinizing effect has recently been put forward by Ohno et al. (1976) (see Sec. II.B). The occasional side-to-side differences in hermaphrodites suggest that phenotypic sex is determined locally, within the reproductive tract. Since secondary sexual characters develop in response to hormones produced by the gonads, the phenotypic sex of the gonads will be of crucial importance, and this presumably depends on the relative numbers of XX and XY cells within each during some critical phase of development. The apparent variation in the degree of dominance of the XY component in XX/XY chimaeras probably merely reflects variation in the frequency with which both XX and XY cells are present in the gonad or its primordium. There is no evidence of variation either in the specific masculinizing capacity of the Y chromosome of different strains, or in the response of XX cells. The small proportion of hermaphrodites (see below) presumably represents the situation where the XY cells in the gonadal primordium are sufficient to prevent normal female development, but not to ensure complete male development.

The embryonic gonad contains two populations of cells of independent origin, the somatic tissue of the genital ridge, and the immigrant population of primordial germ cells. Differentiation of the genital ridge into a male or female gonad occurs even when germ cells are largely or entirely absent for genetic reasons, or are almost all destroyed before entering the genital ridge by treatment of the mother with the drug Busulphan (Merchant, 1975). Although this evidence suggests very strongly that the phenotypic sex of the gonad is determined by the chromosomal sex of the somatic cells rather than by that of the primordial germ cells, the point cannot be regarded as conclusively established, because it has so far proved impossible to ensure that no germ cells enter the ridge. Also, in mouse XX/XY chimaeras, it not infrequently occurs that the somatic tissues of the adult testes are very largely (up to 95%) made up of cells of the XX component (Mintz, 1969; McLaren, 1975a); conversely, Ford et al. (1974) described an adult female XX/XY mouse with a great excess (98%) of XY cells in the ovarian follicles. These observations might suggest that the phenotypic sex of the gonad is unlikely to be determined by the chromosomal sex of the somatic cells; however, it would of course be the somatic cells in the gonadal primordium of the early embryo that played the critical determinative role. The somatic constitution of the adult gonad may not be a reliable guide to the constitution of the gonadal ridge in embryogenesis, because differential cell proliferation of the two component populations of the chimaera is likely to occur. In human XO/XY mosaics, the more masculinized gonad usually shows the higher proportion of XY cells on culture of somatic tissues (Ford, 1970).

2. Secondary Sex Glands

Once the gonad of an XX/XY embryo has begun to differentiate as a testis, the androgens that it produces will cause regression of the Müllerian duct derivatives (female) and development of the Wolffian duct derivatives (male). Some hermaphrodites, however, possess almost complete sets of both Müllerian and Wolffian ducts (Whitten, 1975).

XX cells appear to be just as capable as XY of responding to the androgenic stimulus from a developing testis, because Mintz et al. (1972) described an XX/XY male chimaera in which not only the testis, but also the epididymis and seminal vesicles, consisted largely if not entirely of the XX component. Seminal vesicle fluid contains proteins that differ from one inbred strain to another and that are never found in females; such fluid from another XX/XY male chimaera proved to contain seminal vesicle proteins characteristic of both component strains (Mintz et al., 1972). This elegantly demonstrates both that XX cells can secrete a specifically male protein in response to a male environment, and that the form of the protein is determined by the genotype of the XX cells themselves.

3. Fate of Germ Cells

In *Drosophila*, experiments on germ-line chimaeras have shown that the phenotypic sex of a germ cell is determined entirely by its chromosome constitution (Van Deusen, 1976). In lower vertebrates, on the other hand, the type of gametogenesis (male v. female) that germ cells undergo depends not on their own genotype but on the phenotypic sex of the gonad in which they reside. XX/XY chimaeras provide a means to investigate whether such functional sex reversal is also possible in mammals.

If many of the available XX germ cells differentiated as spermatozoa, or XY germ cells as eggs, the sex ratio of the progeny would depart significantly from unity. On the basis of the sex ratio of the progenies of a large number of individual chimaeras, Mintz (1968) concluded that functional alteration of germ cell sex could not occur in either direction. This conclusion has been amply confirmed for XX germ cells by combined progeny testing and cytologic analysis. "Single-sex" chimaeras (see below) often produce progeny of two genetic types, showing that both components have given rise to functional germ cells; in contrast, XX/XY males give progeny of one type only, and in two such males described by Myst-kowska and Tarkowski (1968) and a further seven from McLaren (1975a), the strain origin of the young was shown to correspond to that of the XY component of the chimaera. Mystkowska and Tarkowski could find no XX cells even among primary spermatocytes in XX/XY males.

Evidence from mice carrying the *sex-reversal (Sxr)* gene suggests that it is the presence of a second X chromosome rather than the absence of a Y that debars XX germ cells from undergoing spermatogenesis. The gene transforms genetic females into phenotypic males: On an XX background almost all the germ cells degenerate, but on an XO background, active spermatogenesis occurs, though the spermatozoa produced are abnormal (Cattanach et al., 1971). It seems that this effect of double X-chromosome dosage is not due to an increased response to androgen: XX males heterozygous not only for *Sxr* but also for *Tfm (testicular feminization)*, an X-linked gene suppressing androgen response (see Sec. III.A), have in effect only a single gene dose of androgen responsiveness but still show widespread germ cell degeneration (Lyon, 1974). The few spermatogenic cells surviving in either *Sxr/+* or *Sxr/+ Tfm/+* testes have proved on examination to be XO in chromosome constitution, having presumably arisen as a result of spontaneous loss of an X chromosome.

The failure of XX germ cells to undergo spermatogenesis prompted an investigation into their fate in male chimaeras. In female mice the germ cells enter meiosis before birth, on the 14th or 15th day of gestation, while in normal males no meiotic stages are seen until several days after birth. Mystkowska and Tar-kowski (1970) reported germ cells in meiotic prophase in the testes of male XX/XY fetuses on the 16th to 17th day of gestation; they failed to reach diakinesis, and by $18\frac{1}{2}$ days they were degenerating (McLaren et al., 1972). The obvious explanation of these meiotic cells is that they are XX germ cells, entering meiosis at the time specified by their genotype. An alternative possibility is that the somatic tissue of the gonad determines the development of the germ cells, so that a patch of XX somatic tissue in an XX/XY testis would drive any germ cells (XX or XY) contained within it into meiosis, while those surrounded by XY somatic tissue continued to divide mitotically. It seems unlikely, however, that the prenatal meiotic cells included any XY germ cells, since their DNA replication pattern proved identical to that of normal female germ cells and they failed to display the prominent sex vesicle characteristic of male germ cells (McLaren et al., 1972). The possibility that XX germ cells only enter meiosis before birth if surrounded by XX somatic tissue cannot be ruled out.

The degeneration of meiotic germ cells before birth in the testis is presumably due to the adverse hormonal environment. A few may survive into the postnatal period, since Mystkowska and Tarkowski (1968) found several oocytes as late as five days after birth in one area of a chimaeric gonad that showed typical testicular organization of the somatic tissue. There is no indication that XX germ cells ever complete even the first meiotic division in the mouse testis.

The situation of XY germ cells in an ovary may be different. Detailed breeding records have so far been reported for only two XX/XY female mice (Ford et al., 1975): Of 23 young, 22 proved to be of the genotype corresponding to the XX component of the chimaera. The remaining animal could have been derived only from the XY component, and proved on chromosomal examination to be of the rare XXY chromosome constitution. The possibility was raised that the presence of a second X chromosome, which in the testis militates against spermatogenesis, might have allowed this germ cell to undergo oogenesis. However, at least one XY oocyte has since been identified at first meiotic metaphase in the ovary of an XX/XY female chimaera (Evans and Ford, personal communication), so it looks as though functional sex reversal of XY germ cells is indeed possible in mammals.

4. Hermaphrodites

True hermaphrodites contain both ovarian and testicular tissue, so that each gonad may be an ovary, a testis, or an ovotestis. In mice, the proportion of XX/XY aggregation chimaeras that develop as hermaphrodites probably depends on the strain combination used. Results published up to 1974 (McLaren, 1976) yield eight hermaphrodites out of a total of 252 chimaeras (3%) if only overt chimaeras from "balanced" combinations are included (7 strain combinations). The two "unbalanced" combinations of Mullen and Whitten (1971) showed no hermaphrodites in a sample of 40 overt chimaeras. As we have seen above, the gonadal primordia of unbalanced chimaeras are probably largely formed from one component only, so disturbed sexual differentiation is unlikely to occur. Those bodies of data in which it is not possible to distinguish overt chimaerism show, as would be expected, a lower incidence of hermaphroditism (10/564 = 1.8%). A recent paper by Whitten (1975) records that 12 hermaphrodites have been identified in about 1200 overt chimaeras (1%); no mention is made of how many of these were from unbalanced combinations.

Hermaphroditism in XX/XY mice is usually of the male type, with male organs on one side of the body and rudimentary female organs on the other. Ovotestes are often seen. Unlike the situation in human hermaphrodites, there seems no tendency for the right gonad to be more masculinized than the left. The male organs may sometimes be functional, as in the animal described by McLaren (1975a) that sired a litter at seven weeks of age but subsequently proved sterile. At 20 months, postmortem examination revealed male organs and an atrophic testis on one side and a rudimentary uterus and ovary on the other.

A high incidence of spontaneous hermaphroditism, both male and female, has been reported in the BALB/c inbred strain of mice, together with a low sex ratio (38% males). Mintz (1971) suggested that the hermaphrodites might be XX/XY chimaeras, since BALB/c embryos cleave slowly and some lose the zona pellucida when still at the morula stage, possibly permitting spontaneous aggregation to occur. Chromosome studies (Whitten, 1975) have since shown, however, that the BALB/c hermaphrodites are mosaics. XO/XY, XO/XYY, and XO/XY/XYY mosaics have all been detected. The disturbed sex ratio is due to the occurrence of XO females. Since the high incidence of hermaphrodites and the low sex ratio are also seen when BALB/c males are outcrossed to females of another strain, but not in the reverse cross, it seems that the BALB/c Y chromosome is particularly susceptible to non-disjunction, both during spermatogenesis (hence the XO females) and during early cleavage (hence the mosaics). No chromosome studies on normal males and females in the BALB/c strain have so far been reported, so it is not possible to estimate what proportion of these spontaneous mosaics develop as hermaphrodites.

Varnum (cited by Whitten, 1975) was able to establish that "selfing" is possible in mammals, by removing the ovary from a lateral hermaphrodite and grafting it to an ovariectomized mouse of the same genotype. The original donor, now bearing only a testis, was successfully mated to the recipient female. Thus both the ovary and the testis in such a hermaphrodite are potentially fertile.

Mosaics arising in man and other animals through loss or non-disjunction of the sex chromosomes (XO/XX, XO/XY, XO/XXY, XO/XXX, etc.) are often associated with abnormalities of sexual development. True hermaphroditism in man, however, is particularly characteristic of the rare XX/XY chromosome constitution. Twenty-two reports of XX/XY individuals in the literature are cited by McLaren (1976). Origin of the XX/XY condition from a single zygote seems unlikely, as it would involve some sequence of events such as an XXY zygote arising through non-disjunction during gametogenesis, followed by two further non-disjunctional mitotic events, giving rise to an XX and an XY cell line, followed by loss of the original XXY line. It seems more likely that they are spontaneous chimaeras. In nine of these XX/XY individuals, chimaerism has been established by the detection of two cell lines, distinguished genetically by a number of different blood-group, biochemical, and chromosomal markers. All show evidence of two paternal contributions, indicating fertilization by two spermatozoa; several also show two genetically distinct contributions from the mother, which could arise either by fertilization of one or other polar body as well as the egg or oocyte, or by precocious loss of the zona pellucida and aggregation of two embryos during cleavage.

Sexual differentiation in XX/XY chimaeras seems strikingly different in man and in the mouse. As we have seen, hermaphroditism in mouse aggregation chimaeras is rare, not more than 3%, which implies that it occurs in not more than 6% of XX/XY individuals. In man, seven of the nine XX/XY cases known to be chimaeric were true hermaphrodites (all male). Although all were identified as a result of ambiguities of their external genitalia or other evidence of sexual abnormality, the difference from the mouse is unlikely to be only one of ascertainment, as the remaining two cases, who were identified in the course of blood-group testing, also proved to show intersexual features. One had been operated on at an early age for an intersexual state of his external genitalia (Zuelzer et al., 1964); the other showed hypospadias (Myhre et al., 1965). There is thus little reason to believe that many normal fertile men will prove to be XX/XY chimaeras. Perhaps in man the mechanism of sex determination is less rigidly canalized than in the mouse, so that the presence of XX as well as XY cells in the gonadal primordium more readily leads to intersexuality.

B. Secondary Chimaeras

When vascular anastomosis is established between neighbouring placentae, blood from each foetus circulates through the other and immunologic tolerance develops. Any cells transferred in the blood may thus persist throughout life. Blood chimaerism originating in this way is the most usual form of spontaneous secondary chimaerism encountered in mammals.

XX/XY chimaerism has occasionally been reported among blood cells of both members of opposite-sexed twin pairs in man, but does not affect sexual development. In marmosets the condition is common, since most pregnancies consist of twins and vascular anastomosis from a very early stage of gestation is the rule, but again no sexual problems result. In cattle, on the other hand, and sometimes in sheep, goats, and pigs, vascular anastomosis between twins of unlike sex leads to the condition known as freemartinism in the female twin, and may be associated with abnormalities in the sexual development of the male partner also. Twinning is rare in the horse, and freemartinism has not been clearly established.

Freemartins in cattle have been recognized since the second century B.C, when M. Terentius Varro (cited by Belloni, 1952) described sterile cows as "female bulls." Columella, in the first century A.D., advised farmers to remove such animals from their herds in favour of cows capable of breeding, or else to break them in for ploughing. The association of freemartins with heterosexual twinning was first appreciated by John Hunter in the 18th century (see Marcum, 1974). The characteristic feature of the freemartin is that the external genitalia are female, while the small gonads resemble testes more than ovaries. Germ cells are present during foetal life, but few if any survive beyond birth. Uterus and oviducts are poorly developed or even absent and some male secondary sex organs (seminal vesicles, vasa deferentia, epididymides) are usually present. Bulls twinned with freemartins may also be affected, and often become more or less sterile in later life.

The effects of the freemartin condition on sexual differentiation of the brain have been examined in sheep by Short (1974). It appears that freemartin ewes may show either male-type or female-type LH discharge patterns in response to oestradiol, and either male or female sexual behaviour, depending presumably on the time and/or extent of the sex reversal of the gonads.

The classical theory of Lillie, that the freemartin female was masculinized during foetal life by sex hormones produced by the testes of the male twin, is now known to be an oversimplification. The regression of the female secondary sexual organs takes place at the same time as the equivalent changes in the male fetus, and hence are indeed thought to be due to some male-derived "Müllerian duct-inhibiting hormone" in the common circulation. The demonstration by Vigier et al. (1976) that no inhibition of the Müllerian ducts or of the gonads has taken place by 60 days of gestation in female calves surgically isolated from their male twins at 37 or 45 days, confirms that the inhibitory effect comes directly from the male foetus. On the other hand the masculinization of the secondary sexual organs in the freemartin occurs considerably later than in the normal male, and may well be caused by androgens coming from her own testicular tissue, exerting a local masculinizing effect on the internal organs but inadequate in amount, time, or distribution to affect the external genitalia. As for the gonad itself, attempts to influence its development experimentally, by the administration of testosterone and other androgens to the female fetus, have proved entirely unsuccessful. Either some as yet unidentified hormone from the male fetus is responsible, or the primary effect is not hormonal, but is due to the presence of XY cells in the fetus.

The ratio of XX to XY leucocytes may vary within wide limits, but shows a strong positive correlation between freemartins and their male twins (Marcum, 1974). Apart from erythrocytes, the only other cell type that might be transferred from one twin to the other would be primordial germ cells. There is embryologic evidence that germ cells may occasionally be found in blood vessels in the course of their migration to the gonadal ridges (Ohno and Gropp, 1965; Jost and Prepin, 1966); on the other hand claims to have identified XX germ cells in the testes of XX/XY bulls (twinned with freemartins) have recently been strongly challenged (Ford and Evans, 1976). No XX cells could be detected among 60 spermatogonia and 903 primary spermatocytes from seven bulls, nor among 27 spermatogonia and 139 spermatocytes from an XX/XY marmoset. XY mitotic cells have occasionally been detected in the gonads of freemartins by Ford and Evans (1976); the nature of these cells is uncertain but it seems unlikely that they are germ cells.

If germ cells do migrate from one twin foetus to another, they should be readily detected by breeding tests on bulls born in single-sex twin pairs. Extensive progeny testing has so far failed to yield any genetic evidence for germ cell exchange, in bulls twinned either with freemartins, or with other bulls of contrasted genetic type. For example, in an experiment in which embryo transfer has been used to produce Hereford bulls twinned with Friesian bulls (Rowson and Newcomb, 1976), the progeny of the Herefords has not yet included any calves whose colour indicates that they were derived from immigrant Friesian germ cells. The unexpected appearance of a characteristic Friesian head pattern in a proportion of the calves (Rowson and Newcomb, personal communication) could be interpreted as an effect of secondary chimaerism upon the developing germ cells (see Sec. III.C.2), but the biologic basis of such an effect is hard to envisage. Claims that an excess of female calves are produced by some bulls twinned with freemartins, as would be expected if XX germ cells were present and undergoing normal spermatogenesis, have not as yet been confirmed (for references, see Ford and Evans, 1976).

Little evidence exists as to how the presence of XY cells in the developing gonad exerts a masculinizing effect. In XX/XY aggregation chimaeras, the relevant cells are almost certainly the somatic cells of the gonad, programmed to develop in a testicular direction. In the freemartin, the only male cells known to be present are red and white blood cells, which do not at first sight appear likely candidates to influence sexual differentiation. However, Ohno et al. (1976) have recently proposed that the male histocompatibility antigen H–Y is "the long-sought product of the testis-determining gene," and is responsible for inducing testicular organization. They claim that H–Y antigen can be detected in the fetal gonads of cattle freemartins, and speculate that XY cells can "disseminate" H–Y antigen, which then coats the XX cells of the gonad and "subverts" development into the male pathway. Against this causal interpretation of the role of male histocompatibility antigen is the variety of evidence suggesting that the antigen actually depends on androgens for its expression (reviewed by Erickson, 1977). Poláčková and Vojtíšková (1968), for example, reported that neonatal castration of male mice allows the adult skin to survive for prolonged periods of time, even permanently, on females of the same strain. The decision as to whether H–Y antigen is indeed the cause of male differentiation, or merely one of its consequences, must await further investigation.

The recent study of Vigier et al. (see above) offers a method for determining the extent to which freemartinism is due to the presence of XY cells in the gonad, rather than to a hormonal influence from the male twin. By the time the male and female foetuses were separated, blood chimaerism was already established, and substantial numbers of XY cells could be detected in the livers of the females. This chimaerism did not in itself lead to inhibition of the Müllerian ducts or gonads. When the study is extended to include the later stages of gestation, it will be possible to see whether chimaerism in the absence of enduring vascular anastomosis is adequate to bring about masculinization of the gonads and secondary sexual organs.

III. Single-Sex Composites

A. X-Inactivation Mosaics

Mosaicism resulting from inactivity of one X chromosome in each somatic cell of an XX female would only be expected to involve characters controlled by X-linked loci. In the mouse, the X chromosome contains a locus that controls the response of target organs to androgens. A mutation at this locus, *testicular feminization (Tfm)*, abolishes the response to testosterone, so that X^{Tfm}/Y males, though their testes are normal, have no other reproductive organs and look like females (Lyon and Hawkes, 1971). X^{Tfm}/X heterozygotes are normal females: Their mosaicism can be demonstrated by the pattern of androgen-dependent enzyme induction in the kidney, but has not been shown to affect sexual differentiation in any way.

Another gene in the mouse, the autosomal dominant *Sex-reversal (Sxr)*, causes XX embryos to develop as phenotypic males (Cattanach et al., 1971). The genetic constitution $Sxr/+X^{Tfm}/X$ thus allows one to study the androgen response mosaicism due to X^{Tfm}/X on a male background, where androgen responsiveness is of course crucial for normal sexual differentiation. Drews et al. (1974) described the various grades of intersexuality seen in the reproductive system of chromosomally XX mice, heterozygous for the *Tfm* gene and sex-reversed to male by *Sxr*, in the presence and absence of an additional gene, O^{hv}, which is thought to affect the expression of *Tfm* by decreasing the proportion of mutant X chromosomes inactivated.

The epididymis provides a particularly informative example of the effects of androgen response mosaicism on sexual differentiation. In X^{Tfm}/X males, the epithelium of the epididymal duct appears clearly mosaic with patches of normally differentiated high columnar cells, in which presumably the wild-type X chromosome is functional, alternating with patches of undifferentiated flat cuboidal cells that have failed to respond to androgens, because their functional X chromosome carries the X^{Tfm} gene (Drews and Alonso-Lozano, 1974). No intermediate cells are seen, so the androgen-sensitive differentiation pattern presumably functions autonomously.

On the other hand the very presence of X^{Tfm} cells in the mosaic males suggests that they are able to respond to the trophic effect of androgens. In X^{Tfm}/Y males the epididymis is totally absent, because the Wolffian duct, made up exclusively of X^{Tfm} cells, regresses. In response to treatment of the mosaic males with testosterone, ^{3}H-thymidine incorporation in the epididymis was stimulated just as much in the androgen-sensitive wild-type as in the androgen-insensitive X^{Tfm} cells (Drews and Drews, 1975). The authors conclude that the stimulus for DNA synthesis must be transferred from one cell population to the other, by metabolic cooperation. The number of X^{Tfm} cells in the mosaic epididymis is much less than the 50% expected, because although they proliferate at the normal rate, they die off much more steadily than do the normal cells.

Sexual behaviour also has been examined in $Sxr/+X^{Tfm}/X$ mosaic males (Ohno et al., 1974). Normal male mice attempt to mount females, and show aggression towards an intruder. Female mice and X^{Tfm}/Y males show neither of these forms of behaviour. Mosaic males appear to have mosaic brains, as the mounting behaviour and the aggression, which are presumably under the control of different centres in the brain, are reduced or abolished independently in different mosaic individuals.

B. Gonosomic Mosaics

In certain mammalian species with unusual chromosomal sex-determining mechanisms, the somatic tissues have a different chromosomal constitution from that of the germ line. Thus in *Microtus oregoni*, the creeping vole, the males have normal XY somatic cells but produce only OY germ cells, while the females have XO somatic cells and XX germ cells (Ohno et al., 1963). In *Myopus schisticolor*, the wood lemming, some females show the normal XX chromosome constitution, while others have XX germ cells but XY somatic cells, presumably carrying a gene that suppresses the male-determining effects of the Y chromosome (Fredga et al., 1976). Mitotic non-disjunction is presumably responsible for the loss of the X and Y chromosomes in the germ lines of the male vole and the XY female lemming respectively and for the doubling of the X chromosome in the germ lines of both the female vole and the XY female lemming.

C. Aggregation Chimaeras

1. Mixed Progenies

When both components of a chimaera are of the same chromosomal sex, XX or XY, both germ cell populations can form functional gametes, and mixed progenies may result. The published information on mixed versus single-type progenies (see McLaren, 1976) gives little indication of the real frequency of mixed progenies from single-sex chimaeras, since often overt chimaeras are not distinguished from non-chimaeric individuals, nor is it possible to identify the XX/XY males, which would not be expected (see above) to produce more than one population of gametes.

Nonetheless, one can confidently state first, that mixed progenies occur from chimaeras of both sexes, so that both male and female germ cells must be derived from more than one progenitor cell in the post-8-cell embryo, and secondly, that by no means all progenies even from single-sex chimaeras are mixed. Female chimaeras produce a higher proportion of mixed progenies than males, presumably because very many fewer females than males are XX/XY in chromosome constitution. Even among known single-sex chimaeras, the incidence of mixed progenies seems to depend on the strain combination: For example, Ford et al. (1975) using one strain combination reported four out of six single-sex chimaeras producing mixed progenies, yet McLaren (1975a) with another combination detected only one out of eight.

Where single progenies prove to be consistently from one of the two germ cell populations, the possibility arises of differential cell proliferation, leading to selection of one population at the expense of the other during the oogonial or spermatogonial period. An example of such germ cell selection in male chimaeras has been reported by Mintz (1968). On the other hand if neither component is at a selective advantage, an excess of single progenies may indicate that one or other cell line is excluded from the germ cell population by chance, perhaps because the two components are not randomly mixed at the time that the germ cells are set aside and the number of progenitor cells is small.

In the strain combination used by McLaren (1975a), the distribution of the two populations of spermatozoa within the gonads of those chimaeras in which they coexisted seemed far from random, judging by the proportions of the two types of progeny in successive litters. In another strain combination, where the two components were more closely related genetically, the distribution of the two types of young showed no departure from randomness.

2. Differentiation of Gametes

The differentiation of gametes in a chimaeric gonad could be influenced by the proximity of cells of contrasting genotype. For example, the shape of mouse spermatozoa varies significantly from one inbred strain to another: Evidence from F_1 hybrids suggested that the phenotype was not determined by the haploid genotype of the spermatozoa, but it was not known whether it reflected the diploid genotype of the spermatogonia, or whether the genetic effect operated through the surrounding somatic cells. These include the Sertoli cells known to form intimate associations with the germ cells during spermatogenesis. Normal males offer no means of deciding between these two possibilities; but in chimaeras, there exist two genetically distinct populations of somatic cells in the testis, so that a germ cell of one genotype will often develop in association with a Sertoli cell of contrasting genotype.

Burgoyne (1975) selected two strains, C3H and C57BL, the spermatozoa of which could be distinguished from one another with 98% accuracy, and examined spermatozoa from five chimaeras between them. The chimaeras had been mated to C57BL females. One male had only C57-like sperm and sired only C57-type offspring; one (an XX/XY) had only C3H-like sperm and sired only F_1-type offspring; the other three had sperm of both phenotypes and produced mixed progenies. The difference between the two sperm populations in the chimaeras was the same as between C3H and C57BL spermatozoa from pure-bred mice of the two strains, and no intermediate forms resembling F_1 hybrid spermatozoa were seen. Enzyme determinations showed that the somatic tissue of the chimaeric testes was derived mostly from the C57BL component, and unlike the breeding results showed no correlation with sperm phenotype. The author concluded that the genetic differences in sperm phenotype must be controlled by genes acting in the germ cells themselves.

An entirely different situation was uncovered by Lyon et al. (1975), in their studies on germ cells in mice carrying the X-linked *testicular feminization* gene. In X^{Tfm}/Y males, spermatogenesis is arrested at the first meiotic division, but in aggregation chimaeras made between X^{Tfm}/Y and normal males, both populations of germ cells produced functional spermatozoa, enabling for the first time X^{Tfm}-bearing eggs from heterozygous females to be fertilized by X^{Tfm}-bearing spermatozoa. Since the X^{Tfm}/Y germ cells were presumably refractory to androgens in the chimaeras as well as in X^{Tfm}/Y males, the determining role of testosterone in spermatogenesis must operate through the somatic cells (probably the Sertoli cells) rather than on the male germ cells directly. In the chimaeric males, an X^{Tfm}/Y germ cell could only complete spermatogenesis if it was associated with a normal XY Sertoli cell. Some degree of intersexuality of the genitalia was seen in the chimaeras, similar to that reported by Drews and his colleagues in $Sxr/+X^{Tfm}/X$ heterozygotes.

Not only the phenotype but also the genotype of a developing gamete could conceivably be modified by the genetically dissimilar cellular environment of the chimaeric gonad. Germ cells differentiate in intimate association with Sertoli cells in the male and with follicle cells in the female; any transfer of genetic information would go undetected as long as germ cells and somatic tissue shared the same genetic constitution. Kanazawa and Imai (1974) claimed that even secondary blood chimaerism, in neonatally thymectomized mice reconstituted with spleen cells from a different inbred strain, had a genetic effect on the male germ cells. Progeny of male chimaeras showed some immunologic properties characteristic of the donor strain, as though some functional genetic material had become integrated into the genome of the germ line during spermatogenesis.

No such effect was detected by McLaren (1975b), in a series of aggregation chimaeras in which the two components carried contrasting alleles at ten genetic loci. Chimaeric males were back-crossed to females of the homozygous recessive component strain; 1851 progeny were born, of which 1092 were from germ cells derived from the multiple recessive component. The progeny segregated cleanly into two types, corresponding to the two populations of germ cells: No intermediates were observed, and no "recombinant" phenotypes. Any somatic contamination of the germ cell genome, at least for those ten loci, must therefore be exceedingly rare if it occurs at all.

IV. Conclusions

Chimaeras and mosaics, whether primary or secondary, spontaneous or induced, offer a variety of ingenious and rewarding approaches to the study of sexual differentiation. In normal embryogenesis, sexual development is rigidly canalized, and errors are few; but the ability to combine in a single composite animal populations of germ cells and/or somatic tissue of different sex chromosome constitutions, or carrying different genes affecting various aspects of sexual differentiation, allows the canalization to be challenged and its underlying mechanisms analysed.

It seems that the presence of even a small proportion of XY somatic cells in the gonadal ridges of the early embryo determines the gonad to develop as a testis rather than an ovary. Intermediate conditions are rare. Secondary sexual development follows the hormonal lead of the gonad, whatever the chromosomal sex of the cells involved. Sex-limited genes function according to the phenotypic sex of their cellular environment. If cells are genetically unable to respond to androgens, their involvement in male differentiation will depend on whether the particular function is autonomous, or can be directed by neighbouring responsive cells.

XX germ cells in a testicular environment enter meiosis before birth, as they would in an ovary, and rarely survive birth. XY germ cells in an ovary, on the other hand, seem able to undergo oogenesis and may give rise to functional eggs. The shape of mouse spermatozoa appears to be controlled by the genotype of the germ line; on the other hand, androgen-resistant germ cells can be stimulated to undergo spermatogenesis, provided that at least some of the somatic tissue in the gonad is capable of responding to androgens. The genetic constitution of the germ cells seems well-buffered against outside influences.

References

Belloni,L.: Valsalva, J Hunter, and Scarpa on The Freemartin. In: Journal of the History of Medicine and Allied Sciences 7, 136—140 (1952)
Burgoyne,P.S.: Sperm phenotype and its relationship to somatic and germ line genotype: a study using mouse aggregation chimeras. Dev. Biol. **44**, 63—76 (1975)

Cattanach, B. M., Pollard, C. E., Hawkes, S. G.: Sex-reversed mice: XX and XO males. Cytogenetics **10**, 318—337 (1971)

Deusen, E. B., Van: Sex determination in germ line chimeras of *Drosophila melanogaster*. J. Embryol. Exp. Morph. **37**, 173—185 (1976)

Drews, U., Alonso-Lozano, V.: X-inactivation pattern in the epididymis of sex-reversed mice heterozygous for testicular feminization. J. Embryol. Exp. Morph. **32**, 217—225 (1974)

Drews, U., Blecher, S. R., Owen, D. A., Ohno, S.: Genetically directed preferential X-activation seen in mice. Cell **1**, 3—8 (1974)

Drews, U., Drews, U.: Metabolic cooperation between *Tfm* and wild-type cells in mosaic mice after induction of DNA synthesis. Cell **6**, 475—479 (1975)

Erickson, R. P.: Androgen-modified expression compared with Y linkage of male specific antigen. Nature (Lond.) **265**, 59—61 (1977)

Ford, C. E.: The cytogenetics of the male germ cells and the testis in mammals. In: Rosenberg, E., Paulsen, C. A. (Eds.): The Human Testis, pp. 139—149. New York-London: Plenum Press 1970

Ford, C. E., Evans, E. P.: Cytogenetic observations on XX/XY chimaeras and a reassessment of the evidence for germ cell chimaerism in heterosexual twin cattle and marmosets. J. Reprod. Fert. **49**, 25—33 (1977)

Ford, C. E., Evans, E. P., Burtenshaw, M. D., Clegg, H., Barnes, R. D., Tuffrey, M.: Marker chromosome analysis of tetraparental AKR↔CBA-T6 mouse chimaeras. Differentiation **2**, 321—333 (1974)

Ford, C. E., Evans, E. P., Burtenshaw, M. D., Clegg, H. M., Tuffrey, M., Barnes, R. D.: A functional "sex-reversed" oocyte in the mouse. Proc. R. Soc. (Lond.) B **190**, 187—197 (1975)

Fredga, K., Gropp, A., Winking, H., Frank, F.: Fertile XX and XY-type females in the wood lemming *Myopus schisticolor*. Nature (Lond.) **261**, 225—227 (1976)

Gartler, S. M., Andina, R., Gant, N.: Ontogeny of X-chromosome inactivation in the female germ line. Exp. Cell Res. **91**, 454—457 (1975)

Jost, A., Prepin, J.: Données sur la migration des cellules germinales primordiales du foetus de veau. Arch. Anat. Microsc. Morph. Exp. **55**, 161—186 (1966)

Kanazawa, K., Imai, A.: Parasexual—sexual hybridization—heritable transformation of germ cells in chimeric mice. Jpn. J. Exp. Med. **44**, 227—234 (1974)

Lyon, M. F.: Sex chromosome activity in germ cells. In: Coutinho, E. M. (Ed.): Physiology and Genetics of Reproduction, Part A. New York: Plenum Press 1974

Lyon, M. F., Glenister, P. H., Lamoreux, M. L.: Normal spermatozoa from androgen-resistant germ cells of chimaeric mice and the role of androgen in spermatogenesis. Nature (Lond.) **258**, 620—622 (1975)

Lyon, M. F., Hawkes, S. G.: X-linked gene for testicular feminization in the mouse. Nature (Lond.) **227**, 1217—1219 (1971)

Marcum, J. B.: The freemartin syndrome. Anim. Br. Abstr. **42**, 227—242 (1974)

McLaren, A.: Germ cell differentiation in artificial chimaeras of mice. In: Beatty, R. A., Gluecksohn-Waelsch, S. (Eds.): The Genetics of the Spermatozoon, pp. 313—323. Edinburgh-New York: Beatty and Gluecksohn-Waelsch 1972

McLaren, A.: Sex chimaerism and germ cell distribution in a series of chimaeric mice. J. Embryol. Exp. Morph. **33**, 205—216 (1975a)

McLaren, A.: The independence of germ-cell genotype from somatic influence in chimaeric mice. Genet. Res. **25**, 83—87 (1975b)

McLaren, A.: Mammalian chimaeras. London: Cambridge University Press 1976

McLaren, A., Chandley, A. C., Kofman-Alfaro, S.: A study of meiotic germ cells in the gonads of foetal mouse chimaeras. J. Embryol. Exp. Morph. **27**, 515—524 (1972)

Merchant, H.: Rat gonadal and ovarian organogenesis with and without germ cells. An ultrastructural study. Dev. Biol. **44**, 1—21 (1975)

Mintz, B.: Hermaphroditism, sex chromosomal mosaicism and germ cell selection in allophenic mice. J. Anim. Sci. **27**, 51—60 (1968)

Mintz, B.: Developmental mechanisms found in allophenic mice with sex chromosomal and pigmentary mosaicism. In: Bergsma, D., McKusick, V. (Eds.): First Conference on the Clinical Delineation of Birth Defects, pp. 11—22. New York: National Foundation 1969

Mintz, B.: Control of embryo implantation and survival. Adv. Biosci. **6**, 317—340 (1971)

Mintz, B., Domon, M., Hungerford, D. A., Morrow, J.: Seminal vesicle formation and specific male protein secretion by female cells in allophenic mice. Science **175**, 657—659 (1972)

Mullen, R. J., Whitten, W. K.: Relationship of genotype and degree of chimerism in coat color to sex ratios and gametogenesis in chimeric mice. J. Exp. Zool. **178**, 165—176 (1971)

Myhre, B. A., Meyer, T., Opitz, J. M., Race, R. R., Sanger, R., Greenwalt, T. J.: Two populations of erythrocytes associated with XX/XY mosaicism. Transfusion **5**, 501—505 (1965)

Mystkowska, E. T., Tarkowski, A. K.: Observations on CBA-p/CBA-T6T6 mouse chimeras. J. Embryol. Exp. Morph. **20**, 33—52 (1968)

Mystkowska, E. T., Tarkowski, A. K.: Behaviour of germ cells and sexual differentiation in late embryonic and early postnatal mouse chimeras. J. Embryol. Exp. Morph. **23**, 395—405 (1970)

Ohno, S., Christian, L. C., Wachtel, S. S., Koo, G. C.: Hormone-like role of H–Y antigen in bovine freemartin gonad. Nature (Lond.) **261**, 597—599 (1976)

Ohno, S., Geller, L. N., Young Lai, E. V.: *Tfm* mutation and masculinization versus feminization of the mouse central nervous system. Cell **3**, 235—242 (1974)

Ohno, S., Gropp, A.: Embryological basis for germ cell chimerism in mammals. Cytogenetics **4**, 251—261 (1965)

Ohno, S., Jainchill, J., Stenius, C.: The creeping vole *(Microtus oregoni)* as a gonosomic mosaic. I. The OY/XY constitution of the male. Cytogenetics **2**, 232—239 (1963)

Poláčková, M., Vojtíšková, M.: Inhibitory effect of early orchidectomy on the expression of the male antigen in mice. Fol. Biol. **14**, 93—100 (1968)

Rowson, L. E. A., Newcomb, R.: A method of determining whether germinal cells migrate in bovine chimeras. 8th Int. Congr. Anim. Repr. Artif. Insem., Abstr. 222 (1976)

Short, R. V.: Sexual differentiation of the brain of the sheep. In: Int. Symp. Sexual Endocrinology of the Perinatal Period, INSERM. **32**, 121—142 (1974)

Tarkowski, A. K.: Mouse chimaeras developed from fused eggs. Nature (Lond.) **190**, 857—860 (1961)

Tettenborn, U., Dofuku, R., Ohno, S.: Noninducible phenotype exhibited by a proportion of female mice heterozygous for the X-linked testicular feminization mutation. Nature (New Biol.) **234**, 37—40 (1971)

Vigier, B., Locatelli, A., Prepin, J., du Mesnil du Buisson, F., Jost, A.: Les premières manifestations du freemartinisme chez le foetus de veau ne dependent pas du chimèrisme chromosomique XX/XY. C. R. Acad. Sci. Paris **282**, 1355—1358 (1976)

Whitten, W. K.: Chromosomal basis for hermaphrodism in mice. In: The Developmental Biology of Reproduction (33rd Symp. Soc. Dev. Biol.), pp. 189—205. New York-London: Academic Press 1975

Zuelzer, W. W., Beattie, K. M., Reisman, L. E.: Generalized unbalanced mosaicism attributable to dispermy and probable fertilization of a polar body. Am. J. Hum. Genet. **16**, 38—51 (1964)

Behavioral Analysis in Drosophila Mosaics

JEFFREY C. HALL

Department of Biology, Brandeis University, Waltham, MA, USA

I. Introduction

Mutations disrupting several different types of behavior in *Drosophila* are currently under intensive investigation (reviewed by Benzer, 1973; Grossfield, 1975; Pak, 1975; Pak and Pinto, 1976). The purpose is to unravel basic features of the development and function of the many parts of the nervous system, through studies of perturbations in nervous system embryology and physiology. To use the fly's visual system merely as an example, one could attempt to analyze this system, through the isolation of mutations showing defects in visually triggered behavior (and many such mutants, to be discussed later, have been found). Given all the background information on the development and anatomy of the photoreceptor cells and neurons in the optic lobes (e.g., Meinertzhagen, 1973), one could ask if a behavior mutant involving the visual system has disrupted cell differentiation or the formation of nerve connections in the developing system. Since much is known about the physiology of vision in *Drosophila* (e.g., Pak, 1975), one could ask if this kind of mutant is defective in a process such as membrane permeability changes, caused by light-induced changes in photopigment conformation. The eventual aim is to learn how the action of particular genes programs into the animal such unknown mechanisms as those having to do with the formation of nerve connections and those involved in phototransduction.

This review discusses the many ways in which a wide variety of behavioral mutants have been analyzed using genetic mosaics. There are several reasons such mosaic studies have been carried out. At an early stage in the analysis of a genetic factor that affects behavior, it is critical to ask where in the animal the mutation exerts its effect. The defective tissues may be in parts of the nervous system that are not directly relevant to the neurologic phenomenon that is under study; or the defective tissues may not be in the nervous system at all. Proper performance of a given behavior requires the completion of several steps in a pathway. Thus, flies mutant for behaviors such as phototaxis, optomotor response, flying, or courtship behavior might be defective in the development *or* the physiology of tissues involved in sensory input, conduction of nerve impulses, or motor output involving specific nerves or muscles; or a mutant might have defects in endocrine tissues. Thus, if one wants to study something like the visual system, by investigating nonphototactic mutants, it is clear that one first has to rule out the possibility that mutant behavior is due to a mere motor defect. Some mutants discovered on the basis of poor phototaxis have, indeed, been found to have normal visual physiology, and have been found (partially through mosaic studies) to have motor defects (these are discussed later). Abnormal behavior could result from defects which can be construed as even more trivial—not specifically involving tissues or functions concerned with the nervous system or associated systems (cf. the many mutations affecting external morphology which have behavioral defects—reviewed by, e.g., Manning, 1965; Grossfield, 1975). A mutant fly, with no external abnormalities, could be generally debilitated because it has poor functioning of a particular feature of metabolism which disrupts nerves and muscles, but also many other tissues (e.g., circulatory, alimentary); metabolism might be defective in tissues, none of which are directly concerned with the control of behavior.

An early part of behavioral mutant analysis, then, must be an objective determination of the *focus* of the mutation (as defined by Hotta and Benzer, 1972): the focus is the anatomic site at which the mutant gene exerts its primary effect. A wealth of techniques are available for constructing mosaic flies, in each of which some tissues are mutant, while others are not; thus, examination and analysis of an array of such mosaics can reveal the focus and indicate rather quickly if it is an interesting focus. Operationally, one examines the behavior of an array of mosaics, scores the genotype of as many tissues as possible in each mosaic, and determines the focus by learning in which part of the animal the expression of the mutant gene is correlated with the behavioral defect.

Several behavior mutants have been analyzed by direct observation of tissues or by physiologic recording, before any mosaic work was carried out. For example, mutants in visually triggered behavior were found to have morphologic abnormalities in the visual system, or defects in visual system physiology (reviewed by Pak, 1975). Here, too, it is essential that these specific defects be analyzed with mosaics. Cell interactions are of obvious and extreme importance in the development and function of the nervous system and associated systems. Therefore, a defect merely observed or recorded in a particular region of the nerve or muscle system might not be the focus. That is, a defect in another tissue—with which the observed one interacts through inductive influences in development, through physiologic communication in postembryonic stages, or even trophic influences in postembryonic stages—could be the real, *primary* focus, and could lead to the initially detected abnormalities. An example of how one could be misled as to the real focus would be a case of abnormal physiology in a mutant fly's optic lobes. Such a defect might, indeed, originate in optic lobe neurons; but, instead, it could be due to defective connectivity or physiology involving photoreceptor cells, which form synapses with cells in the optic lobes.

There are additional, more detailed questions to be asked about behavioral mutants, using mosaic tools. Many behavior mutations affect the behavior of the entire fly—so that a mosaic with some tissues mutant, some not, will be entirely normal or entirely mutant in behavior: The entire mosaic might pass out or not, after a particular treatment (e.g., see later discussion of the "bang-sensitive" mutant). Other kinds of mutations can have the behavioral abnormality scored independently for different parts of the fly: One or more individual legs might show defective behavior in a given mosaic (e.g., see later discussion of the Hyperkinetic mutant). For whole-fly behaviors, it is important to know not only the focus, but also if the left focus *and* the corresponding right focus (e.g., in left and right sides of the bilaterally symmetric brain) must be mutant to give the mutant behavior (defined by Hotta and Benzer, 1972, as "submissive" focus); or if mutant tissue at only one of the homologous foci is sufficient to give mutant behavior (defined by Hotta and Benzer, 1972, as a "domineering" focus). Another question about foci is whether there is a particular, restricted part of the left and/or right side of the fly which must be mutant to give mutant behavior; or, if the focus is "diffuse," such that, for example, any one of several different parts of the nervous system will, if mutant, lead to abnormal behavior. Still another possibility is that there is more than one restricted (non-diffuse) focus for a mutant behavior—such that, for instance, mutant tissue in a particular region of both the head and thoracic nervous system is required to give mutant behavior.

Additional uses of mosaics in behavioral analysis have involved questions on specific interactions between mutant and normal tissues. Finally, the question of autonomy of gene expression can be answered: Is it possible that a genetically normal tissue in a mosaic shows a defect because of the deleterious effects of a circulating substance produced by mutant tissue elsewhere? Conversely, is it the case that genetically mutant tissue in a mosaic (say an eye, which in a completely mutant animal would be defective) will be normal in phenotype, because of the correcting influence of a circulating substance produced by normal tissues elsewhere in the fly? No cases of nonautonomous expression of behavior mutants due to circulating substances have been found, though this is the straw man that seems always to be raised up and knocked down by investigators who analyze behavior mutants with mosaics.

Genetic mosaics have been used in the analysis of many mutant and some nonmutant behaviors in *Drosophila*. Mosaic technology is perhaps most extensive for this organism, but several mosaic studies in other organisms have been performed: of sex-specific behavior in wasps (Whiting, 1932; Clark and Egen, 1975); of reproductive behavior in houseflies (Milani and Rivosecchi, 1954); of female worker behavior in bees (Sakagami and Takahashi, 1956); of genetic retinal degeneration in mice (reviewed by Mintz, 1974) and rat (Mullen and LaVail, 1976); of faulty visual system neuronal connectivity, induced by the albino mutation in mice (Guillery et al., 1973); of genetic muscular dystrophy in mice (Peterson, 1974); and of an "eyeless" mutant in axolotl (Van Deusen, 1973). In these experiments, mosaics were obtained sometimes by sporadic recognition of gynandromorphs in stocks (e.g., Whiting, 1932), or by direct manipulation of tissues of different genotypes (e.g., Van Deusen, 1973; Peterson, 1974). In *Drosophila*, all the mosaic analyses of behavior have been carried out with easily applied genetic tools—employing special strains, chromosomes, or treatments with which the mosaics can be obtained in relatively high frequency. Genetic tools analogous to those in *Drosophila* are sometimes available in other organisms (cf. the wasp gynandromorph-producing strain of Clark and Egen, 1975; and the use of intrinsic genetic mosaicism—X chromosome inactivation—in the analysis of the albino mutation by Guillery et al., 1973). *Drosophila* mosaics studies have frequently had the additional advantage of a large library of mutations which affect external and internal tissues, and which can be used to mark the behaviorally mutant and nonmutant parts of the mosaics.

The examples touched on in this introduction may have made it clear that the following discussion will not really be limited strictly to behavior. Indeed, the analysis by mosaics of "behavioral" mutants in *Drosophila* does and should mean more than that. A major aim of this area of research is to understand the development and functioning of the nervous system. Thus, there are mutants found on the basis of an observable defect in excitable cells on (at least) the periphery of the animal; and these have been shown later to have specific and informative effects on behavior. From the other side, there are the many mutants, found on the basis of a behavioral defect, which now are known to have neurologic defects in the anatomy (even the developmental anatomy) of excitable cells, or defects in the physiologically measurable functioning of such cells. The ideas and techniques of mosaic analysis have frequently had great influence on the matter of behavioral mutants "becoming" neurologic ones.

II. Major Features of Behavioral Mosaic Analysis

The majority of mosaic studies of neurologic mutants in *Drosophila* have involved diplo-X, haplo-X gynandromorphs. As discussed by Hall et al. (1976) and by Janning (this volume), there are several techniques whereby an X chromosome can be somatically lost from a diplo-X zygote, resulting in a sex mosaic for which haplo-X tissues are genetically male, and which will be hemizygous for any X chromosome behavior mutation present on the X which was not lost; the rest of the mosaic will retain both chromosomes. If the mutation is recessive and autonomous, and the X that was lost carried the normal allele, then the haplo-X tissues in the mosaic will be able to express the behavior mutation. If these haplo-X tissues include the focus, the mosaic will have defective behavior.

Data from gynandromorphs (not involving behavior mutants) have been used to generate fate maps of the embryonic blastoderm, showing locations of progenitor cells, mainly in regard to external and internal tissues of the adult (see Janning, this volume). Foci of behavior mutants dealt with in the present review will continually be referred to the basic fate maps (also see Fig. 4 in the present paper). In these maps, the imaginal discs are derived from the more central portions of the left or right half of the blastoderm. The parts of the map referring to the internal tissues that are of principal interest here are areas below (or ventral) to regions for the head and thoracic imaginal discs. The ganglia in the head come from anterior ventral regions; and the thoracic ganglia come from ventral regions in the middle of the fate map, quite far from the ventral midline of the map. These are the same general areas from which the larval nervous system is derived (Poulson, 1950). The adult muscles are believed to come from medial, ventral portions of the fate map, closer to the ventral midline than the sites for thoracic ganglia. These conclusions on muscle landmarks come from the larval fate maps of Poulson (1950—not based on mosaic analysis), and from mosaic studies of mutants that affect thoracic muscles in the adult (discussed later).

The techniques for constructing mosaic fate maps are much the same as those used for studying behavior mutants. The following conditions must be satisfied if these X chromosome-loss techniques, as applied to the mutants, are to be meaningful: (1) Hemizygous mutant flies should be clearly different in behavior from flies heterozygous for the mutation. (2) There must be autonomously expressed "tagging" markers linked to the behavior mutation, and, as discussed by Hall et al. and by Janning (this volume) there are a variety of markers available: both for the cuticle (so that the mosaics can be distinguished from nonmosaic progeny of the cross set up to generate the mosaics) and for internal tissues (so that, after cuticle and behavioral scoring, the distribution of mutant versus nonmutant nervous system tissues can be determined). (3) For the most straightforward statistical analysis of the mosaic data, the behavioral characteristic should be unambiguously scorable at the level of individual flies (i.e., in mutant vs. wild-type individuals); thus, the scoring of a given mosaic as showing mutant versus wild-type behavior will be meaningful. Some methods of behavioral ascertainment—such as phototactic scoring in a "countercurrent" device (Benzer, 1967), or demonstration of shock-avoidance conditioning (Quinn et al., 1974; Spatz et al., 1974)—are performed on populations of flies, and not all wild-type individuals successfully perform the behavior on a given trial. (4) There should be no sexual dimorphism associated with the behavioral mutant, since these kinds of mosaics are mixed as to sex per se, in addition to being mixed as to mutant versus wild-type tissues. For instance, gynandromorphs in which haplo-X tissues were expressing a male-specific mutation affecting courtship behavior would be difficult to analyze unambiguously (as was the case for the courtship song mutant of Schilcher, discussed later).

The limitations just listed are not insurmountable. Although X chromosome neurologic mutants are the only ones usable here, 20% of the genome of *D. melanogaster* is on the X chromosome, so that a fair number of mutations affecting a particular behavior will be (and have been) found on this chromosome. Such mutants can immediately have a mosaic analysis performed, as a first step in their characterization. If the X-linked mutation is dominant, then somatic loss of the usual unstable X chromosomes (i.e., ring-X chromosomes—Hall et al., 1976, Janning, this volume) is not immediately applicable. The dominant, mutant allele could be crossed into a ring-X chromosome, and then the recombinant ring-X could be somatically lost from a diplo-X zygote, where the regular X had the recessive, wild-type allele. Yet these recombinants (requiring double cross-overs) are somewhat tedious to produce, and so it is likely easier to induce somatic loss of a dominant-bearing, standard X chromosome through the use of chromosome destabilizing mutants such as paternal loss *(pal)*, claret-nondisjunctional *(ca^{nd})*, or mitotic-loss-inducer *(mit)* (reviewed by Hall et al., 1976). These chromosome loss-inducing mutants, moreover, allow mosaics with a wide range of male or mutant tissue to be produced. For instance, ca^{nd}-induced mosaics are about half male, half female (as is the case for mosaics from unstable rings); while *mit*-induced mosaics are, on the average, one-eighth male or less.

For autosomal neurologic mutants (e.g., Cosens and Manning, 1969; Harris et al., 1976), somatic loss of a second or third chromosome carrying the dominant allele is invariably lethal to the zygote (reviewed by Wright, 1970). Here, one must resort to the following procedures: (1) A duplication carrying the dominant allele of the behavioral gene could be translocated to the X chromosome, and somatic loss of this X could be accomplished, through the use of one of the destabilizing mutants mentioned above, in zygotes homozygous for the autosomal recessive allele. (2) A small, genetically marked, free duplication carrying the normal allele of the neurologic mutation could be obtained. Some duplications of autosomal material exist (e.g., Y chromosomes carrying small segments of an autosome—see Lindsley and Grell, 1968, Lindsley et al., 1972), and one may be fortunate enough to learn that an autosomal behavior mutant under study is covered by such a duplication. This strategy is not conceptually different from that described in (1); but, whereas (2) is hypothetical, somatic loss of X-linked, autosomal neurologic genes has been accomplished (with respect to the rough-eyed mutants, which have optic lobe abnormalities; and physiologic abnormalities in the ebony mutant— discussed later). (3) Diploid–haploid mosaics—whereby both the second and the third chromosomes are somatically lost—can be produced (though at very low frequencies—reviewed by Hall et al., 1976), and apparently survive to adulthood. Mosaic analysis of behavior would be set up, such that the recessive allele of the autosomal gene is uncovered in the haploid tissues.

It should be noted that behavior mutants with sex-limited expression, e.g., male-specific courtship mutants, can be analyzed in mosaics, by procedures analogous to those just described for autosomal mutants. For instance, somatic loss of a small, marked duplication which carries the normal allele of a courtship-controlling gene, from an entirely haplo-X zygote, would be one way to determine the focus of the mutation-induced defect. This technique would be applicable, and necessary, for not only autosomal courtship mutations, but X-linked mutations as well.

Another technique which can be used for cases where somatic loss of an X chromosome is not feasible is induced mitotic recombination (Becker, this volume). Starting with zygotes heterozygous for, say, a third chromosome recessive neurologic mutant, one can treat the developing animals with ionizing radiation, at a variety of stages. Mitotic segregation can result from an induced cross-over between the centromere and the genetic locus in question, so that clones of homozygous mutant tissue are produced. The later the stage of treatment, the more mutant patches are produced (if the target cells are proliferating), but the smaller the average clone size (because there are of course fewer remaining cell divisions between later stages of induction and the time of the last cell division). One is basically limited to inducing the cross-overs between fairly early embryogenesis (a few hours postfertilization) and fairly late in the third (last) larval instar. Even for the earliest induction stages, the mutant clone sizes are small (e.g., a fraction of the eye surface) relative to those obtained by chromosome loss methods. Mutant clones resulting from somatic cross-overs are recognized by the use of tagging genetic markers, linked to the gene under study. Up to now, nearly all the genes studied by mosaicism resulting from mitotic recombination have had to do with external cuticle phenotypes (reviewed by, e.g., Nöthiger, 1972, Gehring,

1972). Yet some progress has recently been made with respect to this kind of analysis done on genes influencing nerve cells (see below); and there are internal markers now available (Janning, this volume).

III. Cases of Behavioral Mosaic Analysis

A. Simple Foci

1. Visual System Mutants

Of the many neurologic mutants isolated recently, several have involved the visual system and have been studied in gynandromorphs. Hotta and Benzer (1970) examined electroretinograms (ERGs) of mosaics, parts of which were hemizygous for an X-linked mutation that causes an abnormality of the ERG. These mutants had been isolated on the basis of faulty phototaxis and/or optomotor behaviors (see below). *Different parts* of a given mosaic could be examined separately, by, for example, recording the ERG of a left eye with genetically mutant retinula cells, and then recording the ERG of the right eye, with genetically wild-type photoreceptor cells (i.e., heterozygous for the recessive mutation). The ERG mutants studied were (1) tan[1]: light body color; abnormal phototaxis and optomotor response; missing on- and off-transients of the ERG; (2) the BS 18 allele of the no-on-transient-A *(non A)* gene: abnormal phototaxis and optomotor response; missing on- and off-transients; but no known morphologic abnormality; and (3) three alleles of the no-receptor-potential *(norp A)* mutation: abnormal phototaxis and almost no ERG at all—that is, missing photoreceptor potential and transients. Mosaics were selected that had both eyes wild type or both mutant; and several others were chosen with one eye mutant and the other wild type (about 60% of the gynandromorphs tested were of this type—a substantially greater proportion of such cases than one finds among a randomly collected population of such mosaics). Eyes of mixed genotype were not analyzed and would be difficult to interpret, because—as noted by Pak (1975)—the ERG is a crude parameter, concerning the summed electric activity of the visual system. However, in all the mosaics examined, each with an eye of uniform genotype, the genotype of the external eye was predictive of the ERG: 88 eyes tested for tan[1] mosaics, 10 tested for $nonA^{BS\,18}$ mosaics, and 120 tested for *norpA* mosaics. Thus, a genetically mutant eye, with its mutant phenotype, could not be corrected by wild-type tissue elsewhere in the fly.

The conclusion from the above study is either than the foci are the retinula cells (only); or perhaps the retinula cells and a relevant part of the ganglia in the head are always clonally related in these mosaics, so that the focus is the photoreceptor cells and/or some more centrally located neurons. The latter possibly can now be ruled out, since Kankel and Hall (1976) have found from cell lineage analysis involving a nerve cell marker—that the optic lobes and the retinula cells are not infrequently of different genotypes in gynandromorphs, (see Janning, this volume). This frequency is high enough to rule out the possibility that in all 208 eyes—in mosaics with respect to tan[1] and *norpA*—the eye surface failed to be of different genotype from the optic lobes; that is, a more central focus for these mutants is ruled out. Several investigators have argued that the on- and off-transients are lamina (optic lobe) potentials (see the review of Pak, 1975). So, while a focus in the retina is not surprising for *norpA*, such a focus is somewhat unexpected for tan[1]. For it means that a defect in the lamina cells is due to a *primary* defect in cells presynaptic to the optic lobe neurons. This kind of inference is certainly not yet possible for the focus of transient defects in *nonA*. If more than five mosaics were analyzed, it might be found that, for example, a genetically normal eye is sometimes associated with a mutant ERG—and this would imply a focus in the lamina.

There is very little behavioral information on visual system mutant mosaics. Anecdotally, it is known that some of the blind mutants (e.g., tan[1]—Hotta and Benzer, 1970) cause mosaics with one eye mutant, the other wild-type to move against gravity and toward light in a helical path, with the defective eye toward the light (vainly trying to equalize light input to both sides). Homyk (1977) has presented mosaic data on a new allele of tan isolated on the basis of defective flight behavior (Homyk and Sheppard, 1977); the behavioral focus is not simple, and is discussed later.

An X-linked mutant called *P-37*, isolated in the laboratory of W. L. Pak on the basis of poor "fast" phototaxis at low light intensity, has been examined behaviorally in mosaics. Totally mutant flies have suppressed on- and off-ERG transients. Moreover, in optomotor tests, *P-37* flies have very poor response to moving, narrow stripes. Thus, the mutation leads to a defective "high acuity system" (Heisenberg and Götz, 1975). M. Heisenberg (personal communication) has tested the response of 17 genetically normal (*P-37*/ +) eyes, in mosaics having mutant tissue elsewhere (even elsewhere in the head): All such stimulated eyes led to normal optomotor responses. A preliminary conclusion, then, is that the focus for this mutant defect is the retina.

It would be very useful behaviorally to test mosaics constructed with respect to a mutant such as *nonA*. Perhaps it is the case that the ERG defect has a focus in the lamina (see above). There are other defects in this mutant, in that the system for high-acuity optomotor response is almost completely blocked. Even more subtly, tethered $nonA^{H2}$ flies—tested individually for optomotor responses—respond normally to back-to-front movement, but almost not at all to front-to-back movement (Heisenberg, 1972). This last defect seems a likely candidate for having a more central etiology than the (hypothetically) peripheral focus for the defective ERG. Thus, it may be that both ERG and optomotor defects have an optic lobe focus in this mutant; or, it could be that there are different foci for the different behavioral and neurologic abnormalities. The facts and hypotheses concerning *nonA* point up the necessity of carefully defining focal determinations: The analysis depends on what particular defect is being considered, so that different parts of the animal might be affected by a given mutation. Any of these tissues might, if mutant, lead to the same mutant phenotype; or, it could be that mutant tissues at the different foci lead to quite different kinds of defective behavior and physiol-

Table 1. Eye genotype. ERG results from ebony mosaics in *D. melanogaster*. Somatic loss was induced by ca^{nd}, of an X chromosome into which e^+ is inserted (this X is part of $T(1; 3)c152$—see Lindsley and Grell, 1968). Loss of this chromosome uncovered the white eye color mutation (w) and e^{11} as well, which was in homozygous condition on the third chromosomes. ERG's were recorded for individual eyes, to determine if visual system physiology was normal, totally mutant (lacking transient), or partially defective (intermediate)

ERG	Wild-type (heterozygous w^+ and e^+)	Mixed	$w; e^{11}$
Normal	98	24	26
Intermediate	2	4	2
Mutant	0	2	62

ogy. The possibility of multiple-foci will reappear with regard to several other behavior mutants, discussed later. Here, the main warning is that quite different behavioral and physiologic parameters, applied to mosaics constructed for a given mutant, can reveal quite different foci.

Some of the visual system mutants have initially been isolated by way of optomotor defects (reviewed by Heisenberg and Götz, 1975). Some of the mutants are allelic to genes first identified with regard to abnormal phototaxis. The extra-lamina-fiber mutant (*elf*—originally called *H-37*), though, is not an allele of the aforementioned mutants that apparently have foci in the retina. *Elf* flies are defective in the "high-sensitivity system"—and so respond poorly in optomotor tests under low light intensities; also, the mutant has a weak off-transient in its ERG. M. Heisenberg (personal communication) has tested 39 eyes in flies mosaic with respect to *elf*. All the mosaic flies tested have had one eye completely hemizygous for the mutation, and the other eye completely heterozygous. There were at least four cases where the retinula cell genotype (determined by a tagging eye-color marker, w^a) was not predictive of the ERG; and there were cases both with w^a eye, but normal ERG; or w^+ eye and mutant ERG. Thus, the *elf* focus for the ERG defect might be in the optic lobes. The behavioral focus may also be in these tissues.

The third chromosome ebony mutant (*e*) has specific ERG abnormalities. Grossly, it has defective on- and off-transients (reviewed by Pak, 1975). Y. Hotta and R. Hodgetts (unpublished) have analyzed the defect in mosaics, in which haplo-X tissues had lost an X chromosome in which e^+ (and neighboring genes) are inserted. Dr. Hotta has supplied details on the results, which are listed in Table 1. The data reveal that the *e*-induced ERG defect does not have a straight-forward retinular focus. Thus, the *e*-induced defect is in a sense not autonomous to the eye, but the nonautonomy is not due to a presence of a circulating substance, since Hotta and Hodgetts noted several individual mosaics with a mutant ERG in one eye, and a wild-type ERG in the opposite eye. Hotta suggests two possible interpretations of these data: (1) A two-focus model, where one focus is the eye itself, since no eyes with normal genotype produced a mutant ERG, and where the other focus is elsewhere, possibly in an internal tissue. Hotta in fact determined a position on the fate map for the second focus, which came out in a region of the blastoderm between the eye and the proboscis-closest to a posterior corner of the eye (and presumably *not* very near the relatively ventral optic lobe landmarks—cf. Kankel and Hall, 1976). This positioning was consistent with the finding that, among 16 eyes which were nearly all mutant—but had a small patch of wild-type cells in the posterior corner—12 produced a normal ERG, 2 produced near-normal ERGs, and 2 had the "intermediate" result. Note that this second focus is apparently not *at* the posterior corner of the eye, but is merely near it. If the eye itself *or* this second focus is e^+, then a normal ERG will be produced. (2) A second model proposed by Hotta is that of a diffuse focus. Here, there would be two large, homologous blastoderm areas (i.e., on the left and right sides), near the compound eye landmarks, but distinct from them. All of this region in, say, the left side of the blastoderm would have to be mutant to lead to a mutant ERG for the left eye. If the diffuse focus is very close to the eye, a wild-type or mixed-genotype eye would nearly always have at least part of the focus wild-type.

Visual system mutants that cause not only physiologic and behavioral abnormalities, but readily observable morphologic abnormalities as well, have been analyzed with mosaics. The receptor-degeneration-A *(rdgA)* and receptor-degeneration-B *(rdgB)* mutants cause the retinula cells to degenerate, after the adults emerge with normal eye morphology (Hotta and Benzer, 1970; Harris et al., 1976; Harris and Stark, 1977). The outer retinula cells (R_{1-6}) always degenerate in flies expressing any of the various alleles at these loci. The two central retinula cells (R^{7-8}) may also degenerate, depending on the allele. Hotta and Benzer (1970) found that three mutant alleles of the *rdgA* gene are expressed autonomously: An entirely mutant eye in a given mosaic always showed rhabdomere degeneration, whereas eyes heterozygous for a recessive mutant allele were phenotypically normal. Harris et al. (1976) and Harris and Stark (1977) learned that autonomous expression of *rdgA* and *rdgB* mutants extends even to the level of single photoreceptor cells. That is, genetically mutant cells within an eye, or even an ommatidium, of mixed genotype showed mutant phenotype; and any cell heterozygous for one of the X-linked mutations was normal. Finally, the mutant sevenless *(sev)*—which emerges as an adult without R_7—is expressed autonomously, so that in mosaic eyes, totally mutant ommatidia lack R_7, heterozygous ommatidia possess this retinula cell; and, for genetically mixed ommatidia containing R_7, that cell was always found to be heterozygous for the recessive *sev* mutation (Harris et al., 1976). These findings on autonomous expression of the genes causing degeneration of photoreceptor cells did not necessarily knock down a straw man of, say, a focus in the nearby pigment cells, because Mullen and LaVail (1976) found just such a focus in rats mosaic for a retinal degeneration mutant.

Another kind of mosaic analysis has been applied to other mutations that affect eye morphology. Meyerowitz (1977) followed up the preliminary findings of Johannsen (1924), Pilkington (1941), Power (1950), and Pak et al. (1969). These investigators found that certain mutants leading to a disarray of eye facets also disrupt the morphology of the optic lobes. Such phenotypes are seen in the Glued *(Gl)*, rough *(ro)*, and glass *(gl^3)* mutants, all of which are autosomal. *Gl* and *ro* are similar in phenotype, so that not only is there an irregular facet array, but one also finds reduced numbers of retinula cells per ommatidia. The lamina is dramatically abnormal as well, showing very "scrambled wiring." Moreover, there is no external chiasma connecting the lamina to the higher-order optic neurons. These optic lobes not only have abnormal projections of nerves to them, but are also shifted to abnormal positions in the head capsule. The mutants are behaviorally defective in phototaxis (though *ro* is nearly normal here), optomotor response, ERG (abnormal transients), and courtship. The gl^3 mutant has a roughened eye surface, with few facets, no detectable retina cells, a badly scrambled small lamina, and scrambled higher-order optic lobes. Behaviorally, one finds very little photo-tactic response, no optomotor response, and weak courtship. Recordings from gl^3 eyes revealed no ERG at all.

The striking finding on these mutants is that the facet and photoreceptor cell defects appear to induce the defects in the optic lobes. Thus, mitotic recombinants induced in a $Gl/+$ fly lead to homozygous wild-type patches on the eye surface; and, *beneath these patches*, one sees normally arrayed lamina neurons and a normal projection of the lamina to the higher-order lobes. In this same eye, patches of mutant photoreceptor cells still correspond to mutant optic lobes. The same general results are found for *ro* and gl^3, but here, Meyerowitz induced mutant eye-surface patches by irradiating larvae heterozygous for either of these recessive mutations. Thus, mutant optic lobe wiring and mutant projections are found only beneath the externally mutant patches. A critical control was the demonstration that—even for cross-overs induced very early in development, at stages *prior* to those pertaining to the experiments on the mutants—there is no clonal relationship between retina and optic lobe cells. The demonstration that abnormal eye genotype can specify abnormal optic lobe phenotype fails if, from the mosaic experiments, it is simply the case that the genetically mutant patch includes not only eye cells but also optic lobe neurons. Yet, Meyerowitz found that induced patches in the eye—revealed with a simple eye-color marker—do not correspond to marked patches in the optic lobes. The nerve cell marker here was an acid phosphatase-null mutation, linked to the eye-color marker. This was the same internal marker used in gynandromorphs to generate external—internal fate maps (Kankel and Hall, 1976); and the lack of clonal relationship between the eye and optic lobe implied by those maps is confirmed in the control of Meyerowitz. Further, preliminary studies of eye-roughening mutants have made use of chromosome loss mosaics—involving mosaicism produced very early in development, using X chromosomes onto which a given eye-roughening gene had been translocated. Here, mutant patches within the eye always correspond to mutant optic lobe regions; yet, the eye and optic lobes are frequently of different genotype in the gynandromorphs (see above). This reinforces the conclusion that the ordering of cells in the eye is involved in specifying the proper wiring array in the second- and higher-order neurons of the visual system.

2. Locomotor-Physiologic Mutants

A locomotor, neurologic mutant with relatively simple, autonomous foci is the X-linked Hyperkinetic (Hk^1) mutation, studied by Kaplan and his colleagues (e.g., Kaplan and Trout, 1969, 1974; Ikeda and Kaplan, 1974). As for the visual system mutants, mosaics with respect to Hk^1 can be scored for mutant and normal behavior in different parts of a given fly: Mutant behavior here was the vigorous ether-induced leg shaking, like that shown in males homozygous for Hk^1 (in the initial report of this mutant, $Hk^1/+$ legs were noted as showing mild shaking behavior). Leg shaking caused by Hk^1 was found to occur independently for each leg in mosaics (Ikeda and Kaplan, 1970b): Any number of combinations of legs might shake in such a mosaic. Once could have plausibly predicted the very different, all-or-none result, whereby a given mosaic would either have all legs normal, or all of them shaking.

The physiology of the Hyperkinetic abnormality has been analyzed in gynandromorphs. Motor neurons in the thoracic ganglia associated with each leg show ether-induced rhythmic bursts of impulses in totally Hk^1 flies; these impulses are not found in the wild type, and, moreover, are of one or two possible types. In each thoracic ganglion in an Hk^1 mutant, there are abnormal cells which show action potentials lacking a "prepotential." There are also the less common types of mutant cells, for which action potentials are always preceded by a slowly rising depolarization (Ikeda and Kaplan, 1970a). These mutant neurons were called "pacemaker cells." Both kinds of abnormal nerve firings are recessive characters. For mosaics, the mutant patterns of neural activity were always found in the part of the thoracic ganglion corresponding to a leg that showed shaking. Nonshaking legs were shown to have associated motor neurons without the abnormal bursts of impulses (Ikeda and Kaplan, 1970b). This "autonomy" of mutant activity is expected, since cells that fire abnormally will still do so if separated from cephalic and sensory input by surgery (Ikeda and Kaplan, 1970a).

Ikeda and Kaplan (1974) present a model in which the actual focus of the Hk^1 mutation is the "pacemaker" neuron, of which there is at least one per relevant region of the thoracic ganglion (i.e., that region corresponding to a given leg). The more common mutant neurons, with no prepotentials, would then be induced to show their abnormal firing upon stimulation by the cells in which resides the focus. Ikeda and Kaplan (1974) even note the result that—when an appropriate region of the thoracic ganglion is mixed in genotype (part of it hemizygous for Hk^1 and part of it heterozygous), only the genetically mutant tissues show the neurologic abnormality. This report is misleading in that W. D. Kaplan (personal communication) has noted that the mixed genotype of these particular ganglia was only inferred from the physiologic results; that is, there was no independent marking of the genotypes of these cells.

Hotta and Benzer (1972) mapped these simple Hk^1 foci with respect to thoracic cuticular landmarks. They found that there was one focus per leg, on a part of the blastoderm map ventral to areas which give rise to thoracic cuticle—from which the adult thoracic ganglia indeed develop. Hotta and Benzer's results are of course consistent with the conclusions of Ikeda and Kaplan.

It should be noted, parenthetically, that Hk^1 leads to other behavioral defects—even in larvae—besides adult leg shaking (Kaplan and Trout, 1969; Trout and Kaplan, 1970; Burnet et al., 1974). Other Hk^1-induced abnormalities have been studied in mosaics, but these defects do not involve simple foci, and are discussed later.

Several other neurologic mutants affect the behavior of individual legs, and have had separate, simple foci determined for the defects. These are the X-linked temperature-sensitive paralytics: paralytic-temperature-sensitive ($para^{ts}$—Suzuki

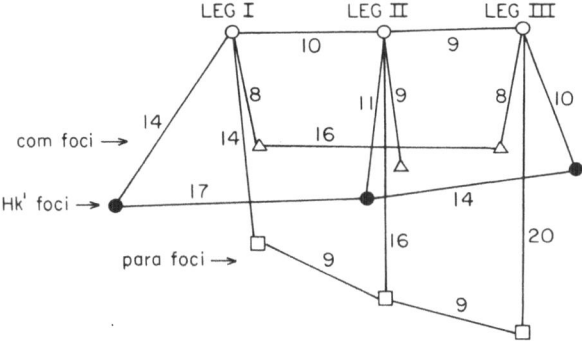

Fig. 1. Foci for behavioral mutants with defects in leg behavior. The diagram is a "fine-structure" map, referring to part of the thorax. Foci for ether-induced leg shaking, influenced by the Hyperkinetic (Hk^1) mutant, are designated by *closed circles* (●) and are mapped with respect to cuticular landmarks (○) referring to the prothoracic leg (*I*), mesothoracic leg (*II*), and the metathoracic leg (*III*). This portion of the map was reported earlier by Hotta and Benzer 1972 (their Fig. 7), who describe the principles and procedures of such behavioral mapping. The same basic strategy was employed for the mapping of foci for leg paralysis induced by the mutant comatose (*com*) at 38° C (△), or by the paralytic-temperature-sensitive mutant (*para*) at 29° C (□). These foci were determined by S. Benzer (unpublished). All numbers in the map are sturt distances (Hotta and Benzer, 1972) designating, for example, the percentage of mosaics in which the leg genotype is different from the behavioral phenotype. The behavioral foci for the different legs are the ones most directly beneath the leg landmarks (*I, II, III*). That is, the focus for *com*-induced paralysis of the prothoracic leg is designated by the triangle, which is in a vertical direction below (and 8 sturts away from) the Leg I landmark

et al., 1971); shibire-temperature-sensitive (*shi^ts*—Grigliatti et al., 1973; Poodry et al., 1973); and comatose (*com*—Benzer, 1973; Siddiqi and Benzer, 1976). For these mutants, the legs become immobile at high temperature (29° C or above, depending on the mutant). Mosaic analysis has shown that individual legs can become paralyzed independently of the behavior or other legs, and this should allow a straightforward determination of the focus for each leg. The analyses here, however, are often complicated by effects of the mutants on other parts of the nervous system besides those regions directly controlling leg movement. Grigliatti et al. (1972) showed, from mosaics constructed with regard to *para^ts*, that externally mutant legs generally showed contractile paralysis at 29° C, whereas heterozygous wild-type legs did not. The abdominal genotype had no effect on the behavior of these mosaics. However, if the external head was all or nearly all mutant, then the mosaic was immobile, but not necessarily with leg paralysis. Thus, flies with mutant head and heterozygous wild-type thorax could not walk, but there was no direct effect on the legs. These data imply that there is a focus in the brain *para^ts*, concerning movement of the whole animal; the focus is not mappable here in a simple fashion (see later discussion of interacting foci). The behavior of individual legs is somewhat independent of the anterior focus, in that a given leg in a mosaic may or may not exhibit the contractile paralysis—whether the anterior focus is mutant or not. S. Benzer (unpublished) mapped leg foci—one per leg—to an area of the blastoderm ventral to thorax cuticle regions (Fig. 1). He was aided by the fact that an otherwise immobile leg in a given $para^{ts}$ mosaic might be induced to show movement by touching the leg or even an antenna. The individual leg foci here likely are in the thoracic ganglia. These foci are, of course, in the same general region as the Hk^1 foci; but, if one overlays the two maps, the $para^{ts}$ foci are in more ventral locations than the Hk^1 foci.

One of the several alleles of $para^{ts}$—number ST 109 (found by S. Benzer)—shows a strikingly different effect with respect to leg behavior in mosaics. For $para^{ts\ 109}$, all six legs are either paralyzed or normal (at 36° C) in a given mosaic. Thus, there could be one focus for this behavioral defect, and it is not mappable in a simple fashion (see below).

The various observations of $para^{ts}$ behavior at high temperature imply that these mutations affect different parts of the nervous system—with respect to a given allele, or in comparing the effects of one allele to those of another. This vague conclusion is not surprising for mutants causing paralytic behavior. These genetic changes could shut down nervous activity generally, through effects on impulse conduction or synaptic transmission. It would perhaps be more unexpected to find that such a mutation affects only a very limited number of nerve cells in very discrete regions of the nervous system. Indeed, physiologic recordings from different parts of the nervous system and at different stages of the life cycle indicate that $para^{ts}$ affects more than one, but definitely not all parts of the nervous system. For example, the adult ERG is normal at high temperature in mutant flies (cf. Suzuki et al., 1971), but propagation of action potentials in larval abdominal nerves is reversibly blocked by temperature treatment (Siddiqi and Benzer, 1976).

The two other paralytic mutants, shi^{ts} and com, also appear to affect several parts of the nervous system. Some of the evidence is from mosaic analysis. Thus, a shi^{ts} or com mosaic may collapse or not at high temperature, irrespective of leg paralysis (L. Hall, unpublished; D. R. Kankel, unpublished). A detailed analysis of the focus for general immobility has not been presented. However, foci for paralysis of individual legs have been mapped for shi^{ts} (L. Hall et al., 1973; Deak, 1976) and for com S. Benzer (unpublished). There is one paralysis focus per leg, and the locations are in the usual place, in blastoderm regions ventral to thoracic cuticular landmarks (Fig. 1). The com foci are in more *dorsal* locations than the Hk^1 or $para^{ts}$ foci, from superimposition of the independently derived maps (Fig. 1). Also, it is the case that "doubly mutant mosaics"—in which somatic loss of an X chromosome uncovers both Hk^1 and com—have certain legs which, for example, do not shake during the ether test and then are paralyzed during the separate high temperature test; so the foci here have been directly separated (S. Benzer, unpublished data). There are data which provide an interesting and apparent confirmation of these results on "directly separated" Hyperkinetic and comatose foci. J. C. Hall and D. R. Kankel (unpublished) have analyzed the genotype of the thoracic ganglia—in appropriately marked flies (cf. Kankel and Hall, 1976), which were mosaic for Hk^1 or com. They learned that the Hk^1 foci are in the more *dorsal* regions of thoracic neuromeres, while the com foci are more ventrally located within such cortical regions of the thoracic ganglia. These "concrete" findings from the internally marked mosaics are wholly consistent with the fact that the "formal" fate mapping of Hk^1 and com foci shows that the former are more *ventrally* located on the map (Fig. 1). This might seem inconsistent with the result that the dorsal portions of the neuromeres contain the *Hk* foci, but it is known that such dorsal regions of the thoracic ganglia neuromeres are, on the fate map, in positions that are *ventral* to the landmarks referring to the (actual) ventral parts of the thoracic ganglia (Kankel and Hall, 1976; Hall, 1977; Flanagan, 1977). In short, the "inversion" of the landmarks in the map for the ventral ganglia is in a sense confirmed by the separate findings on doubly mutant mosaics, and on the internally marked mosaics that were analyzed for Hyperkinetic or comatose behavior.

Interactions among different paralytic mutants have not yet been studied. However, the paralytic foci for *shi^{ts}* or *para^{ts}* could be qualitatively distinguished from the *com* foci, because the first two mutants cause paralysis of the legs at 29° C, whereas the effects of *com* are not revealed until the temperature reaches 38° C. Doubly mutant mosaics could be analyzed on other criteria. For example, in *com* and *para^{ts}* behavior the times required for recovery from paralysis are very different in the two mutants (Siddiqi and Benzer, 1976).

A problem arises in the mapping of *shi^{ts}* or *com* leg paralysis foci. For a given leg, the proportion of all mosaics which have normal leg cuticle (i.e., heterozygous for the tagging marker) but mutant behavior, *or* a marked leg but normal behavior, is used directly to calculate how "far" on the fate map the focus is from the part of the blastoderm from which the leg is derived. These two classes of mosaics are roughly equal for the *Hk^1* mosaics, as they should be for foci which are *points* on the blastoderm—i.e., one point for a leg bristle, and another for the simple focus. Yet the mosaics in regard to *shi^{ts}*- or *com*-induced leg paralysis have substantially more cases of normal leg cuticle, mutant behavior than the reciprocal class (L. Hall et al., 1973; S. Benzer, unpublished). This result is comprehensible if it is the case that either of these two mutants cause rather general abnormalities in nerve cell functioning, and so disrupt many different kinds of nerve cells (interneurons, motor neurons). Thus, any of several different thoracic neurons would, if mutant, not function and lead to leg paralysis. Either a motor neuron or a presynaptic interneuron could thus be considered the focus. The sites responsible for the defects caused by these two mutants are, it seems, not point foci, but are diffuse.

These findings from the mosaics are consistent with physiologic data from *shi^{ts}* and *com*, which indicate that many different parts of the nervous system malfunction at high temperatures (L. Hall et al., 1973; Kelly and Suzuki, 1974; Ikeda et al., 1976; Siddiqi and Benzer, 1976). The nerves affected by these mutants include those in the visual system, because the on- and off-transients of the ERG vanish at high temperature, though the sustained corneal negative component appears to be unaffected. Moreover, the focus for the ERG defect in *shi^{ts}* is not exclusively in the photoreceptor cells, unlike the situation for some of the previously cited mutants of visual system (e.g., tan^1). For *shi^{ts}*, it appears that mutant retinula cells always lead to an absence of transients, but some eyes in mosaics with genetically normal photoreceptor cells have a mutant ERG at high temperature, as if mutant optic lobe cells can also block the transients.

Comatose may lead to more widespread defects than *shi^{ts}*, in that action potentials of larval abdominal nerves in the latter mutants are normal at high temperature but the nerve impulses vanish in *com* larvae under this treatment (Siddiqi and Benzer, 1976).

Shibire may have far more diverse effects on the entire animal than does *com*, because high temperature pulses given to developing *shi*[ts] animals, at a variety of stages, lead to a host of tissue abnormalities (Poodry et al., 1973). These tests have not been carried out on developing *com* animals.

Mosaics were used by Kelly (1974) in conjunction with pharmacologic studies of *shi*[ts]. These mutants are, even at permissive temperatures (22° or lower), more resistant to tetrodotoxin (TTX) than flies from a particular wild-type strain (Oregon-R). TTX is a poison that stops nervous conduction, by selectively blocking the increase in sodium conductance produced by depolarization; i.e., it blocks the "sodium channels" in neuronal membranes. The drug was adminstered to the flies by feeding. Kelly's finding of *shi*[ts]'s relative insensitivity suggested to him that the mutation affects sodium channels. However, the effect is probably indirect, because shibire affects a variety of cell types. Moreover, it should be noted that *shi*[ts] flies are not more resistant than are flies from a variety of other wild-type strains (i.e., aside from the Oregon-R stock which was the source of Kelly's control animals; L. Hall et al., 1978). In any event, Kelly worried that *shi*[ts]'s relative resistance to the drug could be due to general debilitation and poor feeding behavior, in that the mutant flies are by no means entirely normal in behavior at so-called permissive temperatures. Thus, he selected left–right bilateral mosaics that were hemizygous *shi*[ts] on one side, heterozygous on the other. Feeding of TTX to the mosaics resulted in a debilitated heterozygous side, with the opposite side unaffected. Therefore, the drug did seem to have significantly different effects on mutant versus wild-type tissues.

An additional paralytic temperature-sensitive mutant has recently been reported (Søndergaard, 1975). This is the out-cold[ts] mutant, due to a dominant, X-linked mutation (*Ocd*[ts]), which causes paralysis of heterozygous females at low temperatures (18° C); the paralysis is reversible at 25°. Hemizygous mutant males are very debilitated in their behavior and die as young adults (irrespective of the temperature). Søndergaard notes the preliminary mosaic results in which legs are paralyzed independently; the foci appeared to be most closely linked to thoracic cuticular landmarks. The analysis here is one of the few examples of a mosaic study of a dominant, mutant allele. Somatic loss of *Ocd*[ts] in female zygotes was induced by having the mutation-bearing X chromosome inherited from a *ca*[nd] mother.

Mosaic studies of these paralytic, temperature-sensitive mutants have allowed a detailed discrimination of the different effects of the different genes, and also of different alleles at a given locus. The differences in the paralysis, initially studied in entirely mutant animals, do not make the different results from mosaic analysis at all surprising. For instance, there are dramatically different paralysis temperatures, exposure times required for paralysis, and recovery times required for the different mutants (e.g., Siddiqi and Benzer, 1976). The mosaic results also reinforce the necessity for carefully characterizing a given behavioral abnormality, for which a focus is being mapped. Since a paralytic mutant could be expected a priori to affect several portions of the nervous system, it was important carefully to distinguish general immobility from specific defects in the posture and use of the legs.

Table 2. Fate mapping of sluggish mutants in *D. melanogaster*. Seven of these recessive, X-linked recessive mutations were isolated on the basis of poor phototactic and geotactic responses, in the laboratory of S. Benzer. Mutants No. 3 and 4 are allelic. The mutants were found to have normal ERG's, and a positive optomotor response, and thus are apparently "sluggish" mutants that are inexcitable in the countercurrent distribution test of phototaxis (Benzer, 1967). So the faulty phototaxis is not due to a visual system defect. The eighth mutant, *para^{ts}*, was isolated by Suzuki et al. (1971), based on paralysis at 29° C; yet it also has sluggish behavior at 20° C. A. Ghysen (personal communication) has carried out fate-mapping on these mutants, with respect to X-chromosome loss mosaics, and has obtained the following results

Mutant	Blastoderm location (s) of focus (foci) for sluggish behavior	Focus (foci) characteristics	Probable tissue affected
1. EE 85	One focus per leg, ventral to thoracic cuticle landmarks	Simple	Thoracic ganglia
2. KS 153	Same general positions as for EE 85	Simple	Thoracic ganglia
3. EE 54	Anterior, ventral	Submissive	Brain
4. KS 198	Anterior, ventral	Submissive	Brain
5. JM 26	Anterior, ventral	Domineering	Ventral brain
6. KS 140	Anterior, dorsal	Domineering	Brain (?) Trachea (?)
7. BS 35	Between head and thorax cuticle landmarks	Submissive	?
8. para^{ts}	Anterior, ventral	Domineering	Brain

These physiologic mutants allow the construction of formalistic wiring diagrams from mosaics. It could be that a *com* mosaic, with mutant tissue in a particular area of one side of the brain, will have particular thoracic functions absent—displayed in the behavior of a particular leg or wing, or other physiologically measurable defects. Let us assume that there is no mutant tissue in the thorax. Crude wiring diagrams could be constructed, and contralateral control by the brain of a particular thoracic function would be shown, if mutant tissue in one side of the head eliminates that function on the opposite side of the thorax. Levine and Tracey (1973) already generated a wiring diagram concerning the connectivity of the giant motor neuron, which originates in the head and controls a variety of known thoracic muscles. The technique of using mosaics to diagram wiring systems could be tested with Levine and Tracey's background information on this neuron.

A. Ghysen (personal communication) has studied mosaics constructed with respect to several X-linked "sluggish" mutants. These were found initially on the basis of defective phototaxis, but were then shown to have apparently normal physiology of the visual system; and they proved not to move toward light, when excited, because they simply do not move well at all. Two of these mutants were shown to have simple foci, associated with the movement of individual legs (Table 2). That is, the mutants *EE 85* and *KS 153* show "leg weakening" (slow or nonmovement of their legs) when containers of the flies are repeatedly banged down, similar to the treatment received by flies in the "fast phototaxis" countercurrent distribution tests of Benzer (1967). In mosaics constructed with respect to *EE 85* and *KS 153*, mutant tissue in the head or abdominal region was not sufficient to lead to leg weakening. Instead, there were three pairs of independent foci—one pair per left and right pair of legs—analogous to those found for Hk^1-induced shaking of individual legs. Thus, the foci for these two sluggish mutants are likely to appear in the familiar thoracic ganglia region of the blastoderm.

3. Biochemical Mutants

A general approach for the genetic study of neural development and function, different from most of those now being used in *Drosophila*, could involve mutations that disrupt known features of central nervous system (CNS) metabolism. For instance, mutant expression of enzymes concerning the synthesis or hydrolysis of acetylcholine—an important neurotransmitter in the CNS—could be studied in regard to abnormal development, physiology, and behavior. This strategy has been initiated by Hall and Kankel (1976), who have discovered mutations (*Ace*-lethals, on chromosome 3) which eliminate acetylcholinesterase (AChE) activity. Some of the more recently isolated mutations are temperature-sensitive for lethality; and, if allowed to develop to adulthood at permissive (low) temperatures, they are then heat-sensitive for paralysis.

While these AChE-null mutations cause lethality in genetically uniform animals, R. Greenspan and J. C. Hall (unpublished) have generated a series of flies mosaic for nonconditional AChE activity (by inducing somatic loss of an Ace^+-bearing X chromosome, in flies carrying *Ace*-lethals on both their third chromosomes). Histochemical analysis of AChE activity in the nervous systems of such mosaics has revealed that an absence of the enzyme activity in major portions of the optic lobes, of the ventral brain, or of certain ventral regions of the thoracic ganglia is compatible with viability. Yet, an absence of the enzyme activity in the optic lobes is sometimes correlated with deficits in the transients of the ERG. These foci may be thought of as "simple" because the mutant clones lead to local physiologic defects and do not (as far as is yet known) cause severe behavioral abnormalities at the level of the entire organism. Subsequent, more detailed behavioral and physiologic studies of such mosaics may provide information on the physiology of the nervous system (and how it controls various specific behaviors), which will augment the information obtained from the other direction—that is, through physiologic and biochemical investigation of mutants initially found on the basis of defective behavior.

B. Interacting Foci

1. Principles and Procedures

Interacting foci for neurologic mutants are found in cases of abnormal behavior that are scorable only at the level of the entire animal—as opposed to assessing the physiologic phenotype of an eye or the behavioral phenotype of a leg. For interacting foci, the question is not only where is the focus located, but is the focus

Table 3. Data from scoring a left–right pair of landmarks and behavior of mosaic flies

Behavior	Surface landmarks		
	Both sides normal	One side normal, one mutant	Both sides mutant
Normal	a_{11}	a_{10}	a_{00}
Mutant	b_{11}	b_{10}	b_{00}

on only one side of the relevant body region or both sides? This concept could mean that a given behavior mutant has two corresponding, very restricted sites in the left and right sides of a particular tissue, both of which must be genetically mutant to yield the behavioral abnormality—a submissive focus; or only one of which need be mutant to give the abnormality—a domineering focus (Hotta and Benzer, 1972). For example, a neurosecretory cell in the left side of the brain and the homologous right-hand cell could be the focus; and it could be that the mutant phenotype obtains only if both such cells are mutant, because one normal cell could produce enough of a secreted substance to yield a normal phenotype. It is not yet known if the interacting foci to be discussed shortly (several of which have been mapped as points) really consist of such restricted regions of the blastoderm. A substantial amount of mutant tissue (more than one or a few cells) in one or both sides of the brain could be required for mutant behavior.

To carry out the actual mapping of interacting foci—in more definitive terms than scoring the behavior of a few selected mosaics—is difficult. It is not clear what mathematical procedure is the most generally applicable and correct one, if any. Hotta and Benzer (1972) made the first contribution to interactive foci analysis. They generated mosaics which were, on the average, half diplo-X and half haplo-X. In the haplo-X tissues, a recessive behavioral mutant was uncovered. From scoring the genotype of a given pair of cuticle landmarks (e.g., a bristle on the left side of the head and the corresponding bristle on the right side) and from scoring the behavior of the mosaics, the following set of observations is generated (Table 3; cf. Table 4 of Hotta and Benzer, 1972). The symbols a_{11}, b_{11}, etc. are the six observations stemming directly from the landmark and behavioral scoring. It was useful to define two clear-cut behavioral alternatives in this analysis. For example, Hotta and Benzer (1972) were analyzing the interacting foci for the drop-dead *(drd)* mutant. This is an X-linked recessive which causes totally mutant flies to die not long after they emerge as apparently normal adults. Mosaics generated with *drd* were scored as mutant if they developed abnormal symptoms (staggering about) and died within ten days; and they were scored as normal if they were still alive and showed normal behavior at two weeks post eclosion. Now it could have been that the mosaics would show a continuum of adult life spans, with many mosaics being intermediate—e.g., with life spans of two weeks or more, but significantly less than the mean life span for wild-type flies (for whom the fraction of surviving flies does not begin to fall appreciably until 4 to 6 weeks post eclosion—e.g., Hall, 1969; Trout and Kaplan, 1970). An attempt to divide drop-dead behavior, in mosaics, into many subclasses could have ruined the relative simplicity of the following analysis.

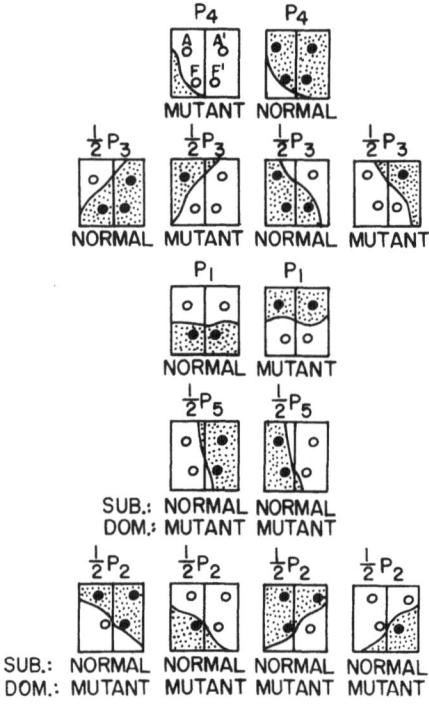

Fig. 2. Various configurations of the mosaic boundary with respect to sites determining a bilaterally homologous pair of surface landmarks (A, A') and a pair of behavioral foci (f, f'). One side of the boundary is covered by normal cells (*shaded*), the other by mutant cells. The probability (e. g., p_1) of each configurations depends on the relative locations of the sites and foci. The behavior expected is given below each diagram. When the boundary falls between the two foci, the resultant behavior depends on whether a mutant focus is submissive (*SUB.*) to a normal one, so that both foci must be mutant to produce mutant behavior; or whether a mutant focus is domineering (*DOM.*) toward a normal one, so that mutant behavior results if either or both foci are mutant. Two of the sixteen theoretically possible combinations are omitted, namely, the one where A and f' are mutant and the one where A' and f are mutant. (These require multiple crossings of the dividing line between the sites and would be expected to be very infrequent, particulary when the sites and foci are close together). Thus, the dividing lines shown here, and their associated probabilities, are taken to be all the possible cases, so that $2p_1 + 2p_2 + 2p_3 + 2p_4 + p_5 = 1.0$ (see Table 4 of Hotta and Benzer, 1972)

Hotta and Benzer reasoned that the mosaics would be in 14 qualitatively different categories, defined by the configuration of the mosaic boundary on the blastoderm. That is, the dividing line between diplo-X and haplo-X tissue would be in a variety of possible orientations with respect to the cuticle landmarks and the interacting foci. The cartoon for this is in Figure 2 (taken from Fig. 9 of Hotta and Benzer, 1972).

The different p values associated with the different mosaic boundary configurations refer to the fact that different categories of dividing line have different chances of occurring, depending on how close a focus is to the cuticular landmark, and how close the left and right homologous foci are to each other. This is, of course, just what one desires to know in order to construct the fate map that locates the focus. A given dividing line category may refer to a different behavior, depending on whether the focus is domineering or submissive, as is indicated in Figure 2.

Hotta and Benzer wrote equations containing the five different p values and the six different observations (Table 3). The solutions derived by Hotta and Benzer were, ultimately, expressions relating the landmark-focus distance or the focus-focus distance to the six observations. The formulas are of course different for submissive or domineering foci. The "distances" here are probabilities between 0 and 1, which are then multiplied by 100 for convenience and defined as "sturt" units (in honor of the originator of the fate mapping concept, A. H. Sturtevant 1929—these sturts are also used by Hotta and Benzer, 1972, to define distances between any two landmarks).

These final equations were used by Hotta and Benzer (1972) to map the interacting foci for the *drd* mutation. The various landmark-focus distances and the interfocus distance led to a determination of the focus, employing the standard fate mapping triangulation procedures (see Janning, this volume). The *drd* focus is in a position on the blastoderm ventral to the sites for head bristles. Thus, the brain is a plausible site for the focus. Indeed, subsequent histologic examination of *drd* flies revealed brain degeneration, the onset of which is correlated with the mutant symptoms that signal impending death (Benzer, 1971).

One might have tried to reach the conclusion of a brain focus from a much quicker handling of the mosaic data. As noted by Hotta and Benzer (1972), only 8% of the mosaics which have entirely normal head cuticle, or all mutant head cuticle, have the opposite behavior; these figures for the thorax and abdomen are, respectively, 16% and 38% (cf. Table 4). Thus, we might have concluded without any equations that the focus is closest to the head (but not "at" the head surface), and thus is likely in the brain. However, the trachea are probably derived from a part of the blastoderm dorsal to the area from which cuticle is derived (Poulson, 1950; Janning, 1975). In addition, the larval ring gland maps in this dorsal, anterior position (Janning, 1976). The quick analysis, therefore, does not distinguish between a possible focus in the brain, the anterior, or in the ring gland (i.e., since part of the adult ring gland complex is derived from the larval ring gland— Bodenstein, 1950). For example, if the trachea were somehow blocked by the action of the mutation, this could lead to brain degeneration and death. The tissue with the histologically observable abnormality—the brain— would therefore not have been the tissue containing the primary focus. It was important to perform the more quantitative focus determination, to show that the *drd* focus not only maps near the head bristle landmarks, but also maps to a position *ventral* to those landmarks.

Hotta and Benzer (1972) decided that the model for a submissive focus was most consistent with the mosaic data. The strategy for arriving at this kind of decision is outlined, in general terms, in Table 4. For the actual *drd* mosaics, a majority had normal phenotype. The equations for the *drd* mosaic analysis then showed that a submissive model was most consistent with the data (Hotta and Benzer, 1972): For instance, an unambiguous triangulation was achieved with the various landmark-to-focus distances. One would thus predict that most bilater-

Table 4. Strategy for deciding on submissive or domineering nature of interacting foci. The strategy can proceed according to the order of questions listed below (see also Table 3 of Hotta and Benzer, 1972)

    A. What is average degree of maleness, per gynandromorph (usually 0.3–0.5)?

    B. What proportion of gynandromorphs have abnormal behavior?

  I. If B. > A., implies domineering focus.

    If B. < A., implies submissive focus.

    C. Among mosaics with major body region (head, thorax, abdomen) of uniform genotype, what proportion of gynandromorphs have "opposite" behavior?

       e.g., of cases with head all normal or all mutant, how many have head normal, behavior mutant; or head mutant, behavior normal?

 II. Body region with smallest proportion of "opposite behaving" mosaics is nearest to focus.

    D. Among mosaics, bilaterally split for genotype of body region nearest focus (e.g., entire left half of head mutant, right half normal, or vice versa), what proportion have abnormal behavior?

III. If D. > .5, implies domineering focus.

    If D. < .5, implies submissive focus.

Conclusion from I. should agree with conclusion from III.

    E. More rigorous decisions on nature of focus.

      1. Original analysis (Hotta and Benzer, 1972).

         Are all A–f and f–f' values, from submissive or domineering model, non-negative? (sometimes a model rejected according to I and III yields negative distances) Do various A–f values from one model or the other, give unambiguous triangulation of focus to a particular region of the fate map?

         Are the various f–f' values, from different landmarks, roughly the same from one model or the other?

      2. Does maximum likelihood (Merriam and Lange, 1974) or minimum Chi-square analysis give consistently low Chi-squares, for all landmarks, from one model or the other?

      3. Does domineering contour mapping (Feitelson and Hall, 1978a, b) yield convergence of "low probability lines" at one region of fate map? If not, does submissive contour mapping yield a convergence?

      4. Using the "focusing" method of Flanagan (1977), can you find by a least-squares procedure the focus's map coordinates which best fit the calculated values for landmark–focus distances? To decide on a domineering vs. submissive focus: Is the probability that a mosaic will not survive, given that a given landmark is mutant, *greater* than the probability that a mosaic will survive, given that the landmark is normal? If so, the focus is domineering; if the reverse holds, the focus is submissive.

---

ally split-head mosaics would be normal in behavior, and Hotta and Benzer (1972) indeed found that almost 80% of them are. All of these behaviorally normal mosaics showed no brain degeneration on either side; but the early-dying mosaics showed brain degeneration on both sides, when they were sectioned soon after they began to show mutant symptoms.

Hotta and Benzer (1972) also used this algebraic procedure for focus determination to map the interacting foci for the mutants with abnormal wing posture, heldup *(hdp)* and upheld *(up)*. These are X-linked factors (then called wings-up A and wings-up B, respectively), which cause the flies to hold their wings in a vertical position and for which mosaics either have both wings held up or both wings in the normal position.

The foci for these mutants are examples of domineering control of abnormal phenotype, in that the solutions to the equations were more consistent with this model (Hotta and Benzer, 1972). This was expected because the majority of mosaics held both wings up. The foci for these two mutants are each far from the head cuticle and thorax cuticle parts of the blastoderm, in a ventral location. This implies a thoracic nervous system or a thoracic muscle focus. In this light, all totally mutant flies or mosaics showing mutant posture show gross abnormalities of muscles in the thorax [also see the later discussion of results of Deak (1977)]. Mapping foci that are far from known landmarks—as in the mapping of these two mutants, for which the fewest sturt units between a landmark and the focus are 20—is probably not very accurate. Thus, whereas muscles are derived from the same general area of the blastoderm as the *hdp* or *up* focus, the current results and analysis do not absolutely rule out a thoracic nerve focus. The abnormal nerves could induce the muscle abnormality, leading to the faulty wing posture, by analogy to the effects of vertebrate muscular dystrophy mutants, which some believe only indirectly affect the muscles (e.g., McComas, 1974). However, the foci for these wing posture mutants are rather close to the ventral midline of the blastoderm, and so the muscle focus is more likely (cf. Poulson, 1950).

There are procedures available—which might supplement or in some cases replace the original ones of Hotta and Benzer—for carrying out the rigorous quantitative positioning of interacting foci (see last section of Table 4). Merriam and Lange (1974) presented a modification of the original analysis. They noted that, for a given pair of surface landmarks (e.g., left and right ocellar bristle) there are more observations than unknowns. That is, one generates a 2×3 table of observations from the landmarks and behavioral scoring (Table 3); yet, there are only five unknown p values, referring to the different dividing lines (Fig. 2). Thus, there is no unique solution to the equations of Hotta and Benzer (1972), which relate the p values to the observations. Merriam and Lange (1974) note that different, equally valid solutions to the equations can result in focus-landmark or interfocus distances that are 40–120% discrepant. If one were to use the various alternative solutions to the equations, then—given the one degree of freedom inherent in Hotta and Benzer's model—rather different map positions for the focus of a given behavioral mutant could result.

A solution to the problem of "too many" observations is a maximum likelihood or minimum Chi-square analysis. These are procedures from which one can find the best p values—the values from which one has the maximum chance of observing the actual data, or p values which yield the smallest overall discrepancy between the data and the expectations generated from the p values. The computer-aided maximum likelihood solution to Hotta and Benzer's model—with respect to *drd* behavior and the ocellar bristles (part of the head cuticle)—yielded landmark-focus *(A–f)* and interfocus *(f–f)* sturt values that were nearly the same as the distances generated by Hotta and Benzer (1972), from their particular solutions to the formulas (cf. Table 5). It must be emphasized that the procedures used by these two sets of authors were not the same, and the similar final values need not have obtained. The *A–f* and *f–f* values here—from both kinds of analyses—came from a submissive model. In this light, Merriam and Lange (1974) provided a useful way to deal with solutions to the formulas, because they com-

Table 5. Fate-map distances in sturts from the focus for the drop-dead mutant to the indicated cuticle landmarks. The Chi-square values (χ^2) are listed for both the submissive (sub) and domineering (dom) models (there is one degree of freedom)

Cuticle structure	Model	Hotta and Benzer (1972)			Merriam and Lange (1974)			J. Hall (unpublished)		
		$A\!-\!f$	$f\!-\!f'$	χ^2	$A\!-\!f$	$f\!-\!f'$	χ^2	$A\!-\!f$	$f\!-\!f'$	χ^2
Ocellar bristle	Sub	15	10	—	15	10	1.5	17	12	4.2
	Dom	—	—	—	25	0	53.0	26	8	38.2
Postvertical bristle	Sub	17	—	—	—	—	—	21	10	9.6
	Dom	—	—	—	—	—	—	24	8	29.6
Palp	Sub	19	—	—	—	—	—	20	13	2.3
	Dom	—	—	—	—	—	—	24	4	16.8
lst	Sub	27	—	—	—	—	—	25	17	0.3
leg	Dom	—	—	—	—	—	—	26	0	24.4
Proboscis	Sub	30	—	—	—	—	—	29	15	1.3
	Dom	—	—	—	—	—	—	28	4	29.0

pared the actual observations (i.e., from the matrix in Table 3) to the expectations from the p values estimated by maximum likelihood. The Chi-square value for the submissive model was very low, allowing one to accept this model for *drd* with a reasonably high probability (although only one pair of landmarks was analyzed). Yet, the Chi-square for the domineering model was massive for the same landmark pair; and this model could be rejected.

I have somewhat extended Merriam and Lange's analysis, by using a computer to generate minimum Chi-square estimates of Hotta and Benzer's p values—with respect to *drd* and, not only the ocellar bristles, but the other landmarks used by Hotta and Benzer (1972) to triangulate the focus (their Fig. 10); Dr. Benzer supplied the needed unpublished data). Table 5 presents a comparison of Hotta and Benzer's distances, with the maximum likelihood generated values (as far as they go), and with the minimum Chi-square sturt distances.

The minimum Chi-square solutions would yield a focus positioning which is essentially the same as that of Hotta and Benzer. Also, it is clear that—for all five landmarks—the submissive model gives the best fits, but perhaps only in a qualitative sense, in that two of the minimum Chi-squares correspond to probabilities that are less than 0.05. It is also noteworthy that the interfocus distances are not too variable (mean of 13.4 sturts $\pm$ SEM 1.2). The $A\!-\!f$ distances should of course be different for different landmarks; yet the real $f\!-\!f'$ is only one value. The similar estimates of $f\!-\!f'$ from the different landmark data seem to buttress the validity of Hotta and Benzer's initial model.

A problem can arise in the application of Hotta and Benzer's model, even when the estimation procedures are employed. If the average mosaic from a given experiment is substantially less than one-half haplo-X, then the symmetry of the model falls apart. Hotta and Benzer (1972) not that if "any part [of a mosaic] is equally likely to occur as mutant or normal ... the probabilities are symmetric around the center." This means that, for instance, the probability of having both sides normal (i.e., both left and right homologous landmarks diplo-X), with both foci normal, is equal to the probability of having both sides mutant, with both foci mutant (this probability is p_4 in Fig. 1). Therefore, the mosaics in classes $a_{11} + b_{11}$ will not be dramatically different in number from the mosaics in classes $a_{00} + b_{00}$ (Table 3). This symmetry allows one to write equations containing only five unknowns. However, several mosaic-generating systems tend to yield mosaics which are, on the average, only one-third haplo-X or less (Hall et al., 1976). Variations in degree of mutant tissue in mosaics is somewhat under one's control, in that particular mosaic-generating mutants (e.g., *pal*, *mit*) may be chosen in order to produce fairly small mutant patches. However, even if one wants half-half mosaics, they might not be forthcoming. Thus, some unstable ring-X crosses give mosaics which are, on the average, only 30–40% haplo-X (e.g., Kankel and Hall, 1976; Feitelson and L. Hall, 1978 b). J. Hall (1977) was forced to use the *pal* method of gynandromorph production in a behavioral study of such mosaics; and the average mosaic here was only one-third male. For behavioral studies involving mosaics that are substantially less than 40% haplo-X, the $a_{11} + b_{11}$ mosaics are much greater in number than the $a_{00} + b_{00}$ mosaics. The overall effect is that the number of p values (cf. Fig. 1) is nine, and this is of course greater than the number of observations from the landmark and behavioral scoring of the mosaics.

Hall (1977) attempted a minimum Chi-square analysis of the focus for male courtship (discussed in more detail later) and found, as expected, that neither a submissive nor a domineering model fit the data. Most of the contributions to the huge Chi-square values—for any given pair of landmarks—came from the fact that there were many more mosaics with both landmarks diplo-X compared to those with both landmarks haplo-X. These mosaic types (equivalent in Hotta and Benzer's model and data) are *not* the ones used critically to diagnose a submissive vs. domineering model; yet their inequality ruins this simplistic application of the interaction foci analysis.

Feitelson and L. Hall (1978 a, b) used a different focus determination procedure, in a study of a mechanical shock-sensitive mutant (bang-sensitive, or *bas*). They applied the "contour mapping" principle, developed by Y. Hotta (unpublished) and Hotta and Benzer (1976). Hotta was first trying to determine the foci of X-linked lethal mutations. Some of these were "cell lethals," in that no mosaics with haplo-X tissues hemizygous for the lethal survived (compare, e.g., Stewart et al., 1972 or Arking, 1975). Other lethals gave surviving mosaics, and Hotta termed these "focal" lethals. In these cases, landmarks with the lowest probabilities of being haplo-X are nearest the lethality focus, which need not of course be a point, but could instead be a relatively diffuse area of the blastoderm—for example, the entire anterior nervous system. If a particular surface landmark has a zero probability of being haplo-X, then this is the focus; but, as expected, none of the 50 lethals analyzed gave such a result. Instead, series of contour lines were obtained. Each line connected landmarks on a standard fate map which had similar

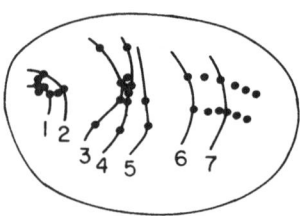

Fig. 3. Hypothetical contour map of lethal mutant focus. The dots are sites on a standard cuticle fate map (e.g., Fig. 5 of Hotta and Benzer, 1972) referring to head (*left portion of map*), thorax (*middle*), and abdomen (*right*) landmarks. The lines connect landmarks with similar probabilities of being haplo-*X*, in mosaics produced with respect to a hypothetical X-linked lethal mutation. Lines with lower (arbitrary) numbers connect landmarks with lower haplo-X probabilities. Thus, the focus here is in the anterior region of the blastoderm—possibly in a region corresponding to the brain

probabilities of being haplo-X. The line referring to the landmarks with the lowest probabilities "converges" on the focus (see Fig. 3). This analysis is similar to that of Bryant and Zornetzer (1973) who—in performing mosaic analysis of X-linked focal lethals—drew circles around surface landmarks on their fate map, each with a radius proportional to the probability that the landmark is haplo-X. Here, small circles are for landmarks nearest the focus, yet these maps are awkward to present and observe.

Feitelson and Hall (1978 b) modified Hotta's model to take into account domineering versus submissive foci. To map the focus of the *bas* mutant, which falls over or "passes out" briefly when subjected to mechanical shock, they calculated (with the aid of the computer) the "maleness average" for each of the different landmarks on the fate map. For the assumption of a *domineering focus*, they considered the mosaics that did *not* give the mutant response to mechanical shock. Consider the maleness-average fraction for a given landmark (e.g., left and right first leg): For the normal behaving mosaics, 2 is added to the numerator for each mosaic that has both left and corresponding right landmark—or *either one of them*—haplo-X. Two is added to the denominator for each mosaic, and the resulting fraction is the maleness average for this particular pair of landmarks.

The rationale for this calculation is based on the fact that few normal-behaving mosaics will have even one member of a landmark pair, which is near the focus, of mutant genotype. This is because only one member of the focus pair need be mutant to cause mutant behavior, and such a mosaic is very likely to have at least one member of the (nearby) landmark pair mutant. The "artificial" addition of 2 to the numerator "forces" the maleness average for this landmark pair to be as high as it could be under the circumstances of being close to the focus. So if there *is* a group of landmark pairs with lower maleness averages than for groups of landmarks elsewhere on the fate map, one can be confident that this convergence of contour lines is consistent with a domineering focus. The result of these calculations with respect to the *bas* mutant was a contour map with landmarks all over the fate map having similar maleness averages. Thus, there was no convergence, and the domineering model would seem not to obtain for this mutant. Apparently, both members of the focus pair need to be mutant to give mutant behavior (submissive focus) so that the mosaics with one or even both members of a landmark pair (near the focus) mutant would frequently have one normal focus. These mosaics would contribute to the maleness average for this landmark pair— and neither it nor any landmark pair would have a dramatically low maleness average. In a subsequent test on the *bas* focus, maleness averages for a submissive model were obtained by adding 2 to the numerator only if both left and right landmarks are haplo-X, and 2 was added to the denominator for each mosaic. Here, some landmarks had maleness averages dramatically *lower* than other groups of landmarks; such landmarks are nearest the focus, because the normal-behaving mosaics had been purposefully selected for the calculations. The resulting contour lines gave a definite convergence, pointing toward a region of the blastoderm beneath the anterior portion of the thoracic cuticle region. This implies a submissive focus in the thoracic ganglia or muscles.

Hall (1977) found that mapping of domineering contours—with respect to early actions in the male courtship sequence—gave a convergence at a particular part of the brain (discussed later). The same major conclusion was obtained when a submissive model was applied—there was a definite convergence at the same area of the fate map as obtained from application of the domineering calculations. The procedures for calculating maleness averages under the two focal possibilities have one-directional applications: if the domineering model works, then the submissive model should as well; but the reverse statement is not true.

The contour mapping procedure thus provides a useful alternative to the more standard interacting foci mapping. Moreover, it allows one to select particular subclasses from the mosaic population, which should not be done in the "dividing line analysis" of Hotta and Benzer (1972). For instance, one might have a mutant which causes the fly to pass out under certain conditions, and then put its entire body into an abnormal posture. Among mosaics, some flies would hypothetically show normal behavior, some would pass out and go into the abnormal posture, and some would only pass out. The focus for passing out could be mapped in the usual way—*if* the mosaics here were roughly half haplo-X, half diplo-X—using data from all mosaics. However, one would want to select the cases which passed out, in order to map the focus for abnormal posture, since the fainting behavior appears to be a prerequisite for the abnormal posture. A contour map would appear to be valid, as applied to this particular subfraction of the mosaics. For the dividing-line model, one could determine sturt distances for the posture focus, on the mosaics that has passed out. This might yield a spurious focus, by analogy to Falk et al.'s (1973) mapping of the larval gonad to a position in the middle of the fate map. Part of the analysis involved choice of only those larval mosaics having a particular posterior structure all or part haplo-X. If cases without the anterior regions (only) haplo-X are thrown out, then this may bias the map. However, since the cases with posterior portions all *or part* haplo-X were chosen, there seems to be an enrichment for mosaics with dividing lines running into the posterior region. This tends artificially to *expand* the distance between a given, real posterior landmark and the gonad landmark. So, instead of the gonad being mapped to a relatively posterior location, its landmark comes out fairly far from the posterior part of the map. Hotta and Benzer (1973) mapped the gonad to a posterior region of the map, by collecting all the larval mosaics they could detect; and moreover, by letting the larvae develop into adults, so that larval tissues could be mapped with respect to each other and with respect to the imaginal discs. The same kind of "distortion" of the map position for a particular tissue, which might be the case in Falk et al.'s study, could occur in behavioral focus mapping, as a result of preselecting only certain kinds of mosaics. This distortion does not necessarily arise in contour mapping.

There is one final procedure presented for the mapping of interacting foci. Flanagan (1977) developed an algebraic method that can be used to determine the focus of mutations which cause the entire fly to be defective. Here, the interfocal distance is a function of the fraction of mosaics which are normal, and the average frequency that landmarks are normal in control mosaics that do not carry the mutation (but which are the type of mosaics used in the mutant study, e.g., those produced by loss of the ring-X). There are separate equations, using these values, for submissive and for domineering foci. The metrics here are easy to determine, as are two of the ones needed for determination of landmark–focus distances: i.e., the probability that a mosaic will survive given that a particular landmark is normal (called q) and the probability that a mosaic will not survive given that the landmark is mutant (q'). In addition, the two equations derived for the calculation of landmark–focus distances (i.e., one equation for a submissive model, one for domineering) each contain their own "unknown quantity" or constant, and these are not trivial to determine, requiring a least-squares procedure. A domineering vs. submissive focus is decided upon by noting that q' is greater than q (for each landmark on the fate map) if the focus is domineering; this has intuitive appeal, and tells you which specific equations to use in calculating distances. Finally, the focus is determined in cartesian coordinates, and placed on a fate map generated with a least-squares procedure for determining landmark locations (using the same kind of coordinates—cf. Flanagan, 1976, and the discussion by Janning, this volume). The analysis has been applied first to an X-linked mutation (doomed, *dmd*), which causes flies to die as pupae or very soon after eclosion (much sooner than in the case of drop-dead). The domineering focus is consistent with a location in the thoracic ganglia, and is confirmed by a rough application of the contour mapping technique (Flanagan, 1977). Histologic examination of newly dead *dmd* flies revealed no neural abnormalities, unlike the case of *drd*'s effects on the brain.

This new "focusing" procedure is in a very general way similar to the algebraic method of Hotta and Benzer (1972), except that quick focusing using Flanagan's method will be difficult, because several key parts of the procedure devolve to the familiar "information available from author on request." However, the new technique is equally applicable to focusing of mutants that cause the entire fly to misbehave or to die; and it does not assume or require that mosaics be half mutant, half normal. The general utility, then, of Flanagan's method is about as broad as in the case of contour mapping, though the latter is more readily applicable for obtaining the general location of a focus. After all, the focusing technique provides merely a formal map location, from which one can only infer the concrete nature of the focus.

2. Cases of Interacting Foci Analysis

Several behavioral foci have been analyzed with mosaics. Some of these cases have been noted already, in order to exemplify the principles and techniques. For the following cases, it is unfortunate that only an anecdotal record exists for the majority of the behaviors. However, more recent studies have made increasing use of the more rigorous quantitative procedures.

a) Locomotor Mutants. The Hyperkinetic mutant, discussed earlier with respect to simple leg-shaking foci, is abnormal in other ways. For example, mutant flies nearly always "jump" when one passes a hand over a container of the flies. This may be only a tentative jump, or it may go so far as to involve a collapse of the fly, with a brief "seizure." Kaplan and Trout (1974) note that the focus for this abnormality induced by Hk^1 appears to be near the head cuticle landmarks. Kaplan has supplied additional (as yet unpublished) details, where he shows that the jump response focus is less than 10 sturts from the head bristle landmarks, which are relatively ventral on the map; the focus is 10–15 sturts from the more dorsal head landmarks; yet the distances to two thoracic landmarks are 20–30 sturts. The mean interfocus distance for the seven head landmarks considered is $27.3 \pm$ SEM 0.9 sturts. Kaplan used the maximum likelihood estimation procedure of Merriam and Lange (1974), and the above distances are all from the domineering model—which, unlike the submissive model, was consistent with the data. The focus here—anterior and ventral on the fate map—suggests a defect in the brain or optic lobes. Levine (1974) learned that the giant motor neurons in flies expressing Hk^1 can be driven by visual input (unlike what is found for wild-type flies). He suggests that the "neurogenetic disorder allowing visual driving of the giant neuron in the mutant is probably acting at or peripheral to the lobula," by analogy to the distal connectivity of the lobular giant neuron in the locust.

W.D. Kaplan (personal communication) has preliminary evidence from double-mutant flies, which is somewhat supportive of a lamina focus for the Hk^1-induced jump response. $Hk^1 \, sev^{LY3}$ double mutants—missing R_7 from each ommatidium in the eye (cf. Harris et al., 1976)—still have the usual jump response. Retina-lamina connections here are changed little (if at all), since R_7 cells do not synapse with lamina neurons. R_{1-6} cells do connect with lamina neurons. These cells do not form or degenerate, respectively, in $Hk^1 \, ora^{JK84}$ or $Hk^1 \, rdgB^{KS222}$ double mutants (cf. Harris et al., 1976). Kaplan has found that there is no jump response in flies of these genotypes; the same result obtains in $Hk^1 \, rdgB^{KS222}$ flies even before the photoreceptor cells show signs of degeneration (though Harris and Stark, 1977, showed that R_{1-6} are never functioning in $rdgB^{KS222}$). Finally, the $Hk^1 \, H\text{-}53$ double mutant has a weak jump response (about one-fifth as many jumps as Hk^1 shows, for a given number of hand-waving tests). $H\text{-}53$ leads to a defective high-sensitivity system and reduced receptor and lamina potentials (i.e., transients) in the ERG (Heisenberg and Götz, 1975). Since these parameters depend on properly functioning R_{1-6} cells and/or the lamina neurons onto which they synapse, the effect of $H\text{-}53$ on the Hk^1-mediated jump response is, again, consistent with a lamina focus for this abnormal behavior. A focus in the retina is definitely ruled out, because the focus by no means maps to the head cuticle. However, it could be that the jump response maps "proximal" to the lamina, in higher-order optic lobes or even in the part of brain involved in pathways that originate from the outer retinula cells. Thus, these double-mutant studies are not a substitute for detailed mosaic analysis.

Hotta and Benzer (1972) mapped the abnormal wing posture foci for *hdp* and *up* to ventral portions of the midblastoderm—thus supposedly to muscle progenitor cells. Deak (1977) has carried the analysis of muscle mutants further. He studied five X chromosome recessive mutations (including *hdp* and *up*), which each lead to abnormal wing posture, poor or no ability to fly, poor or no jumping ability (after a touch to the abdomen), and a variety of fairly gross abnormalities of thoracic muscles. All the behavioral foci (re wing posture, flight, jumping) were domineering, and were in a very ventral blastoderm location—for the mutants *hdp*, *up*, flapwing *(flw)*, and indented-thorax *(int)*. This implies that the behavioral foci are in the muscles. Deak also mapped the muscle abnormalities as "simple" foci, and found that these are in the same very ventral blastoderm regions. Indeed, the "wings-up" phenotype and the flightlessness character—with regard to *hdp*—maps near the indirect flight muscle abnormalities, while the poor jumping ability maps near the tergal depressor of the trochanter (TDT) abnormalities. This is as expected, since these different kinds of muscles are, as noted by the author, required for flight and jumping, respectively.

Roughly the same fate mapping results were found for the other three genes mentioned above. However, a fifth gene—defined by vertical wings *(vtw)* mutations—has strikingly different foci. These mutants hold their wings vertically and lack several of the indirect flight muscles. The mutants cannot fly or jump. Both the muscle abnormalities and the abnormal behaviors map to dorsal blastoderm locations, above the thoracic cuticle landmarks. This is striking, in that no muscles are believed to be derived from this region. Instead, it could be that the trachea come from here (see above), so a defect in these respiratory structures could lead to the muscle and behavioral changes. It should be noted that only *hdp* and *up* have gross TDT abnormalities (Deak, 1977), though all five of the mutants studied here are poor jumpers. Possibly, all really have defective TDTs, but the defect is not observable in sectioned material for some of them. Deak's analysis, while an extension of Hotta and Benzer's study of the muscle mutants, does not prove muscle foci (i.e., for the mutants other than *vtw*). There was no independent muscle marker in this study, only the effects of the mutations themselves on muscle morphology.

Several new mutants with abnormal flying behavior have been systematically isolated by Homyk and Sheppard (1977). Flightless *(fli)* mutants and flight-reduced *(flrd)* mutants have been found, and map to a variety of locations on the X chromosome. Some of these mutants hold their wings in abnormal positions, and may be allelic to certain of the mutants studied by Deak (several of the map positions of the newly induced mutations are similar to those for *flw*, *int*, etc.). Homyk (1977) determined foci for the mutants with defective flight, and found that five *fli* or *flrd* cases map very near the ventral midline of the midblastoderm; these foci are domineering, and no doubt refer to muscle primordia. One focus (for the *fliH*[1] mutant) was submissive, and far enough from the ventral midline that a neural etiology (in the thoracic ganglia) could easily obtain. All of the data here were behavioral, in that no scoring of putatively defective muscles was attempted.

Five of the sluggish mutants, studied recently by A. Ghysen (personal communication) have interesting nonsimple foci (Table 2). Three of these mutants—EE 54, KS 198, and JM 26—are similar in their overall sluggish behavior. They are unexcitable when containers of the flies are banged down, and they move poorly in a phototaxis test when placed in one end of a pair of tubes (with openings juxtaposed), and asked to move toward light. Fate mapping of these mutants revealed that the focus for each is in a relatively anterior position, ventral to head cuticle landmarks; thus, these foci are probably in the brain. The foci for EE 54 and KS 198, which are both submissive, are possibly identical, since these two mutations are allelic. The focus for JM 26 is in a similar blastoderm region, but is domineering. The submissive versus domineering nature of the foci was determined by asking what proportion of the total mosaics were sluggish, and—after a rough mapping of the focus to a major body region—by asking about the behavior of those mosaics bilaterally divided in that region (the more rigorous procedures of Merriam and Lange, 1974 were not used). About 5 to 15% of these mosaics were scored as intermediate in regard to sluggish behavior. Such cases were not included in the fate mapping calculations. The presence of these cases in the mosaic population may mean that the EE 54 focus, for example, is "partially domineering;" or it may mean that the foci here are somewhat diffuse.

Some data have been collected by Ghysen, from JM 26 mosaics which were scored with respect to the internal acid phosphatase marker (used by Kankel and Hall, 1976, and discussed earlier in this review). Nonsluggish mosaics were selected, and these had the genotypes of major head and thoracic ganglia scored after sectioning and histochemical examination. Several such mosaics proved to have entirely mutant thoracic ganglia, or brains containing large mutant clones. In no case, however, was the posterior, ventral subesophageal ganglion mutant so this part of the brain may contain the actual focus.

Two additional sluggish mutants, KS 140 and BS 35, have interacting foci that are somewhat anomalous (Table 2). The KS 140 domineering focus is anterior, but dorsal to the head cuticle landmarks. All parts of the adult anterior CNS mapped by Kankel and Hall (1976) are derived from an anterior but definitely ventral blastoderm region (as confirmed by the computer-aided mapping of Flanagan (1977). The sluggish behavior caused by KS 140 could be due to tracheal blockage or a defective ring gland complex (see above). Or, it is possible that the focus is in the brain, and that either the nervous system map of Kankel and Hall (1976) or the behavioral map of Ghysen is incomplete or in error. Finally there is the focus for the sluggish mutant BS 35. It is a submissive one, between the head and thoracic cuticle landmarks, in a position that is neither dorsal nor ventral (see Fig. 4 for clarification). The only adult tissues known to come from this region are the salivary glands, and they map only roughly to this location (Kankel and Hall, 1976). That the focus for BS 35 does map to this position was in a sense confirmed by Ghysen, who scored the mosaics, independently, on three related behavioral criteria: locomotion (a score was given based on simple observation of the flies), general activity (the measured distance walked down a long, narrow tube, during a specific time interval), and the usual phototaxis test performed on all five of the sluggish mutants discussed here. All three BS 35 foci were in essentially the same position, in the empty region in the middle of the fate map.

The search for flightless mutants discussed earlier (Homyk, 1977; Homyk and Sheppard, 1977) produced "hypoactive" *(hypo)* mutants, in addition to the cases that seemed to have muscular defects. Two *hypo* mutants were mapped in mosaics, in criteria of general inactivity and inexcitability—similar to the behavioral defects analyzed by Ghysen. The *hypoB*[1] and *hypoC*[1] foci are submissive ones, in blastodermal locations that are anterior enough to allow Homyk (1977) to suggest neural defects in the brain, or possibly the anterior thoracic ganglia. These mutants could be allelic to Ghysen's, in terms of the overall behavioral defects, and the results from fate mapping.

b) Physiologic Mutants. The interacting foci for several of the paralytic, temperature-sensitive mutants have been mentioned above, in conjunction with the analysis of the simple leg paralysis foci. Most of the data on the mobility of an entire animal—mosaic with respect to one of these mutants—are neither extensive nor quantitative, and little fate mapping has been performed in order to pinpoint foci other than to say which major body region might contain the focus. Exceptions are the analyses of two alleles of the *para* gene. A. Ghysen (unpublished) has mapped the sluggish behavior shown by the original *para^ts* mutant at 20° C. The focus for this defect would appear to be in the brain (Table 2). Grigliatti et al. (1972) examined a few *para^ts* mosaics, with the only known mutant tissue in the head, and found those with all mutant head cuticle to be completely immobile at 29° C. Thus, there may be a focus for high-temperature immobility in the brain, and it could be the same focus as that for low-temperature sluggishness. However, Ghysen's calculations were consistent with a domineering focus, and Grigliatti et al. (1972) learned that, of the small number of bilaterally split-head cases examined, all were mobile at 29° C. It should be noted here that only one of the seven other sluggish mutants (Table 2) is temperature-sensitive with respect to locomotion, i.e., *KS153*, which is similar to comatose in that it becomes slowly paralyzed when exposed to 38–39° C (A. Ghysen, personal communication). The behavior at high temperature in *KS153* mosaics has not been examined, but do recall that it has clearly mappable defects in leg behavior at lower temperatures. S. Benzer (unpublished) has mapped the all-or-none leg paralysis shown (at high temperature) by *para^ST109* mosaics to be in the same, mysterious region of the blastoderm that contains the focus for the sluggish *BS35* mutant (Table 2)—that is, in the empty region between head and thoracic cuticle landmarks.

For visual mutants, most of the mosaic analysis has had to do with simple foci—that is, based on ERG determinations of individual eyes. Recall that a few of the ERG mutants—*H-37, e,* and *shi^ts*—had foci that are not simply in the retina cells, but could be in the optic lobes or brain. Here, though, the foci are not really interacting, because a focus can be determined separately for each eye. Submissive or domineering foci might be clearly revealed, for these and other visual system mutants, when they are carefully studied for behavioral as well as physiologic defects. Behavioral complexities are just now beginning to be studied in visually mutant mosaics. For instance, the *tan*[101] allele—newly isolated on criteria of flight defects (Homyk and Sheppard, 1977)—has been shown by Homyk (1977) to fail to retract its pro- and meso-thoracic legs in flight. The defect mapped (in mosaics) as a submissive focus in the eye—indeed, the entire eye, because a small patch of normal tissue in one eye was sufficient to allow for normal leg retraction. It would be good to know if other tan alleles show defects in this "complex" behavior, i.e., in addition to their abnormal ERG (which indeed is seen in the new tan[101] mutant).

c) Taste Mutants. Falk and Atidia (1975) have discovered recessive, X-linked mutations which will drink up a sucrose solution which contains 1 M NaCl. Wild-type flies generally avoid such a noxious solution. Two of the mutants (*Lot-94* and *Lot-114*) were allelic, and another was due to a mutation in another gene. The focus for salt tolerance in *Lot-94* was mapped, using the maximum likelihood procedure. The focus was assumed to be submissive, since less than half of the mosaics were mutant (no mention was made of an actual test of a submissive vs. domineering model). The focus was not close to any head, thorax, or abdominal cuticle landmarks; the shortest sturt distance between the focus and a landmark was about 28, with respect to the antenna; the longest distances were about 39 sturts between the focus and the abdominal tergites. It appeared from detailed behavioral testing of the mutant (irrespective of mosaics) as if the labellum were the sensory structure which was "salt-tolerant" in the mutant. Yet the focus for this defect was certainly not close to this head cuticle structure. It is difficult to triangulate a focus for this mutant, since the landmark–focus distances are nearly all the same. Moreover, there is a fair amount of variation in the interfocus distance: $4.7 \pm$ SEM 1.3 sturts. However, the authors conclude that the focus here could be the brain. The distance of the focus to the midline (2.4 sturts) is shorter than would have been predicted from the drop-dead focus (Hotta and Benzer, 1972), or the actual mapping of brain landmarks (Kankel and Hall, 1976)—that is, if a brain focus obtains for *Lot-94*.

d) Circadian Rhythm Mutants. Konopka and Benzer (1971) have isolated three mutants which strikingly alter circadian rhythms of eclosion or locomotor activity. They are all apparently allelic—defining the *per* gene—though two of the mutants alter the periodicity of the rhythms, while a third abolishes the rhythms. Konopka (1972) generated several mosaics with respect to the short-period mutant (*per^s*) and scored them for activity rhythm. No fate mapping was carried out, but he was able to suggest that the focus for this mutant is in the head. Thus, all of 14 mosaics with the external head entirely male had mutant rhythm. Only five of eight mosaics with external head entirely female were normal in rhythm; but the remaining three were arrythmic, not *per^s* in phenotype (the author felt that these three flies were tested at too old an adult age). For cases in which the head was mixed in genotype, an additional phenotypic category was found, i.e., where the fly gave a pattern "resembling the superimposition (sum)" of mutant and normal rhythms. This behavior was never seen in totally mutant or totally wild-type flies. Among 16 "split-head" mosaics (dividing line between mutant and normal cuticle running right down the middle of the head, in an anterior-posterior direction), five had mutant behavior, three were normal, three showed the "sum" phenotype, and five were arrythmic. For additional "mixed head" cases, two were mutant, one normal, one "sum," and two arrythmic. It was clear from this analysis that the thorax and abdominal genotypes were uncorrelated with the behavior. This focus presumably is in the brain, but is not clearly domineering or submissive. The author suggests that rhythm-controlling oscillators can operate independently— in each side of the brain—in some of the mosaics. In other such mixed-head mosaics, oscillators of opposite genotype (or cases where a given oscillator in one side of the brain is of mixed genotype) might be able to interact so that the rhythm control collapses entirely, resulting in arrythmicity.

e) Reproductive Behavior. Reproductive behavior has been examined in mosaics, beginning many decades ago when gynandromorphs in *Drosophila* were first discovered. The results, fragmentary and vague though they were, suggested that genetically male tissue in anterior portions of the fly leads to male courtship behavior (reviewed by Patterson and Stone, 1938; Manning, 1967). Mutations influencing behavior here were not at issue, only genetic markers which indicated which parts of the cuticle were haplo-X. Anterior control of this behavior is not necessarily expected. Such control does obtain for wasp mosaics (Whiting, 1932; Clark and Egen, 1975), but some sex-specific behaviors may be correlated with the genotype of more posterior segments in house flies (Milani and Rivosecchi, 1954) and bees (Sakagami and Takahashi, 1956).

Hotta and Benzer (1976) carried out a more detailed analysis of the early steps of courtship behavior in *Drosophila* mosaics. Mosaics, marked externally as to the distribution of male versus female tissue, were tested for behavior in the presence of females. It was found that the foci for following by males of females and male wing extension at them (see Spieth, 1952, 1974) are essentially inseparable, are domineering, and are near the head cuticle landmarks. This result implies that the brain genotype controls these early courtship stages. A later step in the sequence, attempted copulation, was frequently not performed by mosaics that did follow and show wing extension. Attempted copulation was correlated with the thoracic genotype.

Hall (1977) has analyzed courtship foci further, by using the nervous system cell marker that was applied to cell lineage analysis of the adult nervous system (Kankel and Hall, 1976). Mosaics were tested for behavior with females, then sectioned for histochemical determination of the nervous system genotype. It was found that the focus for male following and wing extension—even bilateral wing extension—is in the more dorsal portion of the supraesophageal ganglion.

A low percentage of Hall's mosaics with all-male brains, or with one side of the brain entirely male, did not show male behavior; but the great majority of such mosaics did. Moreover, none of 163 mosaics with all female brains showed male behavior. Male brain tissue in one side is thus necessary but perhaps not always sufficient for male behavior. The conclusions from these studies of a domineering focus for the control of following and wing extension are analogous to the control, by part of the dorsal brain (mushroom bodies), of courtship song in cricket and grasshopper (deduced from stimulation and ablation experiments— see reviews of Huber, 1965, 1967; Elsner, 1973).

No male-behaving mosaics with entirely female thoracic ganglia attempted copulation (Hall, 1977). While this behavior mapped most closely to the ventral ganglia (by the contour procedure), there was no particular portion of the ganglia for which male tissue was correlated one-to-one with attempted copulation.

Recently, Schilcher and Hall (1978) have mapped the focus related to the wing vibration, which goes along with wing extension in courtship. This *Drosophila* "lovesong" (a short pulse of tone coming at 34 ms intervals) is very separable from wing extension, in the mosaics. Thus, only about two-thirds of the wing-extending mosaics have a normal song, with the remainder equally divided amongst nonsingers and those with a very abnormal sonic output. The focus for the song is also far from the one in the brain associated with wing extension, in that the former is in or very near the mesothoracic part of the thoracic ganglia. The focus here is domineering, in that male tissue in either the left or right mesothoracic neuromere is sufficient to allow for a normal lovesong from both the left and right wings.

Female reproductive behavior has also been examined in mosaics. Nissani (1975), Hotta and Benzer (1976), and Hall (1977) found that whether or not a mosaic will provoke a male to court is strongly correlated with the abdominal genotype. Thus, mosaics with female abdomens nearly always provoke courtship (not necessarily including copulation), whereas those with male abdomens usually do not. The genotype of the head is clearly uncorrelated with provoking, but there may be some involvement of a female thorax in this "sex appeal" (Nissani, 1975; Y. Hotta, unpublished).

If a female or a mosaic is courted it will not necessarily accept copulation (i.e., be receptive), even if the external genitalia are female. W. Harris, R. Rothman, and J. Hall have preliminarily mapped receptivity and—among mosaics selected for female genitals—have found the focus to be in the anterior part of the blastoderm. The criterion for receptivity was the laying of fertile eggs, by mosaics which could lay eggs (see below), and which had been placed with males. Mosaics which on subsequent dissection proved not to have normal reproductive systems were excluded from the analysis. K. White and W. Harris (unpublished) also learned that egg-laying is controlled by the genotype of anterior tissues. That is, a mosaic with female genitalia, and a normal reproductive system, including eggs, may not be able to operate the egg-laying "machinery" unless she has female tissue in anterior (brain?) tissue (note that a female, which has copulated or not, can release eggs). Alternatively, the anterior focus for egg laying may relate to the ring gland's involvement in egg production.

One courtship mutant has been studied in mosaics, the cacaphony mutant (*cac*) of Schilcher (1976a, b; 1977). Males hemizygous for this X chromosome mutation take an inordinately long time to achieve copulation. In addition, they have a specific abnormality of the courtship "song" (which is associated with male courtship wing extension—see review of Bennet-Clark and Ewing, 1970). The song abnormality may not be sufficient for the courtship difficulties, because wingless *cac* males still take longer to achieve copulation than do wingless *cac*[+] males.

Schilcher's mosaic analysis was, as he noted, confounded by the fact that the diplo-X, haplo-X mosaics (the latter tissues being hemizygous for *cac*) were mosaic for sex per se. Of 51 mosaics with heads half-male or more, and thoraces split down the middle (left half male, right half female, or vice versa), 36 displayed a courtship song, consistent with the basic results of Hotta and Benzer (1976) and Hall (1977). All of the songs here—even those produced by vibration of the wing on the heterozygous *cac*, female side of the thorax—were mutant. Since all these mosaics had male tissue in both the head and thorax, the head could contain a domineering focus for wing extension and the abnormal song caused by *cac*; or, the focus for *cac* could be assumed to be a domineering one in the thorax (thoracic ganglia? muscles?). To test these possibilities, Schilcher looked at the songs produced by 14 gynandromorphs with male tissue (hemizygous for the mutation) on at least half the head surface, and nowhere, else on the cuticle. Eight of the songs were mutant and six were "unclassifiable"—the same kind of result found with gynandromorphs that carried only wild-type alleles of *cac*. That is, about 40% of regular gynandromorphs (not carrying *cac*)—which do sing and which have all-female thoraxes—show the abnormal song that is unrelated to *cac*. All gynandromorphs which sang and which had bilaterally split thoraces (half male, half female) had normal wing vibration. These results are of course consistent with the more concrete mapping results of Schilcher and Hall (1978, using the internally marked mosaics), which yielded a mesothoracic male focus for the normal song.

With regard to *cac*, the mosaic results are weakly consistent with the focus being in the head. If it were in the thorax, then some of the singing mosaics with totally *cac*⁺ thorax (on the cuticle), and a "classifiable" song, should have sung normally. One problem is that, though all the mosaics of this type had mutant song, there were only eight gynandromorphs in this category. A more rigorous analysis would involve somatic loss, from a completely male animal, of a small duplication that carries the wild-type allele of *cac*. No such duplication now exists, but Novitski (1963) described a straightforward method for constructing this kind of genetic rearrangement. An alternative procedure, which might work, would be to induce somatic loss of a *cac*⁺-bearing X chromosome, from a diplo-X zygote that is homozygous for the transformer (*tra*) mutation of Sturtevant (1945). Such a mosaic would be uniformly male in phenotype, because XX; *tra/tra* flies are male in external appearance and in behavior. That is, recent quantitative data on the courtship exhibited by such "transformed" males (J. C. Hall and F. von Schilcher, unpublished), indicate regular male-like behavior, on criteria such as frequency of performance of the various courtship steps, fraction of observation period spent actively courting females, duration of copulation, and even the parameters of the lovesong. The *cac*⁺-bearing X chromosome—loss of which would be induced to create the *cac*, non-*cac* mosaics—should carry the duplication for the acid phosphatase gene (Kankel and Hall, 1976). Here, the genotypes of ganglia in the CNS would be scorable, i.e., if the third chromosomes were homozygous for, not only *tra*, but for an *Acph-1*-null allele as well. This technique requires that *cac* be a recessive mutant, that is, in homozygous *tra* flies; and Hall and Schilcher (unpublished) have recently shown this to be so. The general approach here obviates the need for the construction of special chromosomes or chromosome fragments, in regard to mosaic analysis of mutations with male-limited expression.

IV. Conclusions and Prospects

The determination of foci for behavioral mutants will presumably be a routine early step in the analysis of the genetic defects. Mosaic analysis should be done, not just because the technology is readily available, because one wants a general idea as to which tissues are affected by the mutation, and because the fate maps frequently appear to be very elegant. A more serious and conceptual matter is that mosaic analysis can reveal that the tissue thought to be the one primarily affected by the mutation is, instead, indirectly influenced by the primary focus, which is elsewhere. A corollary to this possibility is the notion that multiple defects induced by a particular mutation might, in fact, have a "unifying" etiology—where one kind of cell is abnormal under the influence of the mutation, and this leads to a variety of abnormalities in cells which interact with the primarily affected cells. Thus, the expression of the mutant or normal allele of the gene in the cells that are not the primary focus would be irrelevant.

For behavioral and neurologic mutants in other organisms, one frequently has few pieces of information concerning the actual influence of the gene. For example, there are many mouse mutants—found due to behavioral abnormalities—which derange the development of and eventual morphology of the cerebellum (reviewed by Sidman, 1968; Chung, 1975, 1976). Often, a variety of different cell types end up being intrinsically defective in appearance, or being in abnormal positions within the cerebellum; thus, the connectivity among cells can be quite abnormal. For a mutant such as weaver *(wv)*—in which cerebellar granule cells fail to migrate properly—several other abnormalities apparently follow. It has been hypothesized that a particular class of glial cells guide the migration of the granule cells, and that these glial cells are in a sense the focus (Rakic and Sidman, 1973a, b, c). The absence of migration, induced by the glial cell abnormality, then would lead to the other defects one sees in the cerebellar regions in which the granule cells should have ended up. However, it has also been proposed that the granule cells are the sites of the primary defect in this mutant (Sotelo and Changeux, 1974). The different results from these two laboratories may be influenced by "genetic background" differences, which supposedly lead to different abnormalities in different weaver strains. However, what is needed is a study in which wv versus wv^+ cells are marked in a mosaic cerebellum. Mosaics would be produced by the allophenic techniques reviewed by Gardner (this volume). Then, one would know if the glial cells or the granule cells are the ones primarily defective, and if the abnormalities seen with respect to the other cells are indeed independent of their genotype. A requirement of these studies is to have the glial and granule cells expressing a scorable genetic variant other than weaver itself. Progress in this general area is rapid. Brain mosaics in the mouse have been studied, with regard to the reeler, staggerer, and Purkinje-cell-degeneration mutants (Mullen, 1976, 1977). Through the use of a histochemically detectable enzyme variant, which marked the genotype of some cells in the brain, it was tentatively concluded that the latter two mutants are autonomously expressed in Purkinje cells, but that the abnormalities related to these cells in the reeler may be at least partly due to an interaction with cells or substances not intrinsic to them.

For *Drosophila*, progress has been made mainly with respect to the "gross" fate mapping of a variety of behavioral mutants (Fig. 4). However, many of the foci placed on this map are very tentative, due to very small populations for some of the data and due to the quantitative difficulties which still obtain for properly computing sturt distances or "contour" lines. This review, moreover, is largely a laundry list of many mosaic results, of varying extensiveness and interpretability.

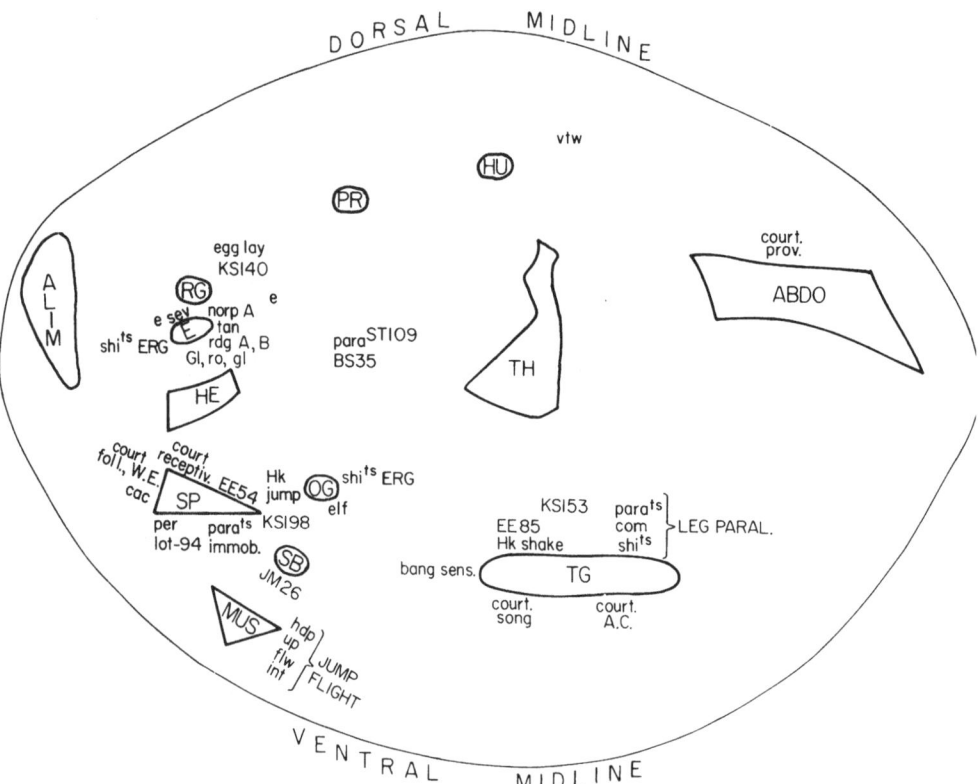

Fig. 4. Fanciful fate map of *Drosophila* behavioral foci. This is a representation of the left or homologous right half of the embryonic blastoderm. The regions leading to various adult tissues are each in outline, with their symbols inside: *ALIM*, alimentary system in thorax; *E*, compound eye; *HE*, head cuticle structures; *RG*, larval ring gland; *SP*, supraesophageal ganglia (dorsal brain); *SB*, subesophageal ganglion (ventral brain); *OG*, optic ganglia; *PR*, proboscis; *HU*, thoracic humeral bristles; *TH*, thoracic cuticle structures; *TG*, fused, ventral thoracic + abdominal ganglia; *MS*, thoracic musculature; *ABDO*, segments of the abdominal cuticle. (Most of these external + internal landmark locations are taken from the maps of Kankel and Hall, 1976. The ring gland site is from Janning, 1976; and the presumptive muscle site is from Hotta and Benzer, 1972, and Deak, 1977). The locations of foci for a variety of behavioral mutants, and the foci for some normal behaviors, are shown. Some of these positionings are very tentative. *P-37*, optomotor defect; *norp A*, absence of ERG; *tan*, ERG transient suppression; *rdg A, B; sev*, morphologic defects in photoreceptor cells; *Gl, ro, gl*, mutant optic lobes, induced by retina morphology mutants; *egg lay*, ability of gynandromorph to release eggs; *court. foll., W.E.*, following, by a gynandromorph, of a female, and wing extension at her; *court. recept.*, willingness of gynandromorph to accept copulation; *cac*, abnormal love song; *per*, abnormal circadian rhythm; *lot-94*, abnormal salt. tolerance; *para^ts immob.*, general immobility due to temperature-sensitive paralytic mutant, and sluggish behavior at low temperature; *Hk jump*, Hyperkinetik [1], visually-triggered jumping; *shi^ts ERG*, temperature-sensitive transient suppression; *e, elf*, ERG transient suppression; *para^ST109*, simultaneous leg paralysis in this particular *para^ts* allele; *EE 54, KS 198, KS 140, JM 26, BS 35*, inexcitability in sluggish mutants; *bang sens.*, sensitivity to mechanical shock; *hdp, up, flw, int, jump, flight*, abnormal jumping, flying in muscle mutants; *vtw*, vertical wings, abnormal jumping and flying in muscle mutant; *Hk shake*, Hyperkinetic, ether-induced leg shaking; *para^ts, com, shi^ts, Ocd^ts leg paral.*, temperature-sensitive leg paralysis; *EE 85, KS 153*, slow, weak leg movement in sluggish mutants; *court. song*, gynandromorph wing vibration at female; *court a. c.*, gynandromorph attempted copulation; *court prov.*, provoking of courtship by gynandromorph

The list does reveal the various modes of mosaic analysis available—ranging from casual observation of a few externally marked gynandromorphs to statistical fate mapping analysis of dozens of mosaics, and then full circle to the kinds of studies involving internal markers, where the behavioral and tissue scoring can perhaps be done on a relatively few cases. It should be mentioned at this point that it is possible to "overquantify" the analysis of interacting foci. Focus determination is hopefully done to suggest which tissue(s) are affected by a mutant, in order to carry out further investigation of what has gone wrong with respect to the focus. It is, to some degree, both unnecessary and impossible to achieve extremely accurate and meaningful focusing—especially for mosaics scored only with cuticular markers. Increasingly detailed, complex, and altered mapping procedures (e.g., Flanagan, 1976, 1977; Feitelson and Hall, 1978a) are certainly not incorrect or useless; but, in the end, they are not a substitute for an eventual study of the tissue made defective by the mutations.

At present, few mosaic studies, even the more quantitatively sophisticated ones, have revealed new principles about the development or functioning of the nervous system. So one resorts to a listing of cases, which at least reveals which techniques are available and which results are possible in the ongoing analysis of behavioral mutants. Most of these results have answered simple questions on tissue specificity of mutant action, and on interactions between mutant and normal tissues. One hopes that such analysis—as it moves beyond a determination of *where* the defect lies, to discovery of *what* has gone wrong at those foci—will be begin to reveal how behaviors are programmed into the nervous system and show how basic neurologic mechanisms work. For now, however, it must be admitted that the kind of information demanded (above) for mouse mutants has not been obtained in *Drosophila*. Indeed, the library of brain-affecting mutants in the fruit fly is not as extensive as that for the mouse. Yet, the conceptual and technical advances in insect mosaic analysis will allow substantial and rapid progress in this area. For example, the detailed and "fine level" mosaic studies of eye-roughening mutants (Meyerowitz, 1977) have revealed that abnormalities of particular neurons can be induced by an abnormal genotype and phenotype in excitable cells presynaptic to the affected neurons. The analysis here depended on—not just the gross fate mapping which one does with mosaics produced very early in development—but on the ability routinely to generate mosaics at later stages, which will have small clones of mutant tissue. Another critical feature of the analysis was the ability reliably to mark neurons of different genotypes. It is also useful to recall that the techniques of induced mitotic recombination allowed this mosaic analysis to be carried out on the expression of autosomal genes. Most of the mosaic techniques described in this review are available only for X-linked mutations.

The conclusions from Meyerowitz's analysis are striking because—not only were primary and secondary foci determined—but the "inductive" influence of the organization of one group of excitable cells on the organization of another group of neurons was demonstrated in a real experimental sense. This is in distinct contrast to the still frequent claims that, for example, growing axons from photoreceptor cells can induce the differentiation and organization of second-order neurons in the developing visual system (e.g., Lopresti et al., 1973; Levinthal,

1974). Here, one is inferring inductive cell interactions from mere observation. The mutant and mosaic techniques have allowed, and will continue to allow real tests of cell interaction hypotheses on nervous system development and functioning.

Thus, in *Drosophila*, one always has the ability first to find out which major body region is affected by any X-linked mutation and to obtain reasonably solid information on which major tissue there (nerve? muscle?) is affected. One can make this initial focus determination with internally scorable mosaics (using the technique of Kankel and Hall, 1976). Therefore, the dreary complications of interacting focus analysis can be avoided. These quantitative complications involved with external marker fate mapping may include, in addition to all those mentioned earlier, the difficulties of defining behavioral phenotypes of the mosaics. Totally mutant or totally wild-type flies may be clearly abnormal and normal in behavior, respectively. However, mosaics may turn out to show more than two phenotypes. For example, the mosaics with respect to the circadian rhythm mutant have a phenotype that is mutant, normal, *or* a "sum" of the two kinds of behavior. Mosaics with respect to the cacophony mutant sing normally, abnormally, or in an "unclassifiable" way. These examples remind us of the sexual behavior results from wasp gynandromorphs (Whiting, 1932, and confirmed by Clark and Egen, 1975). Here, the mosaics have one of at least three different phenotypes: female-like behavior, male-like behavior, or both. It may be possible in these cases to expand the kind of analysis done by Hotta and Benzer (1972). That is, instead of 2×3 tables (cf. Table 3) for behavioral and surface landmark scoring, the raw data would be classified according to entries in 3×3 tables, and a variety of different assumptions tried, on the relationship of different categories of dividing line (Fig. 2) to the various behavioral possibilities.

Another potential problem with cuticular marker techniques of fate mapping is the occasional temptation to use previously drawn landmark fate maps in conjunction with triangulation of as yet unlocalized behavioral foci. This could lead to substantial alterations in focus positioning and could result in serious errors of focus locations. For instance, I mapped the *up*-induced wing posture focus of Hotta and Benzer (1972), using other fate maps made with cuticular markers (reviewed by Hall et al., 1976). The focus always came out in a relatively ventral region of the blastoderm, but, because of the different landmark geometry among the maps, some of the foci here were in more dorsal positions (closer to the nervous system landmarks) than Hotta and Benzer's position.

A corollary to the above warning is that it may be risky to superimpose independently derived fate maps, concerning different behavior mutants, different landmark marking systems (e.g., internal + external vs. external only), or different gynandromorph-generating techniques. As was mentioned earlier, one would have predicted from superimposing maps that the abnormal leg behavior with regard to comatose and Hyperkinetic are different. However, it was much better also to show directly that the foci are separable.

Despite the potential difficulties involved in external landmark fate mapping, the technique of marking nerve cells of different genotypes, in gynandromorphs, has not been widely used. People may continue to shy away from it, because the genetics of the system does not allow the production of mosaics in high frequency (as opposed to the case of ring-X chromosome loss, where mosaics are often $1/4$ to $1/3$ of the originally diplo-X progeny). Another shortcoming of the internal marker system of Kankel and Hall (1976) is that muscle scoring is not possible (because there is virtually no histochemically detectable acid phosphatase activity in wild-type animals). It would be very desirable to confirm the inferences of Hotta and Benzer (1972), Deak (1977), and Homyk (1977) on the presumed muscle foci for the mutants they studied having behavior and wing-posture defects. Here, a gratuitous muscle marker (analogous to the CNS acid phosphatase marker) is needed.

Still another problem with the internally marked mosaics of Kankel and Hall (1976) is that a "fine localization" of nervous system foci may be impossible. This is because of the relatively large clone sizes in mosaic nervous system ganglia. Presumably, there are not many blastoderm cells that will develop into a given ganglion, so that a mosaic with, say, $1/50$ of the dorsal brain being hemizygous for a given mutant cannot be produced. If it could be produced, then one might be able to show that this particular $1/50$ of the ganglion contains the focus. The situations is even worse for the four optic lobes, on each side of the head, which, taken as a unit, are very rarely of mixed genotype. Thus, a lamina versus medulla focus for a visual system mutant cannot be determined easily.

The focus determination, though, does not have to stop with gynandromorph analysis. Small clones of mutant tissue can be generated by inducing mitotic recombination. One is fortunate if the questions being asked have to do with mutant phenotypes that are at least partially expressed at the level of the cuticle (as in the studies of Meyerowitz, 1977). If they are not, then the following approach may be necessary. Suppose that a focus for a behavior mutant has been generally localized to the head ganglia and that there is no cuticle abnormality associated with the mutant. If the mitotic recombinants are induced between the genetic locus in question and the centromere of the relevant chromosome, then a meaningful clone—one which includes brain tissue—is very unlikely to include cuticle areas (which, if mutant, could express a tagging marker such as yellow body color). How will the relevant mosaics be collected for analysis? One is probably forced to apply a behavioral screen to the animals, only some of which will have had a somatic cross-over induced in or near the head ganglia. In other words, one would test the adults, following irradiation of larvae, for the behavioral abnormality associated with the mutation, hope to find some abnormally behaving flies, then section them to see which brain region is genetically mutant. This now begs the question as to how the mutant brain clone will be recognized. Most known behavioral mutants are on the X chromosome (merely because that is where they have been sought). There is no known nerve or muscle cell marker naturally located in X chromosome euchromatin, so that a cross-over between the behavior mutant and the X centromere would also lead to mutant expression of a tagging marker. This problem has now been alleviated, because Harris (1977) has translocated the normal allele of the third chromosome acid phosphatase gene (*Acph-1*) onto the tip of the X chromosome, far from the centromere. If this translocation is the X chromosome which also contains the normal allele of the behavioral mutant, in a female larva heterozygous for the mutant allele, then a cross-over between the behavioral mutation and the centromere can lead to a clone of tissue homozygous for the mutation and lacking the $Acph-1^+$ allele. This clone will be scorable if the third chromosomes are homozygous for an $Acph-1$ null allele.

The above analysis will be very difficult if one must test each putative internal mosaic individually—as opposed to, say, performing a mass phototaxis test (cf. Benzer, 1967) or a temperature-sensitive paralysis test. For instance, one might want to induce haplo-X clones of brain tissue at various times during development, by irradiating developing females heterozygous for a ring-X chromosome and a standard rod-X chromosome (cf. Merriam et al., 1972). The plan would be to determine which small patch of brain tissue must be genetically male in order to allow male courtship behavior—if, indeed, a particular tiny patch of male tissue is sufficient. The problem is that, after collecting adults derived from irradiated earlier stages, an extremely laborious series of courtship observations awaits the investigator. There may be no simple way to "select" for male-behaving mosaics that have male tissue only in a restricted part of the head. (Fertility of the putative mosaics is of course not a useful screening procedure, since male thoracic and abdominal tissue is also required for copulation.)

Mosaics from mitotic recombination can also be used to gain information on neurologic mutants that is of a different kind than all that has been discussed previously. Since the clones of mutant tissue can be induced at different stages of development, it is in principle possible to determine a stage when the formation of the mutant tissue, in the relevant place, no longer leads to a neurologic abnormality. For example, a gene might function during only one stage of larval development, to control some feature of the development of the postembryonic nervous system. After this stage, the gene might not function, so the presence of a mutant allele in the homozygous condition would no longer be relevant.

This sanguine view of the power of such mosaics to provide information on both the place and time of mutant action is overstated. Mitotic recombinants induced at later stages will lead to smaller and smaller patches of mutant tissue. Even if the clone induced at a relatively late stage is within the confines of the focus—and then one finds no expression of the mutation—it could be that this negative result was due to an insufficient mass of mutant tissue. Suppose that the lamina is the focus for an ERG transient-suppressed phenotype, due to a developmental defect in this optic lobe. For relatively early induction of the cross-overs, half-mutant lamina patches might be produced, and this could still result in a noticeable ERG abnormality. For later irradiations, one-eighth lamina patches might be generated, and—whereas the last time of gene action for this case is *after* the stage of irradiation—there is a wild-type ERG because of the small fraction of mutant lamina tissue. So one would conclude that gene action has stopped at a stage which is before the real time. One might try to "separate" the small size of clones induced toward the end of development from the matter of their late induction, by use of the "*Minute* technique" (described by Morata and Ripoll, 1975). Here, non-*Minute* (M^+/M^+) clones, induced in a *Minute* (M^+/M) background, divide faster than the cells of the background genotype, so that "late" clones can be made "large." However, there may be nonspecific behavioral defects in such mosaics due to the prevalence of the *Minute* phenotype. Thus, it is highly desirable to obtain independent evidence on the time of expression of genes which affect the nervous system and behavior. For instance, the effects of a mutation could be turned on and off at different developmental stages, through the relevant treatments of developing animals that carry a conditional allele of the mutation, if any such alleles exist.

Acknowledgements. Several individuals deserve thanks for the many ideas they have contributed to this review, from conversations I have had with them in recent years: Seymour Benzer, Doug Kankel, Martin Heisenberg, Bill Harris, Obaid Siddiqi, Chip Quinn, Ralph Greenspan, Bill Pak, John Merriam, and Linda Hall. I am also grateful to those who supplied unpublished data and techniques: Seymour Benzer, Yoshiki Hotta, Doug Kankel, Bill Kaplan, Bill Harris, Ron Konopka, Florian von Schilcher, Alain Ghysen, Linda Hall, Kalpana White, Lenny Robbins, and Steve Sellers. Finally, there are many who deserve thanks for comments on preliminary drafts of this review: Martin Heisenberg, Bill Kaplan, Jim Haber, Ralph Greenspan, Bob Schlief, Lily Jan, Don Ready, Lee Ehrman, Florian von Schilcher, Bill Harris, and Seymour Benzer.

The author's research, some of which is discussed here, is supported by grants GM-21473 and NS-12346, both from the U.S. Public Health Service, and by funds from a Biomedical Research Support Grant awarded to Brandeis University.

References

Arking, R.: Temperature-sensitive cell-lethal mutant of Drosophila: isolation and characterization. Genetics **80**, 519—537 (1975)

Baker, B. S.: Paternal loss *(pal)*: a meiotic mutant in *Drosophila melanogaster* causing loss of paternal chromosomes. Genetics **80**, 267—296 (1975)

Bennet-Clark, H. C., Ewing, A. W.: The love song of the fruit fly. Sci. Am. **223**, 84—92 (1970)

Benzer, S.: Behavioral mutants of Drosophila isolated by countercurrent distribution. Proc. Natl. Acad. Sci. USA **58**, 1112—1119 (1967)

Benzer, S.: From the gene to behavior. J. Am. Med. Assoc. **218**, 1015—1022 (1971)

Benzer, S.: Genetic dissection of behavior. Sci. Am. **229**, 24—37 (1973)

Bodenstein, D.: The postembryonic development of Drosophila. In: Demerec, M. (Ed.): Biology of Drosophila, pp. 275—367. New York: Wiley 1950

Bryant, P., Zornetzer, M.: Mosaic analysis of lethal mutations in Drosophila. Genetics **75**, 623—637 (1973)

Burnet, B., Connolly, K., Mallinson, M.: Activity and sexual behavior of neurological mutants in *Drosophila melanogaster*. Behav. Genet. **4**, 227—235 (1974)

Chung, S.-H.: Synaptic remodelling in the mutant cerebellum. Nature (Lond.) **257**, 86—87 (1975)

Chung, S.-H.: The brain of the "reeler" mouse. Nature (Lond.) **260**, 14—15 (1976)

Clark, A. M., Egen, R. C.: Behavior of gynandromorphs of the wasp *Habrobracon juglandis*. Dev. Biol. **45**, 251—259 (1975)

Cosens, D. J., Manning, A.: Abnormal electroretinogram from a Drosophila mutant. Nature (Lond.) **224**, 285—287 (1969)

Deak, I. I.: Use of Drosophila mutants to investigate the effect of disuse on the maintenance of muscle. J. Insect. Physiol. **22**, 1159—1165 (1976)

Deak, I. I.: Mutations of *Drosophila melanogaster* that affect muscles. J. Embryol. Exp. Morph. **40**, 35—63 (1977)

Deusen, E., Van: Experimental studies on a mutant gene (*e*) preventing the differentiation of eye and normal hypothalmus primordia in the axolotl. Dev. Biol. **34**, 135—158 (1973)

Elsner, N.: The central nervous control of courtship behavior in the grasshopper *Gomphocerippus rufus* L. (Orthroptera: Acrididae). In: Salánki, J. (Ed.): Neurobiology of Invertebrate Behavior: Mechanisms of Rhythm Regulation, pp. 261—287. Budapest: Acádemiai Kiadó 1973

Falk, R., Atidia, J.: Mutation affecting taste perception in *Drosophila melanogaster*. Nature (Lond.) **254**, 325—326 (1975)

Falk, R., Orevi, N., Menzl, B.: A fate map of larval organs of Drosophila and preblastoderm determination. Nature (New Biol.) **246**, 19—20 (1973)

Feitelson, J. S., Hall, L. M.: The contour technique as a preliminary method of fate mapping complex behaviors in *Drosophila melanogaster*. manuscript in preparation (1978a)

Feitelson, J. S., Hall, L.: Genetic and behavioral analysis of stress-sensitive mutants of *Drosophila melanogaster*. manuscript in preparation (1978b)

Flanagan, J.: A computerized method for making fate maps of *Drosophila*. Dev. Biol. **53**, 142—146 (1976)

Flanagan, J.: Fate mapping the focus of an adult lethal mutation in *Drosophila melanogaster*. Genetics **85**, 587—607 (1977)

Garcia-Bellido, A., Merriam, J. R.: Cell lineage of the imaginal discs in Drosophila gynandromorphs. J. Exp. Zool. **170**, 61—76 (1969)

Gehring, W.: The stability of the determined state in cultures of imaginal disks in Drosophila. In: Ursprung, H., Nöthiger, R. (Eds.): Results and Problems in Cell Differentiation, Vol. V: The Biology of Imaginal Disks, pp. 35—58. Berlin-Heidelberg-New York: Springer 1972

Grigliatti, T. A., Hall, L., Rosenbluth, R., Suzuki, D. T.: Temperature-sensitive mutations in *Drosophila melanogaster*. XIV. A selection of immobile adults. Molec. Gen. Genet. **120**, 107—114 (1973)

Grigliatti, T., Suzuki, D. T., Williamson, R.: Temperature-sensitive mutations in *Drosophila melanogaster*. X. Developmental analysis of the paralytic mutation, para . Dev. Biol. **28**, 352—371 (1972)

Grossfield, J.: Behavioral mutants of Drosophila. In: King, R. C. (Ed.): Handbook of Genetics, Vol. III, pp. 679—702. New York: Plenum Press 1975

Guillery, R. W., Scott, G. L., Cattanach, B. N., Deol, M. S.: Genetic mechanism determining the central visual pathways of mice. Science **179**, 1014—1016 (1973)

Hall, J. C.: Age-dependent enzyme changes in *Drosophila melanogaster*. Exp. Gerontol. **4**, 207—222 (1969)

Hall, J. C.: Portions of the central nervous system controlling reproductive behavior in *Drosophila melanogaster*. Behav. Genet. **7**, 291—312 (1977)

Hall, J. C., Gelbart, W. M., Kankel, D. R.: Mosaics systems. In: Ashburner, M., Novitski, E. (Eds.): Genetics and Biology of Drosophila, Vol. Ia, pp. 265—314. London: Academic Press 1976

Hall, J. C., Kankel, D. R.: Genetics of acetylcholinesterase in *Drosophila melanogaster*. Genetics **83**, 517—535 (1976)

Hall, L., Junker, A. K., Suzuki, D. T.: Mapping the nervous system of Drosophila melanogaster using gynandromorphs of a temperature-sensitive locomotor mutant Shibire (abstr.). Genetics **74**, 105 (1973)

Hall, L. M., Olive, D., Farber, I., Osmond, B. C.: A tetrodotoxin-sensitive mutation in *Drosophila melanogaster*. Submitted to Nature (Lond.) (1978)

Harris, W. A.: The tip of 3R onto the tip of X. Drosophila Info. Service **52**, 68 (1977)

Harris, W. A., Stark, W. S.: Hereditary retinal degeneration in *Drosophila melanogaster*. J. Gen. Physiol. **69**, 261—291 (1977)

Harris, W. A., Stark, W. S., Walker, J. A.: Genetic dissection of the photoreceptor system in the compound eye of *Drosophila melanogaster*. J. Physiol. **256**, 415—439 (1976)

Heisenberg, M.: Comparative behavioral studies on two visual mutants of Drosophila. J. Comp. Physiol. **80**, 119—136 (1972)

Heisenberg, M., Götz, K. G.: The use of mutations for the partial degradation of vision in *Drosophila melanogaster*. J. Comp. Physiol. **98**, 217—241 (1975)

Homyk, T., Jr.: Behavioral mutants of *Drosophila melanogaster*. II. Behavioral analysis and focus mapping. Genetics **87**, 105—128 (1977)

Homyk, T., Jr., Sheppard, D. E.: Behavioral mutants of *Drosophila melanogaster*. I. Isolation and mapping of mutations which decrease flight ability. Genetics **87**, 95—104 (1977)

Hotta, Y., Benzer, S.: Genetic dissection of the Drosophila nervous system by means of mosaics. Proc. Natl. Acad. Sci. (USA) **67**, 1156—1163 (1970)

Hotta, Y., Benzer, S.: Mapping of behaviour in Drosophila mosaics. Nature (Lond.) **240**, 527—535 (1972)

Hotta, Y., Benzer, S.: Mapping of behavior in Drosophila mosaics. In: Ruddle, F. H. (Ed.): Genetic Mechanisms of Development, pp. 129—167. New York: Academic Press 1973

Hotta, Y., Benzer, S.: Courtship in *Drosophila* mosaics: sex-specific foci for sequential action patterns. Proc. Natl. Acad. Sci. USA **73**, 4154—4158 (1976)

Huber, F.: Brain controlled behavior in orthopterans. In: Treherne, J. E., Beament, J. W. L. (Eds.): The Physiology of the Insect Central Nervous System, pp. 233—246. New York: Academic Press 1965

Huber, F.: Central control of movements and behavior in invertebrates. In: Wiersma, C. A. G. (Ed.): Invertebrate Nervous Systems: Their Significance for Mammalian Neurophysiology, pp. 333—351. Chicago: University of Chicago Press 1967

Ikeda, K., Kaplan, W. D.: Patterned neural activity of a mutant *Drosophila melanogaster*. Proc. Natl. Acad. Sci. USA **66**, 765—772 (1970a)

Ikeda, K., Kaplan, W. D.: Unilaterally patterned neural activity of gynandromorphs mosaic for a neurological mutant of *Drosophila melanogaster*. Proc. Natl. Acad. Sci. USA **67**, 1480—1487 (1970b)

Ikeda, K., Kaplan, W. D.: Neurophysiological genetics in *Drosophila melanogaster*. Am. Zool. **14**, 1055—1066 (1974)

Ikeda, K., Ozawa, S., Hagiwara, S.: Synaptic transmission reversibly conditioned by single-gene mutation in *Drosophila melanogaster*. Nature (Lond.) **259**, 489—491 (1976)

Janning, W.: Entwicklungsgenetische Untersuchungen an Gynandern von *Drosophila melanogaster*. III. Einige Beobachtungen an larvalen Gynandern. Verh. Dtsch. Zool. Ges. **1974**, 134—138 (1975)

Janning, W.: Entwicklungsgenetische Untersuchungen an Gynandern von *Drosophila melanogaster*. IV. Vergleich der morphogenetischen Anlagepläne larvaler und imaginaler Strukturen. Wilhelm Roux' Arch. **179**, 349—372 (1976)

Johannsen, O. A.: Eye structure in normal and eye mutant Drosophilas. J. Morphol. **39**, 337—349 (1924)

Kankel, D. R., Hall, J. C.: Fate mapping of nervous system and other internal tissues in genetic mosaics of *Drosophila melanogaster*. Dev. Biol. **48**, 1—24 (1976)

Kaplan, W. D., Trout, W. E.: The behavior of four neurological mutants of Drosophila. Genetics **61**, 399—409 (1969)

Kaplan, W. D., Trout, W. E.: Genetic manipulation of an abnormal jump response in Drosophila. Genetics **77**, 721—739 (1974)

Kelly, L. E.: Temperature-sensitive mutations affecting the regenerative sodium channel in *Drosophila melanogaster*. Nature (Lond.) **248**, 166—168 (1974)

Kelly, L. E., Suzuki, D. T.: The effects of increased temperature on electroretinograms of temperature-snesitive paralysis mutants of *Drosophila melanogaster*. Proc. Natl. Acad. Sci. USA **71**, 4906—4909 (1974)

Konopka, R. J.: Circadian clock mutants of *Drosophila melanogaster*. Ph.D. Dissertation, California Inst. of Technology, Pasadena, California (1972)

Konopka, R. J., Benzer, S.: Clock mutants of *Drosophila melanogaster*. Proc. Natl. Acad. Sci. USA **68**, 2112—2116 (1971)

Levine, J. D.: Giant neuron input in mutant and wild-type Drosophila. J. Comp. Physiol. **93**, 265—285 (1974)

Levine, J., Tracey, D.: Structure and function of the giant motorneuron of *Drosophila melanogaster*. J. Comp. Physiol. **87**, 213—235 (1973)

Levinthal, C.: Neural development in isogenic organisms—a summary report. Am. Zool. **14**, 1051—1054 (1974)

Lindsley, D. L., Grell, E. H.: Genetic variations of *Drosophila melanogaster*. Carnegie Institution of Washington, Publ. No. 627, Washington, D.C. 1968

Lindsley, D. L., Sandler, L., Baker, B. S., Carpenter, A. T. C., Denell, R. E., Hall, J. C., Jacobs, P. A., Miklos, G. L. G., Davis, B. K., Gethman, R. C., Hardy, R. W., Hessler, A., Miller, S. M., Nozawa, H., Parry, D. M., Gould-Somero, M.: Segmental aneuploidy and the genetic gross structure of the Drosophila genome. Genetics **71**, 157—184 (1972)

Lopresti, V., Macagno, E. R., Levinthal, C.: Development of neuronal connections in isogenic organisms: cellular interactions in the development of the optic lamina of Daphnia. Proc. Natl. Acad. Sci. USA **70**, 433—437 (1973)

Manning, A.: Drosophila and the evolution of behavior. Viewpoints in Biology **4**, 125—169 (1965)

Manning, A.: Genes and evolution in insect behavior. In: Hirsch, J. (Ed.): Behavior-Genetic Analysis, pp. 44—60. New York: McGraw-Hill 1967

Mc Comas, A. J.: Neural hypothesis of muscular dystrophy is flourishing. Nature (Lond.) **251**, 569—570 (1974)

Meinertzhagen, I. A.: Development of the compound eye and optic lobe of insects. In: Young, D. (Ed.): Developmental Neurobiology of Arthropods, pp. 51—104. Cambridge: Cambridge University Press 1973

Merriam, J. R., Lange, K.: Maximum likelihood estimates for fate map locations of behavior in Drosophila. Dev. Biol. **38**, 196—201 (1974)

Merriam, J. R., Nöthiger, R., Garcia-Bellido, A.: Are dicentric anaphase bridges found by somatic recombination in X chromosome inversion heterozygotes of *Drosophila melanogaster*? Molec. Gen. Genet. **115**, 294—301 (1972)

Meyerowitz, E. M.: Eye-brain interactions in the development of the *Drosophila melanogaster* optic lobes. Ph.D. Dissertation, Yale Univ., New Haven, Connecticut, USA (1977)

Milani, R., Rivosecchi, L.: Gynandromorphism and intersexuality in *M. domestica*. Dros. Info. Service **28**, 135—136 (1954)

Mintz, B.: Gene control of mammalian differentiation. Ann. Rev. Genet. **8**, 411—470 (1974)

Morata, G., Ripoll: Minutes: mutants of *Drosophila* autonomously affecting cell division rate. Dev. Biol. **42**, 211—221 (1975)

Mullen, R. J.: Genetic dissection of the CNS with mutant-normal mouse and rat chimeras. 6th Annual Meeting of Society for Neuroscience, Toronto, Canada (1976)

Mullen, R. J.: Site of *pcd* gene action and Purkinje cell mosaicism in cerebella of chimaeric mice. Nature (Lond.) **270**, 245—247 (1977)

Mullen, R. J., La Vail, M. M.: Inherited retinal dystrophy: primary defect in pigment epithelium determined with experimental rat chimeras. Science **192**, 799—801 (1976)

Nissani, M.: A new behavioral bioassay for an analysis of sexual attraction and pheromones in insects. J. Exp. Zool. **192**, 271—275 (1975)

Nöthiger, R.: The larval development of imaginal disks. In: Ursprung, H., Nöthiger, R. (Eds.): Results and Problems in Cell Differentiation, Vol. V: The Biology of Imaginal Disks, pp. 1—34. Berlin-Heidelberg-New York: Springer 1972

Novitski, E.: The construction of new chromosomal types in Drosophila melanogaster. In: Burdette, W. J. (Ed.): Methodology in Basic Genetics, pp. 381—389. San Francisco: Holden-Day 1963

Pak, W. L.: Mutations affecting the vision of *Drosophila melanogaster*. In: King, R. C. (Ed.): Handbook of Genetics, Vol. III, pp. 703—733. New York: Plenum Press 1975

Pak, W. L., Grossfield, J., White, N. V.: Nonphototactic mutants in a study of vision of Drosophila. Nature (Lond.) **222**, 351—354 (1969)

Pak, W. L., Pinto, L. H.: Genetic approach to the study of the nervous system. Ann. Rev. Biophys. and Bioengin. **5**, 397—448 (1976)

Patterson, J. T., Stone, W.: Gynandromorphs in *Drosophila melanogaster*. Univ. of Texas Publ. **3825**, 1—67 (1938)

Peterson, A. C.: Chimaera mouse study shows absence of disease in genetically dystrophic muscle. Nature (Lond.) **248**, 561—564 (1974)

Pilkington, R. W.: Facet mutants of Drosophila. Proc. Zool. Soc. Lond., Ser. a **111**, 199—222 (1941)

Poodry, C. A., Hall, L., Suzuki, D. T.: Developmental properties of shibire^[ts1]: a pleiotropic mutation affecting larval and adult locomotion and development. Dev. Biol. **32**, 373—386 (1973)

Poulson, D. F.: Histogenesis, organogenesis, and differentiation in the embryo of *Drosophila melanogaster* Meigen. In: Demerec, M. (Ed.): Biology of Drosophila, pp. 168—274. New York: Wiley 1950

Power, M. E.: The central nervous system of winged but flightless *Drosophila melanogaster*. J. Exp. Zool. **115**, 315—340 (1950)

Quinn, W. G., Harris, W. A., Benzer, S.: Conditioned behavior in *Drosophila melanogaster*. Proc. Natl. Acad. Sci. USA **71**, 708—712 (1974)

Rakic, P., Sidman, R. L.: Weaver mutant mouse cerebellum: defective neurological migration secondary to abnormality of Bergmann glia. Proc. Natl. Acad. Sci. USA **70**, 240—244 (1973a)

Rakic, P., Sidman, R. L.: Sequence of developmental abnormalities leading to granule cell deficit in cerebellar cortex of weaver mutant mice. J. Comp. Neurol. **152**, 103—132 (1973b)

Rakic, R., Sidman, R. L.: Organization of cerebellar cortex secondary to deficit of granule cells in weaver mutant mice. J. Comp. Neurol. **152**, 133—162 (1973c)

Sakagami, S. F. Takahashi, H.: Beobachtungen über die gynandromorphen Honigbienen mit besonderer Berücksichtigung ihrer Handlungen innerhalb des Volkes. Insect Soc. **3**, 513—529 (1956)

Schilcher, F. von: Behavioral and genetic analysis of courtship song in *Drosophila melanogaster*. Ph.D. Dissertation, Univ. Edinburgh, Edinburgh, Scotland (1976a)

Schilcher, F. von: The behavior of *cacophony*, a courtship song mutant in *Drosophila melanogaster*. Behav. Biol. **17**, 187—196 (1976b)

Schilcher, F. von: A mutation which changes courtship song in *Drosophila melanogaster*. Behav. Genet. **7**, 251—259 (1977)

Schilcher, F. von, Hall, J. C.: Neural topography of courtship song in *Drosophila melanogaster*. submitted to J. Comp. Physiol. A (1978)

Siddiqi, O., Benzer, S.: Neurophysiological defects in temperature-sensitive paralytic mutants of *Drosophila melanogaster*. Proc. Natl. Acad. Sci. USA **73**, 3253—3257 (1976)

Sidman, R. L.: Development of interneuronal connections in brains of mutant mice. In: Carlson, F. D. (Ed.): Physiological and Biochemical Aspects of Nervous Integration, pp. 163—193. Englewood Cliffs, M. J.: Prentice-Hall 1968

Søndergaard, L.: A temperature-sensitive mutant of *Drosophila melanogaster*: Out-cold^{ts}. Hereditas **81**, 199—210 (1975)

Sotelo, C., Changeux, J. P.: Bergmann fibers and granular cell migration in the cerebellum of homozygous weaver mutant mouse. Brain Res. **77**, 484—491 (1974)

Spatz, H., Emanns, A., Reichert, H.: Associative learning of *Drosophila melanogaster*. Nature (Lond.) **248**, 359—361 (1974)

Spieth, H. T.: Mating behavior within the genus Drosophila (Diptera). Bull. Am. Mus. Nat. Hist. **99**, 396—474 (1952)

Spieth, H. T.: Courtship behavior in Drosophila. Ann. Rev. Entomol. **19**, 385—406 (1974)

Stewart, M., Murphy, C., Fristrom, J. W.: The recovery and preliminary characterization of X chromosome mutants affecting imaginal discs of *Drosophila melanogaster*. Dev. Biol. **27**, 71—83 (1972)

Sturtevant, A. H.: The claret mutant type of *Drosophila simulans*: a study of chromosomal elimination and cell lineage. Z. Wiss. Zool. **135**, 323—356 (1929)

Sturtevant, A. H.: A gene in *Drosophila melanogaster* that transforms females into males. Genetics **30**, 297—299 (1945)

Suzuki, D. T., Griglaiatti, T., Williamson, R.: Temperature-sensitive mutations in *Drosophila melanogaster*. VII. A mutation (*para^{ts}*) causing reversible adult paralysis. Proc. Natl. Acad. Sci. USA **68**, 890—893 (1976)

Trout, W. E., Kaplan, W. D.: A relationship between longevity, metabolic rate, and activity in shaker mutants of *Drosophila melanogaster*. Exp. Gerontol. **5**, 83—93 (1970)

Trout, W. E., Kaplan, W. D.: Genetic manipulation of motor output in shaker mutants of Drosophila. J. Neurobiol. **4**, 495—512 (1973)

Whiting, P. W.: Reproductive reactions of sex mosaics of a parasitic wasp, *Habrobracon juglandis*. J. Comp. Psychol. **14**, 345—363 (1932)

Wright, T. R. F.: The genetics of embryogenesis in Drosophila. Adv. Genet. **15**, 261—395 (1970)

Subject Index

Results and Problems in Cell Differentiation

A Series of Topical Volumes in Developmental Biology

Editors: W. Beermann, W. Gehring, J.B. Gurdon, F.C. Kafatos, J. Reinert

Volume 1

The Stability of the Differentiated State

Editor: H. Ursprung
1968. 56 figures. XI, 144 pages
ISBN 3-540-04315-2

"This is the first of a series of volumes that will review a few of the central issues in Cell and Developmental Biology, with each author writing on the stated theme with a personal slant. This approach is very successful. The articles have been written for the worker in the field rather than the general reader, and cover a wide variety of cell types, vertebrate and invertebrate, both *in vivo* and *in vitro*. Many of the questions raised are discussed by more than one author – for instance the relationship of differentiation to cessation of DNA synthesis. This gives the reader the advantage of seeing more than one selection of appropriate evidence. On other topics, the authors are pleasingly unanimous – such as the meaningfulness of dedifferentiation. The book is attractively presented, and the editor and publisher should be congratulated." *Quaterly Journal of experimental Physiology*

Volume 2

Origin and Continuity of Cell Organelles

Editors: J. Reinert, H. Ursprung
1971. 135 figures. XIII, 342 pages
ISBN 3-540-05239-9

"It is refreshing to pick up a book entitled *Origin and Continuity of Cell Organelles* and find something more than a discussion of mitochondrial and plastid semiautonomy... Overall, the volume certainly is well worth reading for many different types of biologists." *Science*

Volume 3

Nucleic Acid Hybridization in the Study of Cell Differentiation

Editor: H. Ursprung
1972. 29 figures. XI, 76 pages
ISBN 3-540-05742-0

"...This book is a comprehensive source of knowledge, so that it will be of value to beginners and/or experienced research workers in biochemistry, biology, genetics and experts in other areas of research for gaining more information about new hybridization techniques and interpretation of results." *Neoplasma*

Volume 4

Developmental Studies on Giant Chromosomes

Editor: W. Beermann
1972. 110 figures. XV, 227 pages
ISBN 3-540-05748-X

"This is the fourth volume of a successful series of highly topical volumes containing articles and short monographs on central issues in cell and developmental biology... The present book is dedicated to the memory of Jack Schultz, one of the pioneers in the study of polytene chromosomes after their "rediscovery" in 1933. Since then, so much material on the morphological, genetic, physiological, developmental, and biochemical aspects of giant chromosomes and puffing has been assembled that a coherent presentation is fully justified... This excellent volume should be on the bookshelf of every developmental geneticist and of many others interested in the role of chromomeres as sites of complex, integrated regulatory and informative genetic activities." *Genetica*

Springer-Verlag
Berlin Heidelberg New York

Volume 5
The Biology of Imaginal Disks

Editors: H. Ursprung, R. Nöthiger
1972. 56 figures. XVII, 172 pages
ISBN 3-540-05785-4

"...The main appeal of the book is that it is very much alive, squarely set in the present time and conveying the excitement in the field. Even if you do not read it until next year it will be more up to date than most books published in 1973. Imaginal disks are a model system for developmental biologists who should find the book fascinating and challenging..." *Nature*

Volume 6
W.J. Dickinson, D.T. Sullivan

Gene-Enzyme Systems in Drosophila

1975. 32 figures. XI, 163 pages
ISBN 3-540-06977-1

"This concise volume summarizes a large and diverse body of literature in an extremely useful format. The authors document the methods used, problems attacked, and results obtained for each gene-enzyme system which has been studied in Drosophila to date (1974)...
In all, the authors correlate a large amount of material in an easily readable style focusing on information rather than speculation. This timely book will be useful to cell biologists, geneticists, and especially, developmental biologists."
American Scientist

Volume 7
Cell Cycle and Cell Differentiation

Editors: J. Reinert, H. Holtzer
1975. 92 figures. XI, 331 pages
ISBN 3-540-07069-9

"...Nonetheless, I found the book both stimulating and highly instructive. Although priced beyond the reach of many individuals, I consider it a must for libraries of all institutions with an interest in this very fundamental aspect of biology."
Trends in Biochemical Sciences

Volume 8
Biochemical Differentiation in Insect Glands

Editor: W. Beermann

With contributions by numerous experts
1977. 110 figures, 24 tables. XII, 215 pages
ISBN 3-540-08286-7

The spinning glands and other tissues with secretory functions in Lepidopterons and Dipterons are highly specialized in the production of one or only a few different proteins in huge quantities. Moreover, as these insects can easily be raised and handled in the laboratory they provide a unique possibility to perform cytogenetic and molecular investigations on genomes and to study the formation of giant cells with polyploid and polytene nuclei.

Springer-Verlag
Berlin
Heidelberg
New York